Recent Titles in This Series

165 **Yu. Ilyashenko and S. Yakovenko, Editors,** Concerning the Hilbert 16th Problem
164 **N. N. Uraltseva, Editor,** Nonlinear Evolution Equations
163 **L. A. Bokut′, M. Hazewinkel, and Yu. G. Reshetnyak, Editors,** Third Siberian School "Algebra and Analysis"
162 **S. G. Gindikin, Editor,** Applied Problems of Radon Transform
161 **Katsumi Nomizu, Editor,** Selected Papers on Analysis, Probability, and Statistics
160 **K. Nomizu, Editor,** Selected Papers on Number Theory, Algebraic Geometry, and Differential Geometry
159 **O. A. Ladyzhenskaya, Editor,** Proceedings of the St. Petersburg Mathematical Society, Volume II
158 **A. K. Kelmans, Editor,** Selected Topics in Discrete Mathematics: Proceedings of the Moscow Discrete Mathematics Seminar, 1972–1990
157 **M. Sh. Birman, Editor,** Wave Propagation. Scattering Theory
156 **V. N. Gerasimov, N. G. Nesterenko, and A. I. Valitskas,** Three Papers on Algebras and Their Representations
155 **O. A. Ladyzhenskaya and A. M. Vershik, Editors,** Proceedings of the St. Petersburg Mathematical Society, Volume I
154 **V. A. Artamonov et al.,** Selected Papers in K-Theory
153 **S. G. Gindikin, Editor,** Singularity Theory and Some Problems of Functional Analysis
152 **H. Draškovičová et al.,** Ordered Sets and Lattices II
151 **I. A. Aleksandrov, L. A. Bokut′, and Yu. G. Reshetnyak, Editors,** Second Siberian Winter School "Algebra and Analysis"
150 **S. G. Gindikin, Editor,** Spectral Theory of Operators
149 **V. S. Afraĭmovich et al.,** Thirteen Papers in Algebra, Functional Analysis, Topology, and Probability, Translated from the Russian
148 **A. D. Aleksandrov, O. V. Belegradek, L. A. Bokut′, and Yu. L. Ershov, Editors,** First Siberian Winter School "Algebra and Analysis"
147 **I. G. Bashmakova et al.,** Nine Papers from the International Congress of Mathematicians, 1986
146 **L. A. Aĭzenberg et al.,** Fifteen Papers in Complex Analysis
145 **S. G. Dalalyan et al.,** Eight Papers Translated from the Russian
144 **S. D. Berman et al.,** Thirteen Papers Translated from the Russian
143 **V. A. Belonogov et al.,** Eight Papers Translated from the Russian
142 **M. B. Abalovich et al.,** Ten Papers Translated from the Russian
141 **H. Draškovičová et al.,** Ordered Sets and Lattices
140 **V. I. Bernik et al.,** Eleven Papers Translated from the Russian
139 **A. Ya. Aĭzenshtat et al.,** Nineteen Papers on Algebraic Semigroups
138 **I. V. Kovalishina and V. P. Potapov,** Seven Papers Translated from the Russian
137 **V. I. Arnol′d et al.,** Fourteen Papers Translated from the Russian
136 **L. A. Aksent′ev et al.,** Fourteen Papers Translated from the Russian
135 **S. N. Artemov et al.,** Six Papers in Logic
134 **A. Ya. Aĭzenshtat et al.,** Fourteen Papers Translated from the Russian
133 **R. R. Suncheleev et al.,** Thirteen Papers in Analysis
132 **I. G. Dmitriev et al.,** Thirteen Papers in Algebra
131 **V. A. Zmorovich et al.,** Ten Papers in Analysis
130 **M. M. Lavrent′ev, K. G. Reznitskaya, and V. G. Yakhno,** One-dimensional Inverse Problems of Mathematical Physics
129 **S. Ya. Khavinson,** Two Papers on Extremal Problems in Complex Analysis
128 **I. K. Zhuk et al.,** Thirteen Papers in Algebra and Number Theory
127 **P. L. Shabalin et al.,** Eleven Papers in Analysis

(*Continued in the back of this publication*)

Concerning the Hilbert 16th Problem

American Mathematical Society

TRANSLATIONS

Series 2 • Volume 165

Advances in the Mathematical Sciences — 23

(*Formerly Advances in Soviet Mathematics*)

Concerning the Hilbert 16th Problem

Yu. Ilyashenko
S. Yakovenko
Editors

American Mathematical Society
Providence, Rhode Island

1991 *Mathematics Subject Classification*. Primary 34C05, 34C20.

ABSTRACT. The main subject of the book is related to qualitative properties of vector fields on the plane, in the spirit of the famous Hilbert Sixteenth Problem. Two principal topics are bifurcations of limit cycles of planar vector fields and desingularization of singular points for individual vector fields and for analytic families of such fields.

In addition to presenting important new developments in this area, the book contains an introductory paper which outlines the general context and describes connections between various results obtained in five research papers constituting the volume.

The book can be used by researchers and graduate students working in qualitative theory of ordinary differential equations and dynamical systems.

Library of Congress Card Number 91-640741
ISBN 0-8218-0362-X
ISSN 0065-9290

∞ The paper used in this book is acid-free and falls within the guidelines
established to ensure permanence and durability.
♻ Printed on recycled paper.
This volume was typeset by the authors using $\mathcal{A}\mathcal{M}\mathcal{S}$-TeX,
the American Mathematical Society's TeX macro system.

10 9 8 7 6 5 4 3 2 1   00 99 98 97 96 95

Contents

Amer. Math. Soc. Transl.
(2) Vol. **165**, 1995

Concerning the Hilbert Sixteenth Problem

YU. ILYASHENKO AND S. YAKOVENKO

> Im Anshuß . . . die Frage nach der Maximalzahl und Lage der Poincaré-schen Grenzzyklen für eine Differentialgleichung erster Ordnung und erster Grades von der Form:
>
> $$\frac{dy}{dx} = \frac{Y}{X}$$
>
> wo X, Y ganze rationale Funktionen n-ten Grades in x, y sind.
>
> DAVID HILBERT, *Mathematische Probleme*, 1900

Two principal subjects of this volume are bifurcations of limit cycles of planar vector fields and desingularization of singular points, both for individual vector fields and for analytic families of such fields. These subjects are closely related to the second part of the Hilbert Sixteenth Problem. The goal of this introductory paper is to introduce the general context and outline connections between the various results obtained in the five research papers constituting this volume. We had no intention, however, to give a complete survey of the area: the recent collection of lecture notes [S] covers a broader field and gives a panorama of the current state. Among the notes in [S] the papers [I2], [D2], [R3] and [Rs] are the most close in scope to the subjects treated below.

In this Introduction we refer to the papers constituting the volume, as PAPER 1, . . . , PAPER 5.

§1. The Hilbert problem and finiteness theorems for limit cycles of polynomial vector fields

The shortest (the original) way of formulating this problem is to ask *what is the number and position of Poincaré limit cycles* (isolated periodic solutions) *for a polynomial differential equation* $dy/dx = P(x, y)/Q(x, y)$, *where* P *and* Q *are polynomials of degree* n.

1.1. Different forms of Hilbert's question. The formulation given by Hilbert admits several specifications, described in the following three subsections; the con-

1991 *Mathematics Subject Classification*. Primary 34C05.
The research of the first author was partially supported by the Grant M-98000 from the International Science Foundation.

nections between them are schematically shown on Table 1 (the statements of other problems will appear later).

Recall that a limit cycle of the differential equation

$$\frac{dy}{dx} = \frac{P_n(x, y)}{Q_n(x, y)}, \qquad P_n, Q_n \in \mathbb{R}[x, y], \ \deg P_n, Q_n \leqslant n \tag{1}$$

is a periodic solution which has an annulus-like neighborhood free of other periodic solutions on the (x, y)-plane.

INDIVIDUAL FINITENESS PROBLEM. *Prove that a polynomial differential equation* (1) *may have only a finite number of limit cycles.*

This problem is known also as *Dulac problem* since the pioneering work of Dulac (1923) who claimed to solve it, but gave an erroneous proof.

EXISTENTIAL HILBERT PROBLEM. *Prove that for any finite $n \in \mathbb{N}$ the number of limit cycles is uniformly bounded for all polynomial equations* (1) *of degree $\leqslant n$.*

If we denote

$$H(n) = \left\{ \begin{array}{l} \text{the uniform upper bound for the number of} \\ \text{limit cycles occurring in polynomial differ-} \\ \text{ential equations of degree } \leqslant n \end{array} \right\} \tag{2}$$

then the existential problem consists in proving that

$$H(n) < \infty \quad \text{for any } n \in \mathbb{N}.$$

CONSTRUCTIVE HILBERT PROBLEM. *Give an upper estimate for $H(n)$ or suggest an algorithm for computing such an estimate.*

The solution of this last problem would obviously imply the solution of the previous ones (the individual and the existential versions). However, in the next section we consider other statements, which cannot be reduced to the constructive Hilbert problem.

TABLE 1. Relations between different versions of the Hilbert problem: arrows stand for implications.

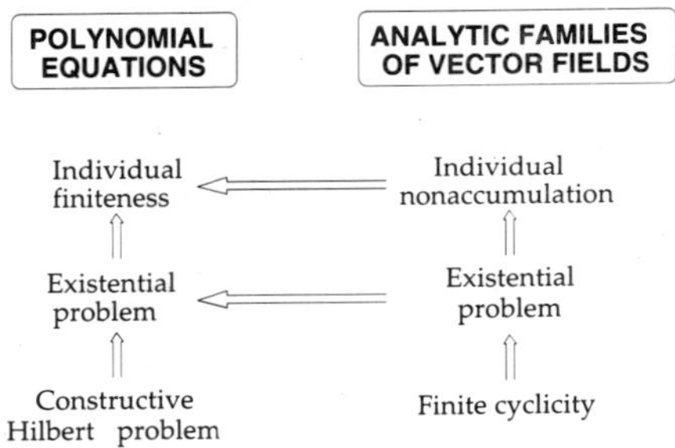

1.2. Nonaccumulation theorem. Of three forms of the Hilbert problem, only the first one (the weakest) is proved. Two independent and rather different proofs were given almost simultaneously by Yu. Ilyashenko [I1] and J. Écalle [É]. The preliminary stages of both proofs include the following preprocessing which was known already to Dulac.

Note that a polynomial differential equation makes sense also at infinity. To make this statement precise, consider the line field on $\mathbb{R}^2$ whose slope at a point (x, y) is $P(x, y)/Q(x, y)$ (we assume that the pair of polynomials has no common factors, thus the points of indeterminacy are isolated; the precise definition of a line field with singularities can be found in PAPER 2). Then this line field extends analytically to the projective compactification $\mathbb{R}P^2 \supset \mathbb{R}^2$: in a neighborhood of the infinite line $\mathbb{R}P^1$ one can multiply the vector field $P_n\,\partial/\partial x + Q_n\,\partial/\partial y$ by a meromorphic nonzero factor in such a way that the resulting vector field, which spans the same line field, would admit an analytic extension onto $\mathbb{R}P^1$ with at most $n+1$ singular points on the infinite line.

To obtain an orientable phase space, one can consider the sphere $\mathbb{S}^2$ together with the canonical covering $\pi\colon \mathbb{S}^2 \to \mathbb{R}P^2$; the pullback of the line field from $\mathbb{R}P^2$ is a symmetric line field on $\mathbb{S}^2$ with isolated singularities. Note that both the sphere and the projective plane are compact.

Assume for a moment that a polynomial vector field possesses an infinite number of limit cycles. By the Poincaré–Bendixson theorem, limit cycles of any differential equation should be "nested" around singular points. Since a polynomial differential equation may have only a finite number of such points, then limit cycles must accumulate (in the sense of Hausdorff metric) to a certain object which consists of some singular points and regular arcs (infinite trajectories) connecting them. Modulo a certain terminological discord, such objects are called *polycycles*. The precise definition that will be used throughout this volume, follows.

DEFINITION. A polycycle of a line field is a cyclically ordered collection of singular points $p_1, p_2, \dots, p_k$ (eventually with repetitions) and arcs (integral curves) connecting them in the specified order: the jth arc connects p_j with p_{j+1} for $j = 1, \dots, k$.

A trivial example of a polycycle is a periodic solution of the differential equation, without singular points and only one arc. Another case of polycycle having no arcs and just one singular point, is also admissible but far from being trivial.

The above arguments reduce the individual finiteness problem to the *nonaccumulation problem*: prove that limit cycles cannot accumulate to a polycycle of a polynomial differential equation. This problem is *semilocal*: its assertion concerns a small neighborhood of a polycycle rather than the whole (x, y)-plane. The following theorem proved by Ilyashenko and Écalle solves the individual finiteness problem (the Dulac problem).

NONACCUMULATION THEOREM. *For any analytic vector field on a two-dimensional real analytic manifold (surface), limit cycles cannot accumulate to a polycycle.*

Thus the nonaccumulation theorem is an analytic rather than algebraic assertion, and it implies the individual finiteness for some analytic differential equations as well.

COROLLARY (individual finiteness theorem for analytic vector fields on the sphere). *An analytic vector field on the two-dimensional sphere $\mathbb{S}^2$ may have only a finite number of limit cycles.*

1.3. Desingularization. The singular points occurring on a polycycle can be of any degree of degeneracy. For instance, the polycycle may consist of just one very degenerate singular point. Nevertheless a procedure is known that reduces the investigation to the case of polycycles with only relatively simple singularities, the *elementary* ones.

DEFINITION. A singular point of a planar differential equation is said to be *elementary*, if the linearization of the equation at this point has at least one nonzero characteristic number (the eigenvalue of the linearization matrix).

The simplest example of a nonelementary singularity is the *cuspidal point*, the singular point with the nonzero nilpotent linearization matrix. Linearization of a vector field at the cuspidal point has the form $\dot{x} = y$, $\dot{y} = 0$.

The procedure of simplification of singular points of a differential equation is known under several names: *desingularization, blowing-up, σ-process, resolution of singularities*. In any case, the idea is to delete a singular point from its small neighborhood and replace it by a one-dimensional curve, a projective line or a circle. For example, to make the *polar blow-up* of the origin, one introduces polar coordinates,

$$(r, \varphi) \xrightarrow{\rho} (x, y) = (r \cos\varphi, r \sin\varphi), \qquad r > 0,\ 0 \leqslant \varphi < 2\pi.$$

The differential equation written in polar coordinates admits an analytic extension to (small) negative values of r and division by a factor of r^ν, where ν is determined by principal terms of the Taylor expansion of the right-hand sides at the origin. After such a division one may consider the system in a narrow annulus $-h < r < h$, $0 \leqslant \varphi < 2\pi$. The entire circle $r = 0$ is the preimage of what formerly was a singular point of the equation, and singularities of the new field on this circle are in some sense simpler then the original singularity at the point $x = y = 0$ on the (x, y)-plane. If necessary, the procedure may be iterated (the new points are in turn blown up) until *all singularities become elementary*. The possibility of blowing up any singular point (satisfying the Łojasiewicz condition in the smooth case or isolated in the analytic category) into elementary singularities is the assertion of Bendixsson–Seidenberg–Dumortier theorem, see [D1], [VdE].

The polar blow-up has some disadvantages. First, it involves trigonometric functions and thus leads to the loss of algebraicity. Second, the points (r, φ) and $(-r, \varphi+\pi)$ correspond to the same point on (x, y)-plane, thus after the resolution the number of singular points is doubled. There exists an algebraic version, the *σ-process*, which operates with polynomial expressions and does not produce twin singularities: from the geometrical point of view it amounts to replacing the annulus around $r = 0$ by the Möbius band, the quotient space of the annulus by the equivalence $(r, \varphi) \sim (-r, \varphi + \pi)$. The central circle becomes then the projective line. These procedures are explained in PAPER 2 and PAPER 3 in more detail.

In any case, when proving the nonaccumulation theorem, one may consider only *elementary polycycles*, that is, polycycles carrying only elementary singularities on some analytic two-dimensional surface.

1.4. Analytic nature of the monodromy map. After this preprocessing one has to study the *Poincaré return map*, or *monodromy* Δ around the elementary polycycle which we denote by γ. This map is defined exactly as in the case of a (nonsingular) periodic orbit: choose a small segment Σ transversal to an arc of the polycycle and let Δ be the map of this segment into itself along solutions of the equation provided that they never leave the small neighborhood of γ. Unlike the case of a periodic orbit, the map Δ is not analytic at the point $p = \gamma \cap \Sigma$ and is usually defined only from one side of p. The problem of investigating analytic properties of the map Δ is very complicated (requires hundreds of pages in both known versions), and as a result one comes to the conclusion that fixed points of the map $\Delta\colon \Sigma \to \Sigma$ cannot accumulate to p.

The analysis carried in [I1] is based on using functional cochains and superexact asymptotic expansions. The main tool in [É] is the theory of resurgent functions and resummability. Currently neither technique allows for any generalization of the proof to the case of vector fields depending on parameters.

§2. Analytic families of vector fields and cyclicity of polycycles

2.1. Universal polynomial family. The natural way to look at *all* polynomial equations (1) at once is to consider them as an *analytic family of line fields on the sphere*; the parameters of this family are coefficients of the polynomials P, Q. The parameter space thus introduced is the Euclidean space with the deleted origin (since the case $P = Q = 0$ does not correspond to a line field), but, in fact, the simultaneous multiplication of both P and Q by a common factor $\lambda \in \mathbb{R}$, $\lambda \neq 0$, does not change the line field. Thus the existential Hilbert problem becomes a particular case of the following global conjecture.

GLOBAL FINITENESS CONJECTURE. *For any analytic family of line fields on the two-dimensional sphere $\mathbb{S}^2$ with a compact finite-dimensional parameter space B, the number of limit cycles for all fields is uniformly bounded over all parameter values.*

In what follows we refer to the family obtained by the compactification of (1) as the *universal polynomial family* of degree n.

2.2. Cyclicity. Suppose that the global finiteness conjecture is wrong for a family of line fields $\alpha(\varepsilon)$, $\varepsilon \in B$. Then, since it is known that each individual field from the family has only a finite number of limit cycles (see §1.2), there must exist an infinite sequence of parameter values $\varepsilon_k \in B$, $k = 1, 2, \dots$, such that the corresponding line fields have monotonously increasing numbers of limit cycles. Since the base B is compact, without loss of generality we may assume that the sequence ε_k converges to a certain point $\varepsilon_* \in B$. Next, since the sphere is compact, the cycles of $\alpha(\varepsilon_k)$ should accumulate to a certain compact subset of the sphere, invariant by the field $\alpha(\varepsilon_*)$. Such sets were introduced by Françoise and Pugh [FP] under the name of *limit periodic sets*. The precise definition looks as follows.

DEFINITION. A subset $\gamma_* \subset \mathbb{S}^2$ is a limit periodic set for the family of line fields $\alpha(\varepsilon)$, $\varepsilon \in B$, at a point ε_* if there exists a sequence of points $\varepsilon_k \to \varepsilon_*$ such that the corresponding line fields $\alpha_k = \alpha(\varepsilon_k)$ have limit cycles γ_k which converge to γ_* in the sense of Hausdorff distance.

The structure of limit periodic sets admits a simple description.

PROPOSITION [FP]. *A limit periodic set either is a polycycle or contains an arc of nonisolated singularities of the field* $\alpha(\varepsilon_*)$.

The following definition introduces an important characteristics of a limit periodic set occurring in a certain family of (line, vector) fields.

DEFINITION. We say that a limit periodic set γ_* occurring in a family of line fields on the sphere for a certain parameter value ε_*, *has cyclicity* $\leqslant \mu$ if there exist neighborhoods U, V, $\mathbb{S}^2 \supseteq U \supset \gamma$, $B \supseteq V \ni \varepsilon_*$ such that for any $\varepsilon \in V$ the field $\alpha(\varepsilon)$ has at most μ limit cycles in U. In other words, the polycycle generates at most μ limit cycles after bifurcation in the family $\alpha(\cdot)$.

The minimal μ (if it exists) is called the *cyclicity* of the limit periodic set, otherwise the cyclicity is said to be infinite.

REMARK. The notions of limit periodic set and cyclicity are defined for families rather then for individual line fields. Still, if we have a polycycle for an analytic individual line field α_*, then cyclicity of such polycycle can sometimes be estimated from above for *any analytic family* $\alpha(\varepsilon)$ unfolding the field α_*. In this case the term *absolute cyclicity* is used.

Returning back to the global finiteness conjecture, we see that the assumption on the unboundedness of the number of limit cycles would lead to a contradiction if the following assertion were proved.

FINITE CYCLICITY CONJECTURE (Roussarie). *Any limit periodic set occurring in an analytic family of line fields on the sphere, has finite cyclicity in this family.*

The above parameter localization procedure can be formulated in the form of an implication.

THEOREM ([R2], see also [Aea]).

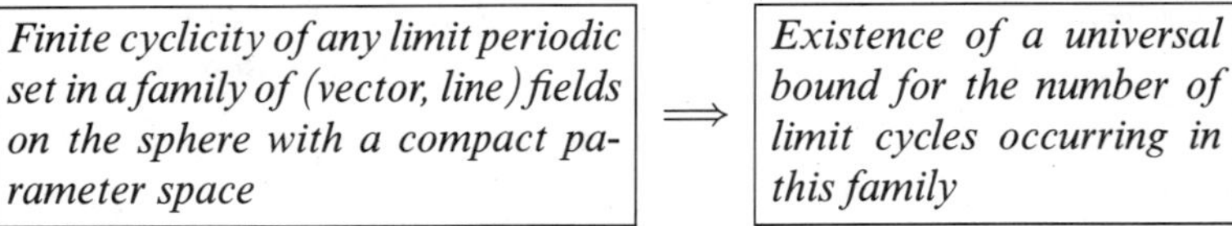

2.3. Examples of cyclicity estimates. There are many different examples of limit periodic sets whose cyclicity is known to be finite or even explicitly computed. For example, a hyperbolic periodic orbit (a trivial polycycle) has absolute cyclicity 1. The simplest nontrivial example of a polycycle (with at least one arc and at least one singularity) is a separatrix loop of a hyperbolic saddle: continuations of stable and unstable invariant curves form a closed loop. If the divergence of the vector field at the saddle point is different from zero, then the loop has absolute cyclicity 1 (Andronov and Leontovich). In those (and in many other cases) the results are valid even for smooth families, and they were in fact established in the classical bifurcation theory. In §3 we give a brief summary of the cyclicity results given *gratis* by that theory and its recent developments.

On the other hand, there are some results that establish finite (not absolute) cyclicity for polycycles occurring in analytic families. In these results no nondegeneracy-type assumptions are made, so they may be applied to polycycles with the identical monodromy.

The simplest case of a periodic orbit was studied in [FP]. The first really nontrivial case of a separatrix loop of a nondegenerate saddle was analyzed by Roussarie [R1] together with the closely related case of cuspidal singular points.

But historically the first case of effective computation of cyclicity that is not absolute, is due to Bautin [B]. Bautin studied bifurcations of limit cycles from an *elliptic singular point*, i.e., the point at which the eigenvalues are complex conjugate, $\sigma \pm i\omega$, $\omega \neq 0$, in the universal family of quadratic vector fields.

BAUTIN THEOREM. *An elliptic singular point has cyclicity* $\leqslant 3$ *in the universal polynomial family of degree* 2.

The original proof is very complicated and involves heavy computations. In PAPER 5 a new proof of this result is suggested. This proof follows to a certain extent the original Bautin's proof, but on the final steps instead of almost incomprehensible manipulations with integrals, simple geometric arguments based on a hidden $\mathbb{Z}_3$-symmetry allow us to arrive to the conclusion.

REMARK. Recently yet another proof of this result was suggested by H. Zołądek [Z]. That proof is based on some rotational symmetry of the problem.

2.4. Quadratic vector fields. Recently an intense attack was launched by Dumortier, Roussarie, and Rousseau [DRR] to solve the existential Hilbert problem for the family of quadratic vector fields, that is to prove that

$$H(2) < \infty.$$

In [DRR], a list of 121 polycycles and degenerate limit periodic sets that may occur after compactification was composed. Out of this list, a substantial number of cases has been analyzed. The principal difficulties in investigating all these cases occur when a polycycle is identical (which means that from one side its neighborhood is filled with closed periodic orbits), or when a limit periodic set with nonisolated singularities occurs. The same applies to investigation of other universal polynomial families, and it is not very likely that the general existential Hilbert problem could be solved without involving some essentially new ideas.

Still there exists a natural way to change settings in the existential Hilbert problem so that those pathologies would be ruled out. This reformulation is known as the *Hilbert–Arnold problem*.

§3. Generic smooth families of vector fields and the Hilbert–Arnold problem

3.1. Generic smooth families of vector fields on the sphere. One might ask why Hilbert had chosen polynomial families as the subject for investigation concerning limit cycles. Perhaps, the reason was that the universal polynomial family is the only constructive family of line fields on the plane which extends to a family of line fields with singularities on the sphere. Towards the second half of this century, the ideology changed and typical smooth objects (vector fields and their families) became a legal subject of consideration.

If we replace in the formulation of the existential Hilbert problem the universal family (1) by a *generic* family of vector fields on the 2-sphere $\mathbb{S}^2$, then at least some of the difficulties mentioned at the end of the previous section disappear.

Recall that for any subset $B \subseteq \mathbb{R}^n$ a function $\varphi: B \to \mathbb{R}$ is said to be smooth if it admits a C^∞-smooth extension to some open neighborhood U of B.

Let $B \subset \mathbb{R}^n$ be a finite-dimensional compact. Then the space of C^∞-smooth families of vector fields $v(\cdot): \mathbb{S}^2 \times B \to T\mathbb{S}^2$ admits the natural topology induced by the metric $d(v_1, v_2) = \sum_k 2^{-k} \|v_1 - v_2\|_k$, where

$$\|v\|_k = \max_{\substack{x \in \mathbb{S}^2,\ \varepsilon \in B,\\ |\alpha|+|\beta|=k}} |D_x^\alpha D_\varepsilon^\beta v(x, \varepsilon)|. \tag{3}$$

Definition. We say that a generic n-parameter family of vector fields on the sphere possesses a certain property $\mathcal{P}$ if this property holds for a residual subset of the total space of all n-parameter families.

The property is said to hold for a generic family (without indicating the number of parameters n explicitly) if it holds for any generic n-parameter family, whatever a finite number n is.

Proposition. *In a generic family only isolated singularities occur, and their multiplicity is bounded over any compact subset in the space of parameters.*

Corollary. *Any limit periodic set occurring in a generic family is a polycycle.*

In general, one may expect that generic families of vector fields in many respects resemble analytic families, but this is an observation rather than a formal claim. Still there are many reasons to believe that the following basic conjecture holds.

Hilbert–Arnold problem. *Prove that in a generic family of vector fields on the sphere $\mathbb{S}^2$ with a compact base B, the number of limit cycles is uniformly bounded.*

The Hilbert–Arnold problem was implicitly formulated in [AI] as a special case of a conjecture stating that for generic n-parameter families of vector fields on the sphere, only a finite number of local bifurcation diagrams can be realized. In fact, the full conjecture as it was formulated in [AI], is wrong: the counterexample is given in Paper 4 below. However, this example does not disprove the Hilbert–Arnold conjecture.

For small $n = 2$ and 3, that is, for few-parameter families of vector fields, the Hilbert–Arnold problem admits investigation by case studies, see §3.2, §3.3 for more information.

Remark. The topology generated by the family of norms $\|\cdot\|_k$, is not the only possible: for example, one may take into consideration only derivatives in the x-variables and disregard those in the parameters ε. Other modifications are also available. In fact, to supply the word *generic* in the formulation of Hilbert–Arnold problem with a precise meaning, is a part of the solution of the problem.

The same localization technique which was used when reducing investigation of analytic families to the study of small neighborhoods of limit periodic sets, works also for smooth families. The difference with the analytic case is in the presence of the natural index n, the number of parameters. It turns out that at least for small n, cyclicity of polycycles occurring in generic n-parameter families, admits an upper estimate in terms of n.

EXAMPLE. In a generic n-parameter family, the maximal multiplicity of a limit cycle does not exceed $n+1$: for example, for a structurally stable vector field only hyperbolic limit cycles occur, in codimension 1 semistable limit cycles of multiplicity 2 may appear, etc. Thus, the cyclicity of a trivial polycycle (without singularities) in a generic n-parameter family, does not exceed $n+1$. For some reasons we exclude trivial (poly)cycles without veritces from consideration when giving the following definition.

DEFINITION. The *bifurcation number* $B(n)$ is the maximal cyclicity of *nontrivial* polycycles occurring in generic n-parameter families.

The definition of the number $B(n)$ does not depend on the choice of the base of the family, but only on its dimension n.

LOCAL HILBERT–ARNOLD PROBLEM (for n-parameter families). *Prove that for any finite n, the number $B(n)$ is finite.*

Solution of this problem would imply solution of the global Hilbert–Arnold problem by virtue of the same compactness arguments as in §2.2.

CONSTRUCTIVE HILBERT–ARNOLD PROBLEM. *Compute explicitly or give an explicit upper estimate for the bifurcation number $B(n)$.*

3.2. Hilbert–Arnold problem for families with small number of parameters. Bifurcations of limit cycles in generic few-parameter families of vector fields were the subject of studies since late thirties. The accumulated information about the simplest bifurcations in generic 1-parameter families can be compressed into a single equality.

THEOREM B_1 (Andronov–Leontovich, 1930s; Hopf, 1940s).

$$B(1) = 1.$$

This result summarizes results of the investigation of three classical bifurcations, separatrix loop of a saddle with nonzero divergence, saddle-node loop, and an elliptic point.

The next number in the series, $B(2)$, has a longer history and the corresponding equality summarizes numerous results.

THEOREM B_2 (Takens, Bogdanov, Leontovich–Cherkas, Mourtada, Grozovskiĭ, early 1970s–1993).

$$B(2) = 2.$$

The proof of this result involves consideration of bifurcations of eight different polycycles which may occur in generic 2-parameter families. Out of this list, four bifurcations were already studied by different authors before the problem of computing $B(2)$ was explicitly formulated, and the cyclicity found to be at most 2. Out of the four remaining cases, three are very simple and correspond to cyclicity 1, and the last case, the *half-apple*, was studied recently by a graduate student T. Grozovskiĭ. It consists of a polycycle with 2 singularities, a nondegenerate saddle, and a saddle-node, and its cyclicity does not exceed 2.

The list of polycycles occurring in dimensions 2 and 3 is given in PAPER 4. A complete investigation of all polycycles occurring in codimension 3, would yield an upper estimate for the bifurcation number $B(3)$, though several cases from this list seem to be very hard (for example, a loop carrying a degenerate cuspidal point). At the same time it is clear that an attempt to obtain a sharp upper estimate for $B(4)$ by a similar case study is hopeless.

3.3. Lips and other ensembles. Investigation of generic 3-parameter families revealed some very simple but surprising facts about bifurcations of limit cycles in such families.

The definition of bifurcation numbers starts from the notion of a polycycle. There might be an alternative approach. For any smooth family $v(x, \varepsilon)$, $x \in \mathbb{S}^2$, $\varepsilon \in B \Subset \mathbb{R}^n$ of vector fields on the 2-sphere one can define an integer-valued counting function $\mathfrak{c}(\cdot) = \mathfrak{c}_v(\cdot)\colon B \to \mathbb{Z}_+$,

$$\mathfrak{c}(\varepsilon) = \text{the number of limit cycles of the field } v(\cdot, \varepsilon) \text{ on } \mathbb{S}^2.$$

This function is defined for all values of the parameter $\varepsilon \in B$ and clearly even for generic families there cannot be any natural bound for $\mathfrak{c}$ in terms of $n = \dim B$, though for such families this function presumably has finite values.

However, if we introduce the *oscillation* function $\mathfrak{o}(\cdot) = \mathfrak{o}_v(\cdot)$ constructed for the family v as

$$\mathfrak{o}(\varepsilon) = \operatorname{osc} \mathfrak{c}(\varepsilon) = \lim_{r \to 0+} \left(\sup_{|\varepsilon' - \varepsilon| < r} \mathfrak{c}(\varepsilon') - \inf_{|\varepsilon' - \varepsilon| < r} \mathfrak{c}(\varepsilon') \right) \geqslant 0,$$

then it turns out that the function $\mathfrak{o}(\cdot)$ admits an upper estimate for all generic 1- and 2-parameter families. Note that the oscillation function is zero for a generic point $\varepsilon \in B$, since generic vector fields are structurally stable.

THEOREM. *For a generic n-parameter family of smooth vector fields on the sphere, the oscillation function does not exceed* 2 *if* $n = 1$, *and is everywhere less or equal to* 3 *for* $n = 2$.

The proof of this theorem is also obtained by studying separate cases: now one has to take into account the possibility of simultaneous formation of several polycycles and multiple limit cycles. It is clear that if two polycycles are disjoint, then their simultaneous occurrence is an event of the codimension equal to the sum of degeneracy codimensions of each polycycle independently, and the oscillation function is equal to the sum of the terms corresponding to bifurcations of each polycycle. In generic 1-parameter families this is the only possibility, and the upper bound equal to 2 appears because a semistable (double) limit cycle may disappear or generate two close hyperbolic limit cycles.

However, in codimension 2 *ensembles* of polycycles appear, that is, graphs formed by several polycycles with a common singularity or an arc. Such ensembles may have degeneracy codimension *strictly smaller* than the sum of codimensions of polycycles constituting them. On the other hand, different polycycles constituting an ensemble, sometimes cannot *simultaneously* generate the maximal number of limit cycles, so that the total number of limit cycles born from an ensemble is only a subadditive function.

EXAMPLE. Consider an ensemble composed by two separatrix loops of the same hyperbolic saddle point with a nonzero divergence. Then this ensemble is a union of *three* polycycles, two simple loops and the eight-shaped contour. Each one of them has cyclicity 1, but their union has cyclicity 2 in a generic 2-parameter family, since the simultaneous generation of limit cycles by all three polycycles is impossible (D. Seregin, E. Malgina, in preparation).

In codimension 2, nine different types of ensembles can occur (see PAPER 4), and their investigation yields the above theorem. One can express the upper bound for the oscillation function $\mathfrak{o}(\cdot)$ in terms of bifurcation numbers similar to the numbers $B(n)$.

DEFINITION. The global bifurcation number $C(n)$ is the upper bound for the number of limit cycles that can be born from *all polycycles* which may simultaneously occur in a generic n-parameter family of vector fields on the sphere.

After introducing this number, an evident estimate holds for the oscillation function,

$$\mathfrak{o}_v(\cdot) \leqslant C(n) \qquad \text{for a generic } n\text{-parameter family } v,$$

and the above theorem can be formulated as follows:

$$C(1) = 2, \qquad C(2) = 3.$$

One might expect that for generic 3-parameter families there also should be such a universal bound, and a problem for estimating the number $C(3)$ should be the next in a row. But the counterexample below shows that it is not the case: although for any generic 3-parameter family the function $\mathfrak{o}(\cdot)$ is locally bounded, a common upper bound for all generic 3-parameter families, does not exist, contrary to the cases $n = 1, 2$. This is explained by a simultaneous occurrence of a continuum of polycycles for isolated values of parameters in generic 3-parameter families of vector fields on the sphere.

The simplest example of a generic 3-parameter family of vector fields on the 2-sphere with a continuum of coexisting polycycles was found by A. Kotova and was called *lips*. Consider two saddle-nodes $S_\pm$ of multiplicity 2 (topologically equivalent to the vector field $x^2\,\partial/\partial x \pm y\,\partial/\partial y$); their simultaneous occurrence is an event of codimension 2. Suppose that the (uniquely defined) trajectory emanating from S_- and the (uniquely defined) trajectory which tends to S_+ are continuations of each other, together forming a heteroclinic orbit η. This means an additional degeneracy of the field, thus the whole picture may occur for isolated values of parameters in a generic 3-parameter family.

Each saddle-node has a parabolic sector entirely filled by trajectories tending to (from) the singular point S_- (resp., S_+). Without increasing the codimension, we may assume that *there exists a trajectory χ_0 passing through interiors of both saddle-nodes*. If there is at least one such trajectory, then all sufficiently close trajectories χ_s, $s \in (\mathbb{R}^1, 0)$, also are biasymptotic to both saddle-nodes, see Figure 1. Finally, we obtain a continuum of polycycles

$$S_+ \to \chi_s \to S_- \to \eta \to S_+, \qquad s \in (\mathbb{R}^1, 0).$$

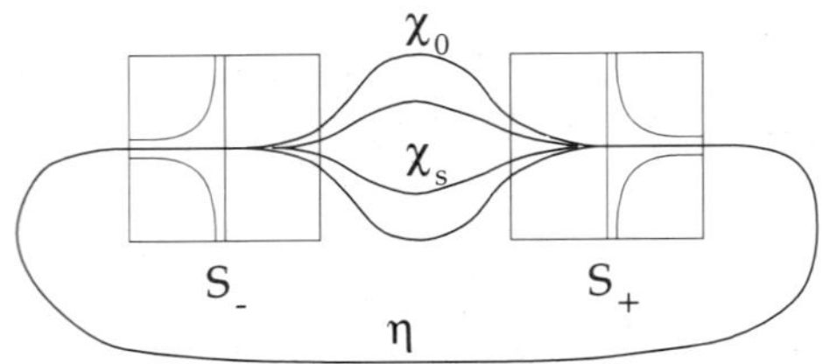

FIGURE 1. Lips.

Bifurcations of lips in generic 3-parameter families are studied in PAPER 4; as a byproduct of this investigation, the *generalized Legendre duality* was constructed.

KOTOVA THEOREM. *For any natural N there exists a 3-parameter family $v(\cdot, \varepsilon)$ of vector fields on the sphere, such that the oscillation function $\mathfrak{o}(\varepsilon)$ constructed for this family, takes a value $> N$ at a certain point, and the same is true for all sufficiently C^k-close 3-parameter families of vector fields, if k is large enough.*

In other words,

$$C(3) = +\infty.$$

COROLLARY (Kotova, Stanzo). *For generic 3-parameter families of smooth vector fields on the sphere, there exists an infinite number of pairwise locally topologically nonequivalent bifurcation diagrams.*

The bifurcation diagram for the lips is constructed in PAPER 4 by V. Stanzo.

The lips are not the only possible "pathology" which may occur in generic 3-parameter families. For example, somewhere "between" the two saddle-nodes, an arbitrary number of nondegenerate saddles may occur without rising the codimension of the whole picture. Those saddles may participate in the creation of polycycles with three singular points, and it is clear that any finite number of such polycycles may coexist. For further details see PAPER 4.

3.4. Elementary polycycles and their finite cyclicity in generic families. The local Hilbert–Arnold problem was solved under an additional assumption that all singular points occurring on a polycycle, are elementary (see the definition in 1.3).

DEFINITION. The *elementary bifurcation number* $E(n)$ is the maximal cyclicity of a nontrivial *elementary* polycycle occurring in a (smooth) generic n-parameter family.

The only nonelementary polycycle that can occur in a 2-generic family, is a cuspidal point whose bifurcations were studied by Bogdanov and Takens, and whose cyclicity was found to be 1. Thus, Theorems B_i, $i = 1, 2$, imply that

$$E(1) = 1, \qquad E(2) = 2.$$

However, the nature of the function $n \mapsto E(n)$ is now understood much better than that of $B(\cdot)$.

THEOREM (Ilyashenko and Yakovenko, 1992). *For any n the elementary bifurcation number $E(n)$ is finite. Moreover, the function $n \mapsto E(n)$ admits a primitive recursive majorant.*

COROLLARY. *The global Hilbert–Arnold problem has a positive solution for families of vector fields in which only elementary singularities occur: any generic family of vector fields on the sphere with a compact finite-dimensional base of parameters and with elementary singular points only, has a uniformly bounded number of limit cycles.*

REMARK. A primitive recursive function is an integer function of a natural argument n, which admits an algorithmically effective computation for any specified value of the argument. The formal definition is given in PAPER 1. In fact, it is very likely that this majorant is elementary, that is, an explicit expression could be written for it.

This theorem was announced in [IY1]. A complete proof of this result is given in PAPER 1. It consists of the four principal steps:

1. C^k*-smooth normalization* of the family near each elementary singularity. The main tool here is provided by the classification theorems from [IY2]. The normal forms are polynomial and integrable. We perform an explicit integration of normal forms in the class of Pfaffian functions introduced by A. Khovanskiĭ [K] and show that the correspondence maps near each singular point in the normalized coordinates can be expressed through elementary transcendental functions which satisfy some algebraic Pfaffian equations. The degree and the total number of these equations can be estimated in terms of n.

2. *"Algebraization"* of the system of equations obtained at the previous step: the reduction procedure suggested in [K] allows us to eliminate transcendental functions from the equations determining fixed points of the monodromy map. After this elimination there appears a system of equations having the form of a *chain map*, a composition of a polynomial map and a jet extension of a generic smooth map.

3. *Gabrielov-type finiteness conditions* are established for a smooth map $F: \mathbb{R}^k \to \mathbb{R}^k$ to have a uniformly bounded number of regular preimages $\#F^{-1}(y)$ when the point y varies over a compact subset of $\mathbb{R}^k$. These conditions are automatically satisfied if a map F is real analytic. We introduce a topological complexity characteristics, the *contiguity number*, in terms of which an upper estimate for the number of preimages can be expressed.

4. *Thom–Boardmann-type construction* allows us to prove that the above finiteness conditions can be expressed in terms of transversality of the jet extension of F to some semialgebraic subsets of the jet space. Moreover, this construction can be generalized to cover *chain maps* of the form $P \circ (j^\ell F)$, where P is a polynomial, and $j^\ell F$ is the ℓ-jet extension of a generic smooth map. This is exactly the class of maps which appear after the Khovanskiĭ elimination procedure (step 2 above). The contiguity number of a chain map is expressed through the integer-valued data (degree of the polynomial P, order of the jet ℓ and dimension of the domain and target spaces).

REMARK. Although the upper estimate obtained in the above theorem is *not family-dependent* (common for all n-parameter families), the established cyclicity is *not absolute*: we require that after the jet extension the family considered as a map

$v: \mathbb{S}^2 \times (\mathbb{R}^n, 0) \to T\mathbb{S}^2$ to be transversal to some algebraic stratified subset in an appropriate jet space.

§4. Parameter desingularization

In 1.2 we explained the role of elementary singularities in proving the nonaccumulation theorem. If something similar could be proved for families rather than for individual vector (or line) fields, then the theorem on finite cyclicity of generic elementary polycycles would imply the Hilbert–Arnold problem in full generality. Unfortunately, such a straightforward approach does not yield immediate results.

4.1. Two approaches to parameter desingularization. From the point of view of analytical geometry, locally a family of (vector, line) fields may be considered as a single (vector, line) field on the *total space of the bundle* $\pi: (\mathbb{R}^2, 0) \times (\mathbb{R}^n, 0) \to (\mathbb{R}^n, 0)$, $\pi(x, y; \varepsilon) = \varepsilon$ which is tangent to the fibers of this bundle. In this section we consider only the case of analytic families.

Isolated singular points occurring in a family of vector fields in such settings correspond to an analytic at most n-dimensional set which intersects the fibers of the bundle π by discrete subsets.

One approach to parameter desingularization, suggested by Z. Denkowska and R. Roussarie [DR], was applied to investigation of families of vector fields that have nonelementary singularity occurring for isolated values of the parameter ε (a typical example is an unfolding of a cuspidal point), which we assumed to be the origin $x = y = 0$, $\varepsilon = 0$. The idea is to blow-up the origin *in the total space of the bundle*, deleting it and pasting an $(n+1)$-dimensional sphere instead. One of the possible ways to do this is to consider a quasihomogeneous mapping

$$\begin{gathered}\sigma: \mathbb{S}^{n+1} \times (\mathbb{R}^1_+, 0) \to (\mathbb{R}^2, 0) \times (\mathbb{R}^n, 0), \\ (\overline{x}, \overline{y}, \overline{\varepsilon}_1, \dots, \overline{\varepsilon}_n; r) \overset{\sigma}{\mapsto} (\overline{x}\, r^{\nu_1}, \overline{y}\, r^{\nu_2}, \overline{\varepsilon}_1\, r^{\mu_1}, \dots, \overline{\varepsilon}_n\, r^{\mu_n})\end{gathered} \tag{4}$$

with the weights ν_i, μ_j chosen in an appropriate way.

After such a procedure the total space becomes not a bundle but rather a singular "foliation", whose fibers have different dimensions; the fiber over a nonzero value of ε is still two-dimensional, while the fiber over $\varepsilon = 0$ is diffeomorphic to the sphere $\mathbb{S}^{n+1}$. The investigation of the pullback of the original family to a neighborhood of the pasted sphere proceeds further in different charts which play different role.

We do not intend to expose the procedure in full detail, referring the reader to the papers [DR], [R3].

The other approach suggested by S. Trifonov is explained in PAPER 2. The main idea is to blow up not just one point in the total space, but rather the entire *singular locus*, the set of all singular points for all fields in the family. A brief explanation of this approach is given in the next section.

4.2. Desingularization after Trifonov. Suppose that in an analytic family of vector fields a singular point depends analytically on the parameters (like in the case when the singularity is nondegenerate). Then without loss of generality one may assume that locally the singularity remains at the origin $x = y = 0$ for all values of the parameters $\varepsilon \in (\mathbb{R}^n, 0)$. Thus we consider a line field on the total space $(\mathbb{R}^2, 0) \times (\mathbb{R}^n, 0)$ parallel to the vertical direction (the second factor), with singular points on the horizontal plane $\{x = y = 0\} \subset (\mathbb{R}^{n+2}, 0)$.

Then the mapping

$$\sigma : \mathbb{S}^1 \times (\mathbb{R}^1_+, 0) \times (\mathbb{R}^n, 0) \to (\mathbb{R}^2, 0) \times (\mathbb{R}^n, 0),$$
$$\sigma(\cos\theta, \sin\theta; \varepsilon) = (r\cos\theta, r\sin\theta; \varepsilon)$$

which in fact is a trivial suspension of the standard polar blow-up, yields a simultaneous blow-up of the singularity at $x = y = 0$ for all vector fields from the family. In a similar way the algebraic σ-process can be applied to all vector fields of the complexified family: as a result, we obtain a new total space which is obtained from the original one by deleting the plane $x = y = 0$ and pasting in the cylinder $\mathbb{C}P^1 \times (\mathbb{C}^n, 0)$.

The real problems begin when the singularity looses its analytic dependence on the parameters. In this case the two basic additional arguments are used.

First, one should consider not all singularities, but only the *essential* ones. To explain the term, note that in analytic families for some exceptional values of the parameters the singular locus may well be nonisolated as a subset of the corresponding fiber. Since the fiber is two-dimensional (and this property will be maintained when iterating the construction, in contrast to the approach of Denkowska and Roussarie), the line field may be extended to the analytic curve of singular points by cancelling nontrivial common factors in the right-hand side of the differential equations. After such cancellation the curve carries only discrete singular points of the extension of the line field. They are called *essential singularities*.

The second idea is to make singular reparametrizations of the family, or what is called in Paper 2 *unfoldings of the base*. The simplest (and in a sense the main one) case of such unfolding is the spherical blow-up of $(\mathbb{R}^n, 0)$ at the origin, as in (4).

Trifonov proved that after an appropriate blow-up of the parameter space, all essential singular points of the pullback of the original family can be placed on a finite number of analytic sections of the foliation. However, the latter looses its locally trivial topological nature: after performing all required steps the fibers become not globally diffeomorphic, though still two-dimensional and locally diffeomorphic.

The final form of the principal result of Paper 2 can be formulated as follows.

Definitions. 1. An *analytic family of two-dimensional surfaces* is a triple $M \xrightarrow{\pi} B$, where M and B are analytic manifolds, $\dim M = \dim B + 2$ and the map π has the constant rank equal to $\dim B$.

2. Let M be a manifold covered by an atlas of charts $\{U_\alpha\}$ and in each chart U_α a vector field v_α is defined. We say that this family of vector fields defines a *line field with singularities on* M, if on each nonempty intersection $U_\alpha \cap U_\beta$ there exist a nonvanishing smooth function $\varphi_{\alpha\beta}$ such that $v_\alpha = \varphi_{\alpha\beta} v_\beta$.

3. An *analytic family of line fields with singularities* is a line field with singularities on M that tangent to all fibers $\pi^{-1}(\varepsilon)$, $\varepsilon \in B$. The family is *proper* if the restriction of π on the singular locus of the family is proper.

Definition. A *simple blow-up* of a two-dimensional surface M is a map $\sigma : \widetilde{M} \to M$ which is biholomorphic except for the preimage $\Sigma \subset \widetilde{M}$ of just one point $p \in M$; the exceptional set Σ is biholomorphic to the projective line $\mathbb{C}P^1$ and the germ $\sigma\colon (\widetilde{M}, \Sigma) \to (M, p)$ is left-right equivalent to the standard blow-up described above.

The last condition means that in a small neighborhood of Σ two local charts can be introduced, (x, u) and (y, v), with the transition functions between them $y = xu$, $v = 1/u$, such that in these charts the map σ has the form

$$(x, u) \mapsto (x, ux), \qquad (y, v) \mapsto (yv, y).$$

A *resolution* of the surface M is an analytic map $\theta : \widetilde{M} \to M$ which is the composition of a finite number of simple blow-ups.

TRIFONOV THEOREM. *For a proper family α of line fields on an analytic family of two-dimensional surfaces, there exists another family of surfaces $\widetilde{M} \xrightarrow{\tilde{\pi}} \widetilde{B}$ and a pair of analytic maps $H\colon \widetilde{M} \to M$, $\rho\colon \widetilde{B} \to B$ such that*:

(1) $\pi \circ H = \rho \circ \tilde{\pi}$;
(2) *the map H restricted on any two-dimensional fiber $\tilde{\pi}^{-1}(\tilde{\varepsilon})$, $\tilde{\varepsilon} \in \widetilde{B}$, is a resolution (a finite composition of simple blowing-ups)*;
(3) *all essential singularities of the pullback family $H^*\alpha$ (defined in the natural way) are elementary.*

REMARKS. 1. The assertion of the theorem holds for both real and complex analytic categories.

2. The new base $\widetilde{B}$ of the new family may be not connected.

4.3. Singular perturbations. Trifonov's theorem claims that after blowing up in a family of analytic line fields, a new family can be constructed in such a way that all *essential* singularities of this new family are elementary. However, this result says nothing about singularities that are not essential. They correspond to the dynamical phenomenon called *singular perturbation* in the theory of differential equations: one needs to study families of, say, vector fields on the plane, which, for certain values of parameters, exhibit a whole curve of nonisolated singularities. A comprehensive discussion on this subject can be found in PAPER 2. The appearance of singular perturbations after parameter desingularization constitutes now the main gap between the theorem on finite cyclicity of elementary polycycles together with the parameter desingularization theorem on one side and the general Hilbert–Arnold problem on the other.

4.4. Complexity of desingularization. When proving the Trifonov theorem on parameter desingularization, one needs effective means to estimate the number of blow-ups necessary to resolve completely an isolated singularity.

Besides, there is a general question: given a germ of vector field at a singular point, how many terms of its Taylor expansion determine completely the topology of the phase portrait of the field? The constructive procedure that answers that question is based on the desingularization technique. The topology of an elementary singular point is determined (in the degenerate case) by the principal term of the restriction of the vector field to the center manifold (curve). Knowing the sequence of steps resolving a nonelementary singularity into elementary ones, one may "glue up" the local phase portraits into the global phase portrait of the original field, provided that the trajectories are not spiralling around the singularity. The latter condition may be guaranteed by assuming that there exists at least one *characteristic orbit*, a trajectory tending to the singularity with a certain limit slope. The procedure of blowing up involves only lower degree Taylor terms, so to estimate the order of the Taylor polynomial which would determine completely the topology of the phase portrait, one needs to control the number of blow-ups.

These matters constitute the subject of PAPER 3 from the present volume. To formulate the main result, recall the principal definition.

DEFINITION. Let $v(x, y) = v_1(x, y)\,\partial/\partial x + v_2(x, y)\,\partial/\partial y$ be the germ of a smooth vector field at the origin. The *multiplicity* of the germ v is the dimension of the local algebra,

$$\mu_0(v) = \dim_{\mathbb{R}} Q_v, \qquad Q_v = \mathbb{R}[[x, y]]/\langle \hat{v}_1, \hat{v}_2\rangle,$$

where $\mathbb{R}[[x, y]]$ is the ring of all formal power series over $\mathbb{R}$ in two variables (x, y), $\hat{v}_i$ are the Taylor series of the coordinate functions v_i, $i = 1, 2$, and $\langle \hat{v}_1, \hat{v}_2\rangle$ is the ideal generated by the two series.

DEFINITION. The jet of a vector field at the singular point is called *topologically sufficient*, if any two vector fields with that same jet are topologically equivalent in a certain small neighborhood of the singular point.

KLEBAN THEOREM. *The order of a topologically sufficient jet for a vector field with a singular point of multiplicity $\mu < \infty$ having a characteristic orbit, does not exceed $2\mu + 2$.*

The existence or absence of the characteristic orbit is a fact that can be also established by analyzing the $(2\mu + 2)$-jet of the vector field.

This theorem gives a quantitative version of the general result by Dumortier [D1]. The proof is based on controlling the number of blow-ups sufficient to resolve a degenerate isolated singularity of multiplicity μ into elementary singularities. Such an approach was suggested by Van den Essen in [VdE].

LEMMA (see PAPER 2). 1. *If the linear part of a vector field v at the singular point is zero, then the sum of multiplicities of all singular points appearing after one blow-up, is strictly less than the multiplicity μ of the original singularity.*

2. *If the linear part is nonzero (and the singular point is still nonelementary), then after no more than $[\frac{\mu}{2}] + 2$ blow-ups it can be resolved into elementary singularities.*

Another result of similar nature, estimating the number of *quasihomogeneous blow-ups* in terms of the multiplicity of the germ of a vector field, was announced recently by M. Pelletier [P].

4.5. Concluding remarks. Concluding this short survey, we would like to return to the general formulation of the Hilbert–Arnold problem. In order to achieve further progress in that direction after proving the theorem on finite cyclicity of elementary polycycles in generic families and the theorem on parameter desingularization, one needs to study bifurcation of limit cycles in singular perturbations. A particular case of a singular perturbation is the family of vector fields

$$\begin{cases} \dot{x} = h(x, y) f_0(x, y) + \sum_{k=1}^{n} \varepsilon_k f_k(x, y, \varepsilon), \\ \dot{y} = h(x, y) g_0(x, y) + \sum_{k=1}^{n} \varepsilon_k g_k(x, y, \varepsilon), \end{cases} \qquad \begin{matrix} (x, y) \in \mathbb{C}^2, \\ \varepsilon = (\varepsilon_1, \dots, \varepsilon_n) \in \mathbb{C}^n, \end{matrix}$$

such that the vector field $f_0\, \partial/\partial x + g_0\, \partial/\partial y$ has only elementary singularities (for example, has no singularities at all). In a more general context the definition of a singular perturbation is given in PAPER 2.

The equation of Van der Pol is a specific example of singularly perturbed vector field, and in that example limit cycles are known to be born.

To specify the problem, consider a *simple cusp*, the cuspidal singular point of maximal nondegeneracy: its linear part $y\, \partial/\partial x$ is nilpotent and nonlinear terms are generic. It is known that after three blow-up steps such point is resolved into elementary ones. Thus applying the technique of parameter desingularization, one may try to prove an analog of the result obtained in PAPER 1. Denote by $EC(n)$ the maximal cyclicity of a polycycle carrying only elementary singularities and simple cusps and occurring in generic n-parameter families.

CONJECTURE. *The number $EC(n)$ is finite for any $n < +\infty$ and the function $n \mapsto EC(n)$ admits a primitive recursive majorant.*

Proving this conjecture would be a first step towards obtaining the complete solution of the Hilbert–Arnold problem.

References

[Aea] V. I. Arnold, M. I. Vishik, Yu. S. Ilyashenko, A. S. Kalashnikov, V. A. Kondratyev, S. N. Kruzhkov, E. M. Landis, V. M. Millionshchikov, O. A. Oleinik, A. F. Filippov, and M. A. Shubin, *Some unsolved problems from the theory of differential equations ant mathematical physics*, problems of Yu. S. Ilyashenko, Uspekhi Math. Nauk **44** (1989), no. 4, 191–202; English transl. in Russian Math. Surveys **44** (1989).

[AI] V. I. Arnold and Yu. S. Ilyashenko, *Ordinary differential equations* Dynamical systems, I, Itogi Nauki. Sovremennye Problemy Matematiki: Fundamental ′ nye Napravleniya., vol. 1, VINITI, Moscow, 1985; English transl., Encyclopaedia of Math. Sci., vol. 1, Springer-Verlag, Heidelberg, 1988.

[B] N. N. Bautin, *On the number of limit cycles appearing with variations of coefficients from an equilibrium state of the type of a focus or a center*, Mat. Sb. (N. S.) **30** (1952), 181–196; English Transl., Amer. Math. Soc. Transl., 1954; Reprinted in: Stability and Dynamical Systems, Amer. Math. Soc. Transl. Series 1, vol. 5, Providence, RI, 1962, pp. 396–413; *Du nombre de cycles limites naissant en cas de variation des coefficients d'un état d'équilibre du type foyer ou ventre*, first announcement of the result, Soviet Math. Doklady **24** (1939), no. 100, 669–672. (Russian)

[DR] Z. Denkowska and R. Roussarie, *A method of desingularization for analytic two-dimensional vector field families*, Bol. Soc. Brasil. Mat. **22** (1991), no. 1, 93–126.

[D1] F. Dumortier, *Singularities of vector fields on the plane*, J. Differential Equations **23** (1977), 53–106.

[D2] ———, *Techniques in the theory of local bifurcations: blow-up, normal forms, nilpotent bifurcations, singular perturbations*, [S], pp. 19–74.

[DRR] F. Dumortier, R. Roussarie, and C. Rousseau, *Hilbert 16th problem for quadratic vector fields*, Preprint (1991), to appear in J. Differential Equations;———, *Elementary graphics of cyclicity.* 1, 2, Preprint (1993).

[É] J. Écalle, *Introduction aux fonctions analysables and preuve constructive de la conjecture de Dulac*, Hermann, Paris, 1992.

[FP] J. P. Françoise and C. C. Pugh, *Keeping track of limit cycles*, J. Differential Equations **65** (1986), 139–157.

[I1] Yu. Ilyashenko, *Finiteness theorems for limit cycles*, Amer. Math. Soc., Providence, RI, 1991.

[I2] ———, *Local dynamics and nonlocal bifurcations*, [S], pp. 279–320.

[IY1] Yu. Ilyashenko and S. Yakovenko, *Cyclicité finie des polycycles élémentaires*, C. R. Acad. Sci. Paris, Sér. I **316** (1993), 1081–1086.

[IY2] ———, *Finitely-smooth normal forms for local families of diffeomorphisms and vector fields*, Russian Math. Surveys **46** (1991), no. 1, 1–43.

[K] A. Khovanskiĭ, *Fewnomials*, Amer. Math. Soc., Providence, RI, 1991.

[P] M. Pelletier, *Éclatements quasi-homogènes*, C. R. Acad. Sci. Paris Série I **315** (1992), 1407–1411.

[R1] R. Roussarie, *Cyclicité finie des lacets et des points cuspidaux*, Nonlinearity **2** (1989), 73–117.

[R2] ———, *A note on finite cyclicity and Hilbert's 16th problem*, Dynamical systems (Valparaiso, 1986) (R. Bamón, R. Labarca, and J. Palis, eds.), Lecture Notes Math., vol. 1331, Springer-Verlag, Berlin and New York, 1988, pp. 161–188.

[R3] ———, *Techniques in the theory of local bifurcations: cyclicity and desingularization*, [S], pp. 347–382.

[Rs] Ch. Rousseau, *Bifurcation methods in polynomial systems*, [S], pp. 383–428.

[S] Dana Schlomiuk (ed.), *Bifurcations and periodic orbits of vector fields*, NATO ASI Series C (Mathematical and Physical Sciences), vol. 408, Kluwer, Dordrecht, Boston, London, 1993.

[VdE] A. Van den Essen, *Reduction of singularities of the equation* $A\,dy = B\,dx$, Equations differentielles et systemes de Pfaff dans le champ complexe (R. Gerard and J.-P. Ramis, eds.), Lecture Notes Math., vol. 712, Springer-Verlag, Berlin and New York, 1979, pp. 44-60.

[Z] H. Zołądek [Kh. Zholondek], *Quadratic systems with center and their perturbations*, J. Differential Equations **109** (1994), no. 2, 223–273.

Translated by S. YAKOVENKO

YU. I.: MOSCOW STATE UNIVERSITY AND MOSCOW MATHEMATICAL INSTITUTE, MOSCOW, RUSSIA
E-mail address: yuilyashenko@glas.apc.org

S. YA.: THE WEIZMANN INSTITUTE OF SCIENCE, REHOVOT, ISRAEL
E-mail address: yakov@wisdom.weizmann.ac.il

Amer. Math. Soc. Transl.
(2) Vol. **165**, 1995

Finite Cyclicity of Elementary Polycycles in Generic Families

YU. ILYASHENKO AND S. YAKOVENKO

ABSTRACT.
We prove that a polycycle (separatrix polygon) occurring in a *generic* finite-parameter family of smooth vector fields on the real plane, can generate at most a finite number of limit cycles within the family, provided that all the vertices of the polycycle are elementary (have at least one nonzero eigenvalue of the linearization matrix).

This number admits an effective upper bound depending only on the number of parameters of the family. As a byproduct, we establish some results concerning the number of nondegenerate preimages of a point for maps of the form $P \circ f : (\mathbb{R}^n, 0) \to \mathbb{R}^n$, where P is a vector polynomial, and f a generic smooth germ.

The principal result of the paper implies that a generic smooth finite-parameter family of vector fields on the 2-sphere with a compact parameter space admits a *uniform upper estimate* for the number of limit cycles, provided that all singular points occurring in this family are elementary. This is a particular case of the *Hilbert–Arnold problem*.

0. Introduction

0.1. Formulation of results. The main goal of the paper is to estimate the number of limit cycles that appear after generic perturbation of an elementary separatrix polygon on the real plane. All the necessary definitions are given in the introductory paper to this volume.

THEOREM I. *An elementary polycycle occurring in a generic smooth finite parameter family of vector fields can generate in this family at most a finite number of limit cycles.*

THEOREM II. *For any natural k, the number of limit cycles which can be born from an elementary polycycle occurring in a generic k-parameter family, admits an upper estimate by a certain number $E(k) < \infty$ which depends only on the number of parameters.*

The function $k \mapsto E(k)$ is an effectively computable primitive recursive function.

The exact meaning of the term *generic,* which occurs in the formulation of the

1991 *Mathematics Subject Classification.* Primary 34C05, 34C23.
The research of the first author was partially supported by the Grant M-98000 from the International Science Foundation.

two theorems will be explained later: it is slightly different for Theorem I and for Theorem II (for the latter case the genericity assumptions are somewhat more restrictive), so that formally Theorem II does not imply Theorem I.

As an auxiliary tool for the investigation of bifurcations of polycycles we establish the following result which belongs entirely to the realm of singularity theory.

THEOREM III. *Let* $P = (P_1, \dots, P_m)\colon \mathbb{R}^n \to \mathbb{R}^m$ *be a vector polynomial of degree* $d = \deg P = \max_j \deg P_j$ *in* n *variables. Then for a generic germ* $f\colon (\mathbb{R}^m, 0) \to (\mathbb{R}^n, 0)$ *the number of small regular preimages of points in the target space,*

$$\limsup_{r\to 0^+} \sup_{y\in\mathbb{R}^m} \#\{x \in \mathbb{R}^m : |x| < r,\ F(x) = a,\ \det F_*(x) \neq 0\}$$

by the composite map F *between the spaces of the same dimension,*

$$F = P \circ f\colon (\mathbb{R}^m, 0) \to \mathbb{R}^m,$$

is uniformly bounded by a certain effectively computable number $B(n, m, d)$ *depending only on the degree of the polynomial* P *and the numbers* n, m *of variables; the function* $(n, m, d) \mapsto B(n, m, d)$ *is a primitive recursive function of its natural arguments.*

In the same manner as before, the term *generic* in this theorem can be made precise: in this case one may find the smoothness order $\ell \in \mathbb{N}$ determined by n, m, such that the set of germs for which the assertion of Theorem III holds, is open in C^ℓ-topology and dense in $C^{\ell-1}$-topology.

The accurate definition of a primitive recursive function is given below, but the most important in the formulation of Theorems II and III is the constructive and algorithmic nature of the upper estimate, compared to the purely existential form of Theorem I. One should mention here the recent result of A. Mourtada [Mou], establishing a similar constructive nature of the upper estimate, but for polycycles satisfying much stronger restrictions.

The notion of a primitive recursive function which occurs in the formulation of Theorem II, is defined as in [Man, pp. 181–183]. The formal (inductive) definition looks as follows.

DEFINITION 0.1. A function $\zeta : \mathbb{Z}_+^p \to \mathbb{Z}_+^q$ is called *primitive recursive* if it can be obtained from the *basic functions*

$$x \mapsto x + 1, \quad (x_1, \dots, x_p) \mapsto 1 = \text{const}, \quad (x_1, \dots, x_p) \mapsto x_i,$$

in a finite number of steps; on each step one may use only the basic functions and functions obtained on the previous steps and perform the following operations (assuming that the numbers of arguments match the formulas):

- composition (substitution) $\zeta, \eta \rightsquigarrow \zeta \circ \eta$;
- juxtaposition $\zeta, \eta \rightsquigarrow (\zeta, \eta)$ (a function whose components are ζ and η);
- recursion $\zeta, \eta \rightsquigarrow \xi$, where the function ξ is defined recursively as

$$\xi(x_1, \dots, x_p, 1) = \zeta(x_1, \dots, x_p),$$
$$\xi(x_1, \dots, x_p, k + 1) = \eta(x_1, \dots, x_p, k, \xi(x_1, \dots, x_p, k)), \qquad k > 1.$$

For practical means a primitive recursive function is to be understood as an integer-valued function which can be effectively computed using some explicit iterative algorithms. In fact, the latest achievements in the constructive algebraic geometry suggest that the estimate $E(k)$ can be explicitly expressed in a closed form.

The case of a polycycle consisting of only one elementary singular point with no arcs at all, is well known. Being elementary, such a singular point can generate limit cycles only if it is a slow focus, that is, the linearization matrix has a pair of two imaginary eigenvalues. This bifurcation was investigated by F. Takens, and the result is the following: *a slow focus occurring in a generic k-parameter family, may generate no more than k limit cycles.*

Thus we may restrict ourselves to the investigation of polycycles containing at least one arc. From now on we will implicitly assume that integer variables enumerating vertices of the polycycle, are considered modulo n, the number of arcs. Among the vertices there can be repetitions, while all the arcs γ_i must be distinct.

The results of this paper were announced in [IY3]; this publication contains the detailed proofs. For convenience of the reader, we outline below the main ideas and constructions.

0.2. Four stages of the proof. The proofs of both Theorem I and Theorem II consists of four relatively independent steps.

Step 1. *Normal forms for local families of vector fields and their integration.* In §1 we use normal forms to establish an explicit form for the correspondence maps near singular points on the polycycle and show that these maps satisfy Pfaffian equations with algebraic coefficients depending smoothly on the parameters of the family. As a result, we obtain a *basic system* of equations for determination of limit cycles.

Step 2. *Khovanskiĭ reduction procedure.* In §2 we describe in detail a part of the procedure, suggested by A. Khovanskiĭ, for the investigation of systems of equations that involve functions satisfying Pfaffian equations. In §3 we make a formal reduction from the basic system to a *mixed functional–Pfaffian system of equations.* After application of the Khovanskiĭ reduction technique to the latter, we obtain several *chain maps*, the maps of the form

$$x \mapsto P(x, f(x), f'(x), \dots, f^{(n)}(x)),$$

P a vector polynomial of some known degree, f a generic smooth function,

and the problem of estimating the number of limit cycles reduces to estimating the number of nondegenerate preimages of points by the chain maps.

Step 3. *Gabrielov-type finiteness condition.* In §5 we establish certain conditions under which a smooth map between the spaces of the same dimension admits an upper estimate for the number of nondegenerate preimages (in this case we say that the map possesses the *Gabrielov property*). These conditions are automatically satisfied when the map is real analytic (the case investigated by A. Gabrielov). It turns out that the natural framework for introducing these conditions is that of stratified sets and mappings. All necessary definitions, including the *contiguity number* of a stratified variety, a measure of the topological complexity of the latter, is given in §4. As a result, we introduce the notions of *nice* and *locally nice* maps and show that (locally) nice maps possess the Gabrielov property.

Step 4. Universal criminant sets. The finiteness condition introduced in §5, is of an inductive nature: the class of nice maps is defined in intrinsic terms. In §6 we show that in fact the (local) niceness can be expressed in terms of transversality to some universal subsets of appropriate jet spaces. As a result, we prove that for an open and dense subset of smooth vector-functions f the chain map $P \circ f$ with a polynomial exterior part P is nice (which proves Theorem III above), and the same is true if:

(1) instead of f the jet extension $j^s f$ is substituted into the chain map, or
(2) some components of the map f do not depend on some coordinates of the source space (in this case the Cartesian genericity introduced in §6 should replace the usual genericity).

The latter case exactly corresponds to the composite map that appears after the second step. This observation concludes the proof of the theorems.

The detailed description of separate steps follows.

0.3. Normal forms of local families and their integration. The first step is to write down finite differentiable normal forms for smooth local families of vector fields in small neighborhoods of singular points (in the Cartesian product of the phase space and the space of the parameters). The list of normal forms was obtained in [IY1]. This list has the following main property: all the normal forms are polynomial and integrable. The smoothness of the normalizing transformation is as high as necessary: the smaller the neighborhood, the greater the degree of differentiability of the conjugacies taking the family into the normal form.

We summarize results from [IY1] in §1 below and perform an explicit integration of the normal forms from the list. Precisely, in a small neighborhood of a singular point we choose two small segments transversal to the vector field for all values of the parameters, and investigate the correspondence maps, associated with these pairs of transversal segments (the maps taking a point of the first segment into the first intersection of the corresponding phase curve with the second segment whenever this intersection point is defined, see Figure 1).

For all polynomial families from the list of normal forms (Table 1), and for a special choice of the transversals, the correspondence maps can be explicitly computed. This computation shows that these maps are *singular*: their domains do not contain any entire neighborhood of the origin, and some of their derivatives tend to infinity on the boundaries of the domains. On the other hand, they satisfy certain simple Pfaffian equations with polynomial coefficients differentiably depending on the parameters.

The simplest example is deformation of a nonresonant saddle. Any local unfolding of a saddle point with an irrational ratio of the eigenvalues is finitely differentiably equivalent to the linear family

$$\dot{x} = x\,, \qquad \dot{y} = -\lambda(\varepsilon)\, y.$$

If one takes the two transversal segments to be parallel to the coordinate axes with the local coordinates being restrictions of the coordinates on the plane, $\xi =$

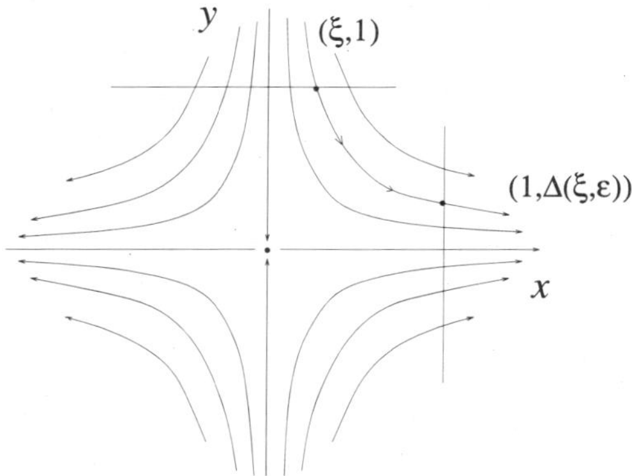

FIGURE 1. Correspondence map for a hyperbolic sector.

$x|_{\{y=1\}}$, $\eta = y|_{\{x=1\}}$ (see Figure 1), then the correspondence map takes the form

$$\eta = \Delta(\xi, \varepsilon) = \xi^{\lambda(\varepsilon)}, \qquad \xi > 0.$$

Returning back to the natural (though less rigorous) notation $(\xi, \eta) \mapsto (x, y)$, we see that the graph $y = x^{\lambda(\varepsilon)}$ of the correspondence map Δ is an integral curve for the Pfaffian differential equation

$$x\,dy - \lambda y\,dx = 0, \qquad \lambda = \lambda(\varepsilon),$$

which is linear in the phase variables x, y and quadratic if we treat the parameter λ as an additional independent variable. Recall that $\lambda(\cdot)$ is a smooth function of the original parameters of the family. In other cases the formulas become much more complicated, but the above two features still hold.

0.4. Singular–regular systems determining the number of limit cycles. Each singular point on the polycycle may be endowed with so the called *normalizing coordinates* in its sufficiently small neighborhood (and by rescaling these coordinates one may ensure that the unit square belongs to this neighborhood). Take standard transversals for the normal form (the sides of the unit square which are intersected by the arcs of the polycycle) and consider their preimages on the initial phase plane. Thus each singular point on the polycycle turns out to be separated from the rest of the polycycle by two transversals (segments), see Figure 2. Since the polycycle is ordered by the direction of the flow on it, one of the transversals becomes the "entrance to", while the other is the "exit from" the neighborhood of the singular point: phase curves enter the neighborhood through one of the transversals and leave through the other. Moreover, each transversal is endowed with a certain *normalizing C^r-smooth chart* on it, coming from the coordinate functions for the normal form, see §1.2 for the precise description. Denote the chart on the "entrance" transversal near the jth singularity by x_j, and that on the corresponding "exit" transversal by y_j. When referring to charts on transversals, we shall always have those in mind.

Written in the above charts, the correspondence maps Δ_j belong to a certain *canonical* list of functions, transcendental and irregular themselves, but satisfying

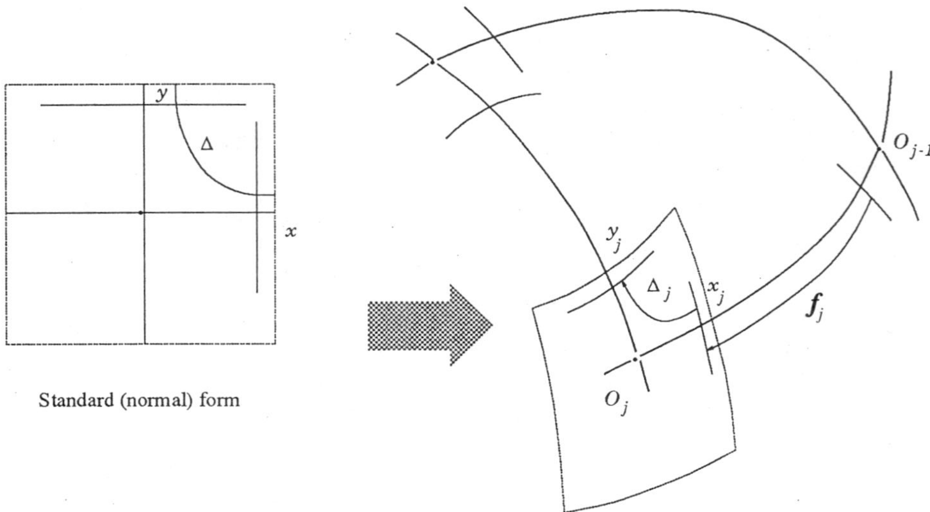

FIGURE 2. Construction of entrance and exit transversals.

certain Pfaffian equations of the form

$$\omega_j = 0\,, \qquad \omega_j = A_j(x_j\,,\,y_j\,,\,\varepsilon)\,dx_j + B_j(x_j\,,\,y_j\,,\,\varepsilon)\,dy_j\,, \tag{0.1}$$

with polynomial (in $x_j\,,\,y_j$) coefficients depending differentiably on the parameters on the family.

Together with these singular maps the correspondence maps associated with regular arcs of the polycycle arise. They have the same smoothness as the charts on the transversals. According to the cyclic ordering of the singular points and the arcs, these maps take points of jth "exit" transversal to points of the $(j+1)$ th "entrance" one. Denote these maps by f_j: in the charts $y_j\,,\,x_{j+1}$ they become (scalar) functions, which we denote by the same symbol,

$$f_j(\cdot\,,\,\varepsilon)\colon y_j \mapsto x_{j+1} = f_j(y_j\,,\,\varepsilon).$$

The monodromy map (otherwise called first return map) of the polycycle may be decomposed into the chain of singular maps Δ_j and regular maps f_j of the total length $2n$, where n is the number of arcs on the polycycle. Limit cycles correspond to fixed points of the monodromy. But instead of writing an equation for fixed points of this monodromy map, we replace it by a system of $2n$ equations, which will be called the (preliminary) *basic system*:

$$\begin{cases} y_j \;\;\;= \Delta_j(x_j\,,\,\varepsilon)\,, & j = 1\,,\,\dots\,,\,n\,, \\ x_{j+1} = f_j(y_j\,,\,\varepsilon)\,, & j = 1\,,\,\dots\,,\,n \pmod n). \end{cases} \tag{0.2}$$

Here x_j and y_j are coordinates of the intersection of a limit cycle with the transversals. This system involves C^r-smooth functions f_j which are regular (with respect to both the variables and the parameters), and the maps Δ_j which are explicitly known (modulo the reparametrization $\varepsilon \mapsto \lambda(\varepsilon)$), but essentially singular. The problem is to estimate the number of solutions to the system uniformly over all sufficiently small values of the parameters.

As soon as the system (0.2) is written, the problem of estimating the number of limit cycles becomes purely analytical, though differential equations will appear once again within the framework of the Khovanskiĭ reduction procedure.

0.5. The Khovanskiĭ reduction process. The system (0.2) is difficult to analyze because of the presence of irregular functions Δ_j. The key idea of the second step is to replace the corresponding (functional) equations in the system (0.2) by the Pfaffian (differential) equations of the form (0.1). As a result we obtain the *mixed functional–Pfaffian system of equations* of the form

$$(0.3)\qquad \begin{cases} \omega_j = 0, \\ F_j(x, y, \varepsilon) = 0, \end{cases} \qquad j = 1, \dots, n,$$

$$\omega_j = A_j\, dx_j + B_j\, dy_j, \qquad F_j(x, y, \varepsilon) = x_{j+1} - f_j(y_j, \varepsilon),$$

$$(x, y) = (x_1, y_1, \dots, x_n, y_n) \in (\mathbb{R}^{2n}, 0), \qquad \varepsilon \in (\mathbb{R}^k, 0),$$

where ω_j are the Pfaffian forms (0.1). This system is to be interpreted as follows: one has to take an integral manifold Γ for the Pfaffian equations of the system (0.3) and then compute its intersection with the level set $F^{-1}(0)$, where $F\colon (\mathbb{R}^{2n}, 0) \to \mathbb{R}^n$ is the map with the coordinate functions F_j. To estimate the number of isolated solutions to (0.2), one needs to estimate the number of isolated points in the intersection. It turns out that *it is sufficient to analyze only transversal intersections* of Γ with a generic fiber $F^{-1}(b)$ for b sufficiently close to the origin in $\mathbb{R}^n$. Since the integral manifold and the level set have complementary dimensions, a transversal intersection always consists of isolated points, which we call *nondegenerate solutions* to the system (0.3). What we are interested in is the upper estimate for their number, uniform over all the integral manifolds Γ and all sufficiently small values of the parameters.

The procedure suggested by A. Khovanskiĭ [Kh] permits us to replace a mixed functional–Pfaffian system of the form (0.3) by two systems of similar form, but containing $n-1$ Pfaffian equations and $n+1$ functional ones: the number of nondegenerate solutions to the initial system is estimated from above by the sum of the numbers of nondegenerate solutions to these two auxiliary systems.

The idea of the Khovanskiĭ construction is illustrated by the following example: if Γ is a boundary of a domain in $\mathbb{R}^n$ and at the same time an integral manifold to the Pfaffian equation $\omega = 0$, then the form ω takes (eventually after multiplication by -1) positive values on any vector on the boundary, which points outward. So if there is a curve (not necessarily connected or compact, but smooth), then between any two successive intersections of this curve with Γ there must be a *point of contact*, at which the curve is tangent to the hyperplane $\omega = 0$. The tangency condition can be expressed in an explicit algebraic way through coefficients of the form ω and the equations defining the curve, see §2 for details. Therefore, we can estimate the overall number of transversal intersections by the number of points of contact plus the number of noncompact components of the curve.

Returning to the general case, we point out that the reduction process is constructive: the additional functional equation is obtained by algebraic operations and differentiation applied to the functions F_j and coefficients of 1-forms. Iterating this construction, we end up with a collection of systems having no Pfaffian equations at all. These systems have the following form:

$$\mathbf{p}(\boldsymbol{x}, \varepsilon, \mathbf{f}(\boldsymbol{x}, \varepsilon), \dots, \mathbf{f}^{(n)}(\boldsymbol{x}, \varepsilon)) = 0. \tag{0.4}$$

Here $\mathbf{p}$ is a vector polynomial taking its values in $\mathbb{R}^{2n}$, $\boldsymbol{x} = (x, y) \in \mathbb{R}^{2n}$, $\mathbf{f} = (f_1, \dots, f_n)$, and $\mathbf{f}^{(\ell)} = (f_1^{(\ell)}, \dots, f_n^{(\ell)})$, where $f_j^{(\ell)} = (\partial^\ell f_j / \partial y_j^\ell)$, $\ell = 1, \dots, n$.

The polynomials $\mathbf{p}$ for all the systems of the form (0.4) appearing after the elimination, depend only on the types and the order of singular points on the polycycle. Therefore only a finite number of such systems can arise, and it turns out that their degree admits an upper estimate by an explicitly computable elementary function of the number of parameters k of the original family.

The last two stages of the proof deal with systems only of the type (0.4). In order to get rid of the parameters, we consider them as new independent variables. Denote $\mathbf{x} = (\boldsymbol{x}, \varepsilon) \in (\mathbb{R}^{2n+k}, 0)$ and let $\mathbf{P} = (\mathbf{p}, \mathrm{id})$ be the polynomial constructed from $\mathbf{p}$ by adding new components identical in the parameters ε.

Then one may reformulate the problem as follows: *given a vector polynomial map* $\mathbf{P}$, *estimate the number of nondegenerate solutions of the system*

$$\begin{aligned} &(\mathbf{P} \circ j^n \mathbf{f})(\mathbf{x}) = \mathbf{y}, && |\mathbf{x}| < r, \\ &\mathbf{x}, \mathbf{y} \in (\mathbb{R}^{2n+k}, 0), && \mathbf{y} = (0, \dots, 0, \varepsilon_1, \dots, \varepsilon_k), \end{aligned} \tag{0.5}$$

for a generic smooth vector-valued function $\mathbf{f}$ (the symbol $j^n\mathbf{f}$ stands for the n-jet extension of the function $\mathbf{f}$), *uniformly over all values of the parameter* $\mathbf{y}$. The map occurring in the left-hand side of the formula, is the *chain map*, whose interior part $\mathbf{f}$ is finitely differentiable (and in a some sense generic), while the exterior term $\mathbf{p}$ is polynomial.

0.6. Gabrielov property of smooth maps. The following theorem is one of the basic facts of real analytic geometry.

Gabrielov theorem [G]. *If M is a compact analytic set in the real space, and $g: M \to \mathbb{R}^n$ a real analytic map, then the number of the connected components of the preimages $g^{-1}(a)$ is bounded from above uniformly over $a \in \mathbb{R}^n$.*

For brevity we say that a map between two spaces of the same dimension possesses the Gabrielov property if it admits a uniform upper bound for the number of preimages of regular points.

If in the system (0.5) the function $\mathbf{f}$ is real analytic, then the chain map possesses the Gabrielov property. Indeed, the dimension of the source and the target spaces in (0.5) coincide, therefore the regular preimages will be discrete sets of points.

In general, the Gabrielov theorem fails for smooth maps: one may easily construct counterexamples with unbounded (or even infinite) number of regular preimages. We find certain sufficient local conditions on a smooth map $\mathbf{F}: M^n \to \mathbb{R}^n$

guaranteeing that the number of regular preimages in $\mathbf{F}^{-1}(a)$ is uniformly bounded for all values of $a \in \mathbb{R}^n$. These conditions are obtained in §5 in an attempt to modify the original proof of the Gabrielov theorem [G] for finitely differentiable maps. For reasons to be clarified later, we consider the category of stratified sets and their smooth maps.

The principal idea in estimating the number of nondegenerate preimages for a map $\mathbf{F}\colon M^n \to \mathbb{R}^n$ is to consider the critical locus C of the map, that is, the set of points in the source space, at which the rank of $\mathbf{F}$ is not maximal. Distinct nondegenerate preimages of a variable point in the target space can coalesce or disappear only on such a subset. Suppose for a moment that the critical locus were a smooth submanifold of codimension 1, and consider a projection π of the target space $\mathbb{R}^n$ onto a generic linear subspace of codimension 1. Since we are dealing only with nondegenerate preimages, both the initial value of $\mathbf{b}$ and its projection $\mathbf{b}' = \pi(\mathbf{b})$ may be chosen as regular values of $\mathbf{F}$ and $\pi \circ \mathbf{F}$ respectively.

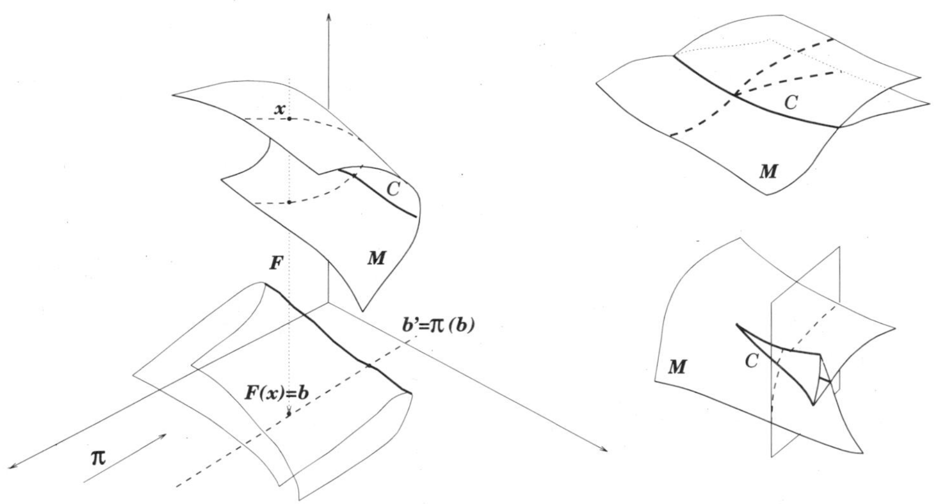

FIGURE 3. Bifurcation of preimages on smooth and stratified sets.

If the variable point $\mathbf{b}$ moves along the line $\pi^{-1}(\mathbf{b}')$ in the target space, then the number of nondegenerate preimages is a piecewise constant integer function of the point, and the discontinuities of this function occur only on $\mathbf{F}(C)$, see Figure 3 (left).

The number of confluent preimages for a generic case is 2, because the standard singularity of a map is the fold [AVG], and for $|\mathbf{b}|$ sufficiently big there are no preimages at all (we assume that the domain of the map $\mathbf{F}$ is compact). Thus the number of nondegenerate preimages of a point for the map $\mathbf{F}$ does not exceed the number of nondegenerate preimages for the restriction of the map $\pi \circ \mathbf{F}$ to C, since for an integer-valued function vanishing at infinity the maximal value does not exceed one half of the sum of absolute values of jumps. But the dimensions of both the source and the target spaces for the map $\pi \circ \mathbf{F}|_C\colon C \to \mathbb{R}^{n-1}$ are smaller than before. So if the inductive application of the above construction is possible up to dimension zero, then the Gabrielov property holds for the initial map $\mathbf{F}$. Indeed, a map of zero-dimensional compact manifold into $\mathbb{R}^0 = \{0\}$ by definition possesses the Gabrielov property.

Unfortunately, the critical locus for a chain map $\mathbf{F} = \mathbf{P} \circ (j^n\mathbf{f})$ might become a nonsmooth manifold even for a generic map $\mathbf{f}$, because we do not make any assumptions about the polynomial $\mathbf{P}$ (which could be quite degenerate). But for a generic $\mathbf{f}$ the critical locus is always a *stratified submanifold*, as it will be explained in 0.7, and the above construction may be modified in such a way that the case of stratified submanifolds would be covered as well. The number of coalescing preimages must not necessarily be equal to 2 anymore (see the examples on Figure 3, right), but this number still admits an effective upper estimate in terms of a special topological characteristic introduced below, the *contiguity number*. Using this idea, we give sufficient conditions for a smooth map between two stratified manifolds of the same dimension to possess the Gabrielov property. These conditions are of inductive nature and constructive: knowing the contiguity numbers of all stratified manifolds occurring in the construction, one may give an upper estimate for the number of regular preimages.

In the last section we prove that in the generic case the chain map admits iteration of the above construction in the category of stratified submanifolds, thus the finiteness property holds for such maps.

0.7. Universal criminant sets and a version of Thom–Boardmann construction. The conditions obtained in §5 hold for a generic function $g\colon M^n \to \mathbb{R}^n$. Moreover, for a generic function g the composition $\mathbf{P} \circ g$ (with a polynomial exterior part) also satisfies these conditions provided that the dimensions of the source and target manifolds for $\mathbf{P}$ and g are appropriate: $g\colon M^n \to \mathbb{R}^N$, $\mathbf{P}\colon \mathbb{R}^N \to \mathbb{R}^n$. Finally, we prove that the same is true with respect to the chain map of the form (0.5) as well.

To prove these claims, for each polynomial $\mathbf{P}$ occurring in chain map (0.5) we construct in §6 a certain family of universal algebraic subsets Σ^s in the corresponding jet spaces $J^s(\mathbb{R}^n, \mathbb{R}^N)$, $s = 0, 1, \dots, n$, such that if the jet extensions of the interior part of a chain map are transversal to those universal subsets, then the chain map itself possesses the Gabrielov property. The construction of the algebraic subsets Σ^s is similar to that of the universal Thom–Boardmann sets known in singularity theory.

To clarify the idea of constructing the universal sets, we start with a more simple case of chain maps of the form $F = \mathbf{P} \circ g$. The critical locus of the map F consists of points at which $\det F_* = 0$. Computing the determinant, we obtain the condition of the form

$$R\left(x, g_i(x), \frac{\partial g_i}{\partial x_j}(x)\right) = 0,$$

where R is a certain polynomial (determined by the vector polynomial $\mathbf{P}$), and g_i stand for the coordinate functions of the map g. Such an equation can be interpreted in the following manner: in the space of 1-jets of maps $\{g\colon M^n \to \mathbb{R}^N\}$ there is an algebraic surface Σ^1 defined by the equation $R = 0$ (note that the arguments of the polynomial R are the natural coordinates on the jet space), and the critical locus C in the Gabrielov construction is exactly $(j^1 g)^{-1}(\Sigma^1)$. Being algebraic, the surface Σ admits a stratification by virtue of the Whitney theorem. Therefore if the 1-jet extension of the map g is *transversal* to Σ^1, then the preimage $(j^1 g)^{-1}(\Sigma^1)$ is also a stratified variety. Thus transversality to Σ^1 guarantees that the first step in the above construction can be implemented, reducing the dimension by 1.

To construct the second set Σ^2, we take a regular (smooth) point on Σ^1 and find a local polynomial representation of the latter set as the zero locus for a polynomial R_1 with almost everywhere nonvanishing differential (the initial polynomial R could be very degenerate, since we do not impose any nondegeneracy conditions on the polynomial P). Then the critical locus for F is locally defined by the equation

$$R_1\left(x\,,\, g_i(x)\,,\, \tfrac{\partial g_i}{\partial x_j}(x)\right) = 0\,,$$

and we should restrict to this set the map whose $n-1$ components are $\mathbf{P}_2 \circ g\,, \ldots, \mathbf{P}_n \circ g$. The restriction would have a critical point if the Jacobian $n \times n$-matrix of all n functions would be degenerate. Since R_1 already involves first derivatives of the functions g_i, such a Jacobian would be a polynomial combination of derivatives of g_i up to the second order, and the set Σ^2 is defined as the corresponding subset in the space of 2-jets. Again transversality to Σ^2 guarantees that the second critical locus can be stratified, etc.

Actually the construction is somewhat more complicated: the main difficulty is to replace a system of polynomial equations determining the universal criminant sets Σ^k by another system which would give a regular representation of the same sets so for any smooth point of Σ^k one could choose a local polynomial representation with linear independent differentials. In other words, one needs to replace the ideal spanned by original polynomials determining Σ^k by its *real radical*, retaining control over degrees of generators. Disregarding this circumstance, the above exposition captures essentially all the basic features of the process. Obviously if the interior part of the chain map is a jet itself, or if its components do not depend on some of the variables, then the algebraic nature of all the operations still holds, though the sets themselves are constructed as subvarieties of the *Cartesian jet space* (see below). What is important is that the transversality theorem in the strong form holds for such space as well (O. Shelkovnikov, see §6).

This procedure provides us with explicit estimates of the contiguity numbers of algebraic varieties $\Sigma^1, \Sigma^2 \ldots$, and, as will be proved in §4, their transversal preimages have the same contiguity numbers. Therefore we get an explicit description of all factors in the recurrent formula arising from the Gabrielov-type construction. Actually the result proved in §§4–6 is a generalization Theorem III from 0.1.

0.8. Proof of the main theorems. Now, after all the four steps were explained, the sequence of implications proving the two main theorems, looks as follows:

(1) The jet extension of a generic map $\mathbf{f}$ is transversal to all universal criminant sets constructed in 0.7, therefore

(2) the inductive process of the reduction of dimension, described in 0.6, can be (effectively) carried on until the zero-dimensional case, thus proving that the correspondent chain map admits an (effective) estimate for the number of nondegenerate preimages, therefore

(3) the mixed functional–Pfaffian system (0.3) has a uniformly bounded number of nondegenerate solutions, therefore
(4) the singular-regular system of equations (0.2) admits an (effective) upper estimate for the number of isolated solutions close to the origin, therefore
(5) the number of limit cycles appearing after perturbation of a polycycle, admits an (effective) upper bound depending only on the number of parameters in the initial family determined by the functions $\mathbf{f}$.

IMPORTANT REMARK. When performing reductions, we introduce different integer parameters, such as the number of equations, dimensions of phase and parameter spaces, upper estimates of degrees of polynomials, etc. The reader should always bear in mind that *all the integer values occurring in our considerations admit an effective upper estimate by some primitive recursive functions of the basic parameter, namely the number k of independent parameters of the initial family*, even if we do not explicitly specify such an estimate. This is an implication of the effective algorithmic nature of all constructions used in the reductions.

We say about *effectively computable* estimates if they can be obtained in the class of primitive recursive functions, and use the term *explicit estimate* if an expression in a closed form can be obtained.

0.9. Technical remarks. In this Introduction we outlined the general scheme of arguments proving Theorems I and II. In fact, our constructions are somewhat more complicated. To avoid eventual confusion we briefly mention here the principal differences between the exposition in the Introduction and in the main body of the paper.

• We consider the parameters of normal forms as additional parameters, forgetting that they are in fact smooth functions of the original parameters ε. Clearly, by doing that we cannot increase the number of solutions of all systems of equations.

• When one applies the Khovanskiĭ reduction to a functional–Pfaffian system (0.3) restricted to a ball of radius r, the value r enters explicitly into the construction. As a result, the left-hand side of the system (0.4) becomes dependent on r, though in a polynomial way. To overcome this difficulty, we need to consider r as an additional variable of the vector polynomial $\mathbf{p}$.

• The Pfaffian equation for the correspondence map associated with the deformation of the $(n:m)$-resonant saddle is replaced by three equations involving two additional variables z, w. This is done in order to avoid occurrence of terms x^m, y^n in the resulting formulas. Otherwise we would be unable to control the degree of the vector polynomial $\mathbf{p}$.

• When proving Theorems I and II, we reduce them in a number of steps to different assertions which will be proved later. In order to make the logical structure of the paper more transparent, we finish the formulations of such theorems by the sign $\lhd$ meaning that the proof will appear later. The end of the proof is marked by the usual sign $\square$.

0.10. Acknowledgments. The authors are grateful to A. Khovanskiĭ for numerous discussions, which essentially contributed to §2, and to A. Gabrielov for the opportunity to look through his thesis. O. Shelkovnikov proved the Cartesian transversality theorem. A. Shcherbakov and A. Nabutovskiĭ helped us to learn the theory of stratified subsets, and Y. Yomdin suggested to refer to Thom's isotopy lemma when introducing the contiguity number. J. Mather sent us a copy of his unpublished manuscript containing another proof of the results related to constructions involving stratified sets. M.-F. Coste-Roy, E. Becker, and R. Neuhaus explained to us the new results concerning effective computation of real radicals in the ring of polynomials $\mathbb{R}[x_1, \dots, x_n]$. We appreciate very much their help and support.

§1. Normal forms for local families and their applications

In this section we begin construction of the functional–Pfaffian system whose number of solutions majorizes the number of limit cycles.

1.1. Local families and polynomial normal forms. A local family of planar vector fields is the germ of a map

$$v\colon (\mathbb{R}^2, 0) \times (\mathbb{R}^k, 0) \to (\mathbb{R}^2, 0), \qquad (x, y, \varepsilon) \mapsto v(x, y, \varepsilon).$$

A C^r-smooth conjugacy between two local families v and w of the above form is a map

$$H\colon (\mathbb{R}^2, 0) \times (\mathbb{R}^k, 0) \to (\mathbb{R}^2, 0), \qquad (x, y, \varepsilon) \mapsto H(x, y, \varepsilon),$$

such that

$$H_* v(x, y, \varepsilon) = w(H(x, y, \varepsilon), \varepsilon),$$

where H_* stands for the Jacobian matrix with respect to the variables x, y (this definition does not yet allow for reparameterization of a local family). Two families are finitely differentiably equivalent if for any $r < \infty$ there exists a C^r-conjugacy between them. The two families v, w are orbitally equivalent if there exists the germ of a nonvanishing function $\varphi\colon (\mathbb{R}^2, 0) \times (\mathbb{R}^k, 0) \to \mathbb{R}^1$ such that v is equivalent to $\varphi \cdot w$.

To allow for a reparameterization of local families, we say that a family $v(\cdot, \varepsilon)$ is induced from another family $w(\cdot, \lambda)$, $\lambda \in (\mathbb{R}^m, 0)$, if $v(\cdot, \varepsilon) = w(\cdot, \lambda(\varepsilon))$, where $\lambda(\varepsilon)$ is the germ of a smooth map $(\mathbb{R}^k, 0) \to (\mathbb{R}^m, 0)$. The number of new parameters m may be different from k.

Assume that the family $w(\cdot, \lambda)$ is *global* (i.e., the expression $w(x, y, \lambda)$ makes sense for all $(x, y, \lambda) \in \mathbb{R}^{m+2}$); this happens in particular when w is *polynomial* in all its arguments. Restricting the parameters λ to a small neighborhood of a certain point $(0, 0, \mathbf{c}) \in \mathbb{R}^2 \times \mathbb{R}^m$, we obtain the *localization* of the global family w, which formally becomes a local family after the parallel translation $\lambda \mapsto \lambda - \mathbf{c}$.

DEFINITION 1.0. 1. A local family $v = v(\cdot, \lambda)$ is a finitely smooth orbital versal unfolding (in short, versal unfolding) of the germ $v(\cdot, 0)$ if any other local family unfolding this germ is finitely differentiably orbitally equivalent to a family induced from v.

2. A polynomial family $w(\cdot, \lambda)$, $\lambda \in \mathbb{R}^m$ is a *global finitely smooth orbital versal unfolding* (in short, global versal unfolding) for a certain class of local families of vector fields if any local family from this class is finitely differentiably orbitally equivalent to a local family induced from some localization of w.

To investigate a versal unfolding means to investigate at the same time all smooth local finite-parameter families that unfold the same germ $v(\cdot, 0)$. The main result describing versal unfoldings of germs of elementary singularities on the plane is given by the following theorem.

THEOREM 1.1 [IY1]. *Suppose that a generic finite-parameter family of smooth vector fields on the plane possesses an elementary singular point for a certain value of the parameters. If this point has at least one hyperbolic sector, than the family is finitely differentiably orbitally equivalent to a family induced from some localization of one of the families given in the second column of Table* 1. □

TABLE 1. Unfoldings of elementary singular points on the plane.

Type	Normal forms	Correspondence maps	Pfaffian equations
S_0	$\begin{cases}\dot{x} = x, \\ \dot{y} = -\lambda y,\end{cases}$ $\lambda = \lambda_0 \in \mathbb{R}^1$	$y = x^{\lambda}$, $x > 0,\ y > 0$	$x\,dy - \lambda y\,dx = 0$
S_μ	$\begin{cases}\dot{x} = x\left(\frac{n}{m} + P_\mu(u, \lambda)\right) \\ \dot{y} = -y,\end{cases}$ $u = u(x, y) = x^m y^n$, $P_\mu(u, \lambda) = \pm u^\mu(1 + \lambda_\mu u^\mu) + W_{\mu-1}(u, \lambda)$, $\lambda = (\lambda_0, \dots, \lambda_\mu)$	$0 = m \ln y + \int_{x^m}^{y^n} \frac{du}{uP_\mu(u, \lambda)}$ $x > 0,\ y > 0$	$y\,P_\mu(y^n, \lambda)\,dx - \left(\frac{n}{m} + P_\mu(y^n, \lambda)\right) \times x\,P_\mu(x^m, \lambda)\,dy = 0$
D_μ^c	$\begin{cases}\dot{x} = Q_\mu(x, \lambda), \\ \dot{y} = -y,\end{cases}$ $Q_\mu(x, \lambda) = \pm x^{\mu+1}(1 + \lambda_\mu x^\mu) + W_{\mu-1}(x, \lambda)$, $\lambda = (\lambda_0, \dots, \lambda_\mu)$	$y = C(\lambda)\,x$, $C = \int_{-1}^{1} \frac{du}{Q_\mu(u, \lambda)}$ $x, y \in \mathbb{R}$	$x\,dy - y\,dx = 0$
D_μ^h	The same as above	$0 = \ln y + \int_x^1 \frac{du}{Q_\mu(u, \lambda)}$, $y > 0,\ x \in \mathbb{R}$	$Q_\mu(x, \lambda)\,dy - y\,dx = 0$

In what follows we use the following notation for different types of singular points and their normal forms (the subscript indicates the degree of degeneracy of a singularity):

S_0 — Nonresonant saddle;

S_μ — Resonant saddle whose quotient equation (see (1.0) below) has the singular point of multiplicity $\mu + 1$ at the origin, $\mu \geqslant 1$; if we want to specify explicitly the resonance between the eigenvalues, we use the extended notation $S_\mu^{(n\,:\,m)}$ assuming that the natural numbers m, n are relatively prime;

D_μ — Degenerate elementary singular point of multiplicity μ.

Everywhere below $W_{\mu-1}(z, \lambda) = \lambda_0 + \lambda_1 z + \cdots + \lambda_{\mu-1} z^{\mu-1}$ is a Weierstrass polynomial of degree $\mu - 1$.

Remarks to Theorem 1.1 and Table 1. 1. Normal forms from Table 1 are global versal unfoldings of the germs from the corresponding classes. This is a reformulation of Theorem 1.1 in terms introduced in Definition 1.0.

2. There are other types of elementary points on the plane. Some of them also admit finite-parameter versal unfoldings, other have functional moduli even for C^1-smooth classification, see [IY2]. Table 1 contains the list of unfoldings only for those types of singular points that can occur on a polycycle with at least one arc.

3. The natural domains for the standard correspondence maps $y = \Delta(x, \lambda)$ given in the third column of Table 1 do not always coincide with those for which the implicit formulas (obtained by integration of normal forms, see below) make sense. For example, in the case D_μ^c the natural requirement is that there are no zeros of the polynomial Q_μ on the segment $[-1, 1]$, while in the case D_μ^h the correspondence map is defined for values (x, λ) such that there are no roots of the polynomial $Q_\mu(\cdot, \lambda)$ on the segment $[x, 1]$, and in both cases the natural domain of the correspondence maps coincides with the domain where the implicit formulas make sense. On the other hand, the correspondence map for the case S_μ is defined for all sufficiently small positive x, while the formula from the third column in Table 1 makes sense only if $P_\mu(x^m, \lambda) \neq 0$.

4. The localization of the universal families from Table 1 always satisfies the following additional condition: the vector $\mathbf{c} \in \mathbb{R}^{\mu+1}$ has all coordinates except for the last one equal to zero. This is a way to say that in the reparametrization mentioned in Definition 1.0 the parameters of the versal unfoldings are smooth functions of the original parameters of the local family,

$$\lambda = \lambda(\varepsilon) = (\lambda_0(\varepsilon), \ldots, \lambda_\mu(\varepsilon)),$$

and the map $\varepsilon \mapsto \lambda(\varepsilon)$ has the following properties: it is smooth and all the values $\lambda_j(0)$ except for the last one $\lambda_\mu(0)$, vanish, while the last one $\lambda_\mu(0)$ is the formal invariant of the germ $v(\cdot, 0)$.

In particular this means that the global versal unfolding for a nonresonant saddle (which has only one parameter) should be localized at a point $\mathbf{c} \in \mathbb{R}^1$ which is the hyperbolicity ratio of the nonperturbed germ $v(\cdot, 0)$.

1.2. Correspondence maps. In the third column of the table the correspondence maps $y = \Delta(x, \lambda)$ for the polynomial normal forms are given. They are implicitly defined by the equations relating x to y, these equations depending explicitly on the parameters λ and thus implicitly on the original parameters ε. In each case we explain the choice of segments transversal to the phase curves of the family, and the local charts on these segments.

Singularities of the type D_μ give rise to two types of correspondence maps. Recall that for each singular point on the polycycle two arcs are defined, one of them having this point as an ω-limit set, the other having it as the α-limit. In the saddle case all such arcs (separatrices) play the same role, while in the degenerate case there are two invariant curves of a different nature, the *hyperbolic* invariant curve tangent to the eigenvector with a nonzero hyperbolic eigenvalue, and the *center curve* (one-dimensional center manifold) tangent to the eigenvector with the zero eigenvalue. According to this observation, the following two subcases should be distinguished.

D_μ^h — One of the arcs belongs to the hyperbolic curve, while the other is a part of the center manifold (curve). The superscript h stands for "hyperbolic", since the polycycle includes the hyperbolic arc of the singular point.

D_μ^c — Both arcs are parts of the center manifold. In this case the correspondence map is not defined for the unperturbed system, but after perturbation the singularity may disappear into the complex domain (the index μ must be odd), and limit cycles pass near the center manifold. This case is referred to as D_μ^c, c stands for "center".

We study all four possible cases of correspondence maps one by one, integrating explicitly the normal forms from the second column of Table 1 and deducing the Pfaffian equations that appear in the fourth column.

Nonresonant saddle. For the nonresonant saddle, the computation of the correspondence map is straightforward: all integral curves are given by the formula $y(x) = Cx^{-\lambda}$, $C \in \mathbb{R}$. Therefore, the solution passing through a point $(x, 1)$ on the entrance transversal $y = 1$ intersects the exit transversal $x = 1$ at the point $(1, x^\lambda)$.

Resonant saddle. In this case the computations are similar to those in [I]. Namely, the resonance monomial $u = x^m y^n$ (see Table 1) satisfies the *quotient equation*

$$\dot{u} = mu\, P_\mu(u, \lambda). \tag{1.0}$$

We choose the transversals as before. Then the time required to get from a point $(x, 1)$ on the entrance transversal, to the point $(1, y)$ on the exit transversal, $y = \Delta(x, \lambda)$, is equal to $-\ln y$ for $x > 0$, because the equation for the y-coordinate in the normal form is separated.

During the same time, the value of the variable u changes from x^m at the entrance point to y^n at the exit point. Since the quotient equation for u is autonomous (therefore integrable), this time is equal to the integral below, and the equality

$$\int_{x^m}^{y^n} \frac{du}{mu\, P_\mu(u, \lambda)} = -\ln y,$$

implicitly determines the function $y = \Delta(x, \lambda)$.

Degenerate elementary singularity. For $\mu > 0$ odd, two types of correspondence maps near singularities of the type D_μ are possible.

If the entrance transversal is $x = -1$ and the exit one is $x = +1$, then the correspondence map (whenever defined) is linear, because the normal form is linear in y. The corresponding type is denoted by D^c_μ; to obtain an explicit formula for the coefficient $C(\lambda)$ of the linear map, we need to find the time elapsing from the start on $x = -1$ until the trajectory arrives to the transversal $x = +1$, and then integrate the linear autonomous equation $\dot{y} = -1$ on this time interval.

Otherwise, the map taking points on the entrance transversal $y = 1$ to points on the exit transversal $x = 1$ is always defined on a certain semi-interval depending on the parameters. The way of computing the correspondence map is analogous to that for resonant saddles: the variables in the normal form are separated. Using this fact, one can write the equation

$$-\ln y = \int_x^1 \frac{du}{Q_\mu(x, \lambda)},$$

which is equivalent to the expression given in Table 1. The singular map associated with this case is of the type D^h_μ.

1.3. Pfaffian equations for correspondence maps. All correspondence maps from the third column of Table 1 satisfy some simple Pfaffian equations given in the fourth column. To prove this, we proceed as follows.

Case S_0. The function $y = x^\lambda$ satisfies the Pfaffian equation

$$x\,dy - \lambda y\,dx = 0.$$

Case S_μ. If one writes the function

$$F(x, y) = m \ln y + \int_{x^m}^{y^n} \frac{du}{uP_\mu(u, \lambda)},$$

then its differential vanishes on the graph of the correspondence map, because the function itself vanishes identically on the graph. Therefore, on the tangent bundle to the graph the differential dF vanishes,

$$0 = dF(x, y) = \frac{m\,dy}{y} - \frac{m\,dx}{x\,P_\mu(x^m, \lambda)} + \frac{n\,dy}{y\,P_\mu(y^n, \lambda)}.$$

The last identity implies that the graph is the integral curve of the Pfaffian equation $\widetilde{\omega} = 0$, where

$$\widetilde{\omega} = x\,P_\mu(x^m, \lambda)(n/m + P_\mu(y^n, \lambda))\,dy - y\,P_\mu(y^n, \lambda)\,dx.$$

Case D^h_μ. The similar computation applied to the function

$$F(x, y) = y + \exp \int_x^1 \frac{du}{P_\mu(u, \lambda)}$$

yields the identity

$$dy + \frac{y\,dx}{P_\mu(x, \lambda)} = 0$$

on the graph of the correspondence map. This identity is equivalent to the polynomial Pfaffian equation given in Table 1.

Case D^c_μ. This is the simplest case: any linear function satisfies the Euler-type Pfaffian equation

$$x\,dy - y\,dx = 0.$$

Thus we proved that each correspondence map, being irregular as they are, satisfies some simple Pfaffian equation. The most important property of these equations is their *algebraicity*: the coefficients of the Pfaffian forms are polynomial both in the corresponding phase variables x, y and in the parameters of the versal unfoldings λ entering as coefficients of the polynomials P, Q.

Actually we will use somewhat different forms of Pfaffian equations, because of additional topological properties required by the Khovanskiĭ reduction procedure, and we will consider them on slightly different domains. The new equations differ inessentially from the ones listed in the fourth column of the table, and remain to be algebraic.

1.4. Basic system. Here we describe the system of equations which will be analyzed from now on. Assume that a polycycle occurs in a generic k-parameter family of vector fields, and all the vertices of the polycycle are elementary.

Then the number n of vertices does not exceed k. Moreover, one can claim that each vertex is of one of the types S_{μ_j}, $\mu_j \geqslant 0$, or $D^{h,c}_{\mu_j}$, $\mu_j \geqslant 1$, and $\sum \mu_j \leqslant k$. Indeed, the codimension of each type of singularity is μ_j, therefore by the transversality theorem all more degenerate cases can be avoided by slightly perturbing the family of vector fields. The first inequality for the number of vertices is also almost evident. If an arc joining jth and $(j+1)$ st vertices is formed by two hyperbolic invariant curves for the corresponding vertices, then such a phenomenon occurs in codimension 1. If the arc starts (or ends) in a parabolic sector of an elementary singularity, then such a heteroclinic connection is itself a codimension 0 phenomenon, but in this case the degree of degeneracy of the corresponding vertex must be at least 1. Summing up these (independent) codimensions, we conclude that too many (more than k) heteroclinic connections cannot occur in a generic k-parameter family.

Next, we proceed with introducing the normalizing C^ℓ-smooth local coordinates near each elementary vertex, as this is described above (the exact order of smoothness will be specified later on). Then a pair of C^ℓ-smooth transversals can be chosen near each vertex and endowed with local C^ℓ-smooth charts x_j, y_j in such a way that the correspondence map taking a point with a coordinate x_j on the "entrance" transversal to a point with the coordinate y_j on the "exit" transversal, will be of one of the standard types listed in Table 1.

More precisely, for each vertex $j = 1, \dots, n$, Theorem 1.1 yields the *localization point* $\mathbf{c}_j = (0, \dots, 0, c_j) \in \mathbb{R}^{\mu_j+1}$, where $c_j \in \mathbb{R}^1$ is the formal invariant of the unperturbed singular point, and also if the jth vertex is a resonant saddle, then the rational hyperbolicity ratio $n : m$ is explicitly specified.

Denote by $\Delta_{l,\mu}(x, \lambda)$ the correspondence map for each of the four types of singularities from Table 1, $l = S_0$, S_μ, D^c_μ or D^h_μ, with the corresponding index $\mu \in \mathbb{N}$ (by definition, $\mu = 0$ for $l = S_0$). In case S_μ with $\mu > 0$ we consider the relatively prime pair of natural numbers as an additional parameter of the corresponding map, so in this case the rigorous notation would be $\Delta_{S_\mu,\mu}(x, \lambda, n, m)$.

DEFINITION 1.1. 1. The *unspecified basic system* for determination of limit cycles occurring in k-parameter families of vector fields is the system of n regular and n singular functional equations in $2n$ variables x_j, y_j, depending on parameters λ^j, n_j, m_j, ε,

$$\begin{cases} y_j = \Delta_{l_j, \mu_j}(x_j, \lambda^j [, n_j, m_j]), & \lambda^j \in \mathbb{R}^{\mu_j+1}, \\ x_{j+1} = f_j(y_j, \varepsilon), & \varepsilon \in (\mathbb{R}^k, 0). \end{cases}$$

$$(1.1) \qquad j = 1, \ldots, n \mod (n), \quad l_j \in \left\{S_0, S_\mu, D_\mu^c, D_\mu^h\right\},$$

$$n_j, m_j \in \mathbb{N}, \quad \mu_j \in \mathbb{Z}_+, \quad \sum \mu_j \leqslant k, \quad n \leqslant k,$$

Δ depends on n_j, m_j only if $l_j = S_\mu$ with $\mu > 0$.

2. A *specified* basic system is one of a finite number of unspecified basic systems together with an explicit indication of *specification*, which by definition is the collection consisting of:

(1) localization points $\mathbf{c}_j = (0, \ldots, 0, c_j) \in \mathbb{R}^{\mu_j+1}$; in particular this means that hyperbolicity ratios of all nonresonant saddles are explicitly given;
(2) hyperbolicity ratios $n_j : m_j$ for all resonant saddles;
(3) smooth functions $f_j(x, \varepsilon)$ depending on parameters ε defined in some open neighborhoods $(\mathbb{R}^{k+1}, 0)_j$; we assume that $f_j(0, 0) = 0$;
(4) *characteristic size*, that is, the value $r > 0$ which determines the *domain* of the specified basic system as follows:

$$(x, y) \in I_r = \{|x_j| < r, \ |y_j| < r, \quad j = 1, \ldots, n\} \subset \mathbb{R}^{2n};$$
$$(\lambda, \varepsilon) \in B_r = \left\{\|\lambda^j - \mathbf{c}_j\| < r, \ \|\varepsilon\| < r\right\} \subset \mathbb{R}^{k+\mu_1+\cdots+\mu_n},$$

where λ is the tuple of *all parameters* of all normal forms from Table 1, $\lambda = (\lambda^1, \ldots, \lambda^n)$; the characteristic size must be so small that all functions f_j were defined for the corresponding values of their arguments.

NOTES TO DEFINITION 1.1, RELATED DEFINITIONS. 1. Speaking less formally, the unspecified basic system is a system of equations depending on parameters with the slots for the functions f_j and for the resonant hyperbolicity ratios $n_j : m_j$ left open. There is only a finite number of unspecified basic systems, each being completely characterized by the string of discrete data

$$(1.2) \qquad \mathcal{T} = (l_1, \mu_1, \ldots, l_n, \mu_n)$$

subject to the restriction $n \leqslant k$, $\sum \mu_j \leqslant k$. We call the data $\mathcal{T}$ the *combinatorial type of the unspecified basic system.* In §§2, 3 we introduce the language of Pfaffian manifolds, and in this language the notion of an unspecified basic system reduces to the notion of a Pfaffian manifold.

2. The specification supplies all the remaining information which is necessary to transform the basic system into a system of parameter-depending functional

equations so that one could speak about the number of solutions for all appropriate values of parameters λ, ε, n_j, m_j. Formally, we consider a specification as a tuple

$$\begin{aligned}\mathcal{S} = \{r, c_1, \dots, c_n, n_{j_1}, m_{j_1}, \dots, m_{j_s}, n_{j_s} \text{ (for all resonant saddles)};\\ f_1(\cdot, \cdot), \dots, f_n(\cdot, \cdot)\},\\ r > 0,\ c_j \in \mathbb{R}^1,\ m_{j_\alpha}, n_{j_\alpha} \in \mathbb{N},\ f_j \in \mathbb{C}^\ell(\mathbb{R}^{k+1}, 0).\end{aligned} \tag{1.3}$$

The string

$$\mathcal{S}_a = (c_1, \dots, c_n, \dots, m_{j_\alpha}, n_{j_\alpha}, \dots) \in \mathbb{R}^{n+2s}$$

will be referred to as the *algebraic part of the specification* (for reasons to be clarified later), while the string of functions

$$\mathcal{S}_f = \mathbf{f} = (f_1, \dots, f_n)$$

is called the *functional part* of the specification. The value r was earlier introduced as the characteristic size. Slightly abusing the language, we say that the functions f_j are defined in the domain $I_r \times B_r$, though in fact each of them depends only on one variable x_j and on the part of the parameters $(\varepsilon, \lambda) \in B_r$. Later we will introduce a more rigorous language of Cartesian functions, see §3.

3. When referring to a certain unspecified basic system, we will identify it with the corresponding type $\mathcal{T}$, while the specified system will be referred to as $(\mathcal{T}, \mathcal{S})$, or $(\mathcal{T}, \mathcal{S}_a, \mathcal{S}_f; r)$, or $(\mathcal{T}, \mathcal{S}_a, \mathbf{f}; r)$, whichever form would be more convenient at any moment.

4. The preliminary basic system as it appeared in (0.2), can be obtained from the corresponding specified basic system by adding to the latter the equations determining the reparametrization from Theorem 1.1,

$$\lambda^j = \lambda^j(\varepsilon) = (\lambda_0^j(\varepsilon), \dots, \lambda_{\mu_j}^j(\varepsilon)), \qquad j = 1, \dots, n,$$

subject to the requirements $\lambda^j(0) = \mathbf{c}_j$. Clearly, if we drop out those additional equations and consider λ^j as free parameters, then the number of solutions of the preliminary basic system would not exceed that for specified basic system.

After all these notions (or rather the language) being introduced, we can formulate the problem of estimating cyclicity of elementary polycycles occurring in generic k-parameter families in the following form.

THEOREM 1.2 (Existential finiteness theorem for basic systems in codimension k). *For any type $\mathcal{T}$ of unspecified basic system and any choice of the algebraic part $\mathcal{S}_a$ and the characteristic size r_0 one can choose the order of smoothness ℓ_0 and an open dense subset $\mathfrak{F} = \mathfrak{F}_{\mathcal{T}, \mathcal{S}_a, r_0}$ in the space of C^{ℓ_0}-smooth functions $C^{\ell_0}(I_{r_0} \times B_{r_0}, \mathbb{R}^n)$ such that for every $\mathbf{f} = (f_1, \dots, f_n) \in \mathfrak{F}$ the number of isolated solutions $\mathfrak{B}(\mathcal{T}, \mathcal{S}_a, \mathbf{f}; r_0)$ to the specified basic system $(\mathcal{T}, \mathcal{S}_a, \mathbf{f})$ in the domain I_{r_0} is uniformly bounded over all parameter values $(\lambda, \varepsilon) \in B_{r_0}$:*

$$\begin{aligned}&\mathfrak{B}(\mathcal{T}, \mathcal{S}_a, \mathbf{f}; r_0)\\ &\quad= \sup_{(\varepsilon, \lambda) \in B_{r_0}} \#\{(x, y) \text{ satisfying } (1.1),\ (x, y) \in I_{r_0}\} < \infty.\end{aligned} \tag{1.4}$$

THEOREM 1.3 (Constructive finiteness theorem for basic systems in codimension k). *In the notations of the previous theorem one can find another order of smoothness and a narrower subspace* $\widetilde{\mathfrak{F}} = \widetilde{\mathfrak{F}}_{\mathcal{T}, \mathcal{S}_a, r_0} \subset C^\ell(I_{r_0} \times B_{r_0}, \mathbb{R}^n)$, *open in* C^ℓ*-topology and* $C^{\ell-1}$*-dense, such that for any* $\mathbf{f} \in \widetilde{\mathfrak{F}}$ *the number of isolated solutions of the specified basic system in a family of domains of decreasing size admits an effective upper bound*:

$$\limsup_{r \to 0^+} \mathfrak{B}(\mathcal{T}, \mathcal{S}_a, \mathbf{f}; r) \leqslant E(k),$$

where $E(k)$ *is a primitive recursive function of the integer argument* k. *The necessary order of smoothness* ℓ *can be also effectively estimated from above.*

The choice of the subspace $\widetilde{\mathfrak{F}}$ *depends on the algebraic part of the specialization, while the upper estimate* $E(k)$ *does not, being uniform for any choice of the specialization (as soon as* $\mathbf{f}$ *remains in* $\widetilde{\mathfrak{F}}$*) and any type* $\mathcal{T}$ *of the basic system in codimension* k.

REMARK. The genericity-type assumptions appearing in the formulation of the above theorems need to be clarified, since the functions $\mathbf{f}$ constituting the functional part of the specialization depend only on parts of coordinates. We say that a subset $\mathfrak{F} \subset C^\ell$ is dense if by a C^ℓ-small variation of the coordinate functions f_j *preserving their dependence only on the same variables and parameters*, each vector-function $\mathbf{f} = (f_1, \ldots, f_n)$ can be put into $\mathfrak{F}$. In the same manner, we say that $\mathfrak{F}$ is open if such variations, provided they are sufficiently C^ℓ-small, cannot move any element $\mathbf{f} \in \mathfrak{F}$ from $\mathfrak{F}$. More formally these notions are introduced in §3 under the name of Cartesian genericity.

These two theorems evidently imply Theorems I and II respectively. The rest of the paper is devoted to their proof.

§2. Khovanskiĭ reduction procedure

In this section we outline the method of reducing a functional–Pfaffian system to a functional one. In its full generality, this construction is described in the book [Kh].

2.1. Pfaffian systems and their separating solutions. Let M be a smooth orientable n-dimensional manifold, not necessarily compact or connected, and ω a smooth 1-form on it.

DEFINITION 2.1. A codimension 1 smooth submanifold $\Gamma \subset M$ is the separating solution for the Pfaffian equation $\omega = 0$, if

(1) Γ is the integral manifold, that is, the restriction of ω to the tangent bundle of Γ is identically zero:

$$\forall x \in \Gamma, \ \forall v \in T_x\Gamma \qquad \omega(v) = 0;$$

(2) Γ does not pass through singular points of ω:

$$\forall x \in \Gamma \quad \omega(x)|_{T_xM} \neq 0;$$

(3) Γ is the boundary of a domain $D \subseteq M$ and the co-orientation induced on Γ by ω coincides with its co-orientation as the boundary. In other words, on any vector pointing outward from D, the form is positive.

Let now $\omega_1, \dots, \omega_k$ be an ordered k-tuple of smooth 1-forms on M. Consider the system of Pfaffian equations

$$\omega_1 = 0, \dots, \omega_k = 0. \tag{2.1}$$

DEFINITION 2.2. A submanifold Γ is a separating solution for the system of Pfaffian equations if there exists an increasing chain of smooth submanifolds,

$$\Gamma = \Gamma_k \subset \Gamma_{k-1} \subset \cdots \subset \Gamma_1 \subset \Gamma_0 = M \tag{2.2}$$

such that for any $j = 1, \dots, k$, Γ_j is the separating solution for the Pfaffian equation on Γ_{j-1} defined by the restriction of the form ω_j to Γ_{j-1}.

Now we add functional equations, considering a Pfaffian system (2.1) together with a smooth map $F \colon M \to \mathbb{R}^{n-k}$ so that the total number of equations is equal to the dimension of M.

DEFINITION 2.3. Let $\Omega = (\omega_1, \dots, \omega_k) \in (\Lambda^1(M))^k$ be a k-tuple of 1-forms, and $F \colon M \to \mathbb{R}^{n-k}$ a smooth map. A solution to the *mixed functional–Pfaffian system*

$$\Omega = 0, \quad F = a, \qquad a \in \mathbb{R}^{n-k} \tag{2.3}$$

is a pair (Γ, L_a), where $L_a \subseteq M$ is the preimage $F^{-1}(a)$, Γ a separating solution for the Pfaffian system $\Omega = 0$, and the intersection $\Gamma \cap L_a$ is nonempty. The solution is *regular* if a is the regular value for the restriction of the map F to Γ.

If (Γ, L_a) is a regular solution, then the intersection $\Gamma \cap L_a$ consists of isolated points.

DEFINITION 2.4. The *Khovanskiĭ number* $\mathcal{K}(\Omega, F)$ for the mixed system (2.3) is the upper bound for the cardinalities $\#(\Gamma \cap L_a)$ over all regular solutions of the system.

REMARKS. 1. The interior and the exterior topologies on Γ are equivalent, because Γ is the separating solution.

2. The Khovanskiĭ number is also defined if $k = 0$, i.e., there are no Pfaffian equations at all. In this case one can define formally $\Gamma = M$, and $\mathcal{K}(\varnothing, F)$ is equal to the upper bound of the cardinality of preimages $\#F^{-1}(a)$ of *regular values* for the map $F \colon M \to \mathbb{R}^n$.

3. The notation (2.3) is slightly misleading, because the point a plays a role of indeterminate variable in the definition of the Khovanskiĭ number. So formally a functional–Pfaffian system of the form (2.3) is the pair (Ω, F) of a Pfaffian system and a smooth map with the total number of equations (both Pfaffian and functional) being equal to the dimension of the manifold M: $\#\Omega + \#F = k + (n-k) = n = \dim M$. The notation (2.3) looks more traditionally, so it will be sometimes used as well.

4. If we want to stress in the notation the phase space M of the functional–Pfaffian system, we use the notation $\mathcal{K}_M(\Omega, F)$. Usually this is necessary when both F and Ω are defined on the Euclidean space $\mathbb{R}^n$, while we are interested only in solutions belonging to some (open) ball.

The main idea of the reduction principle is to estimate the Khovanskiĭ number for a given mixed system by the Khovanskiĭ numbers of some auxiliary systems *containing a smaller number of Pfaffian equations.*

2.2. Reduction principle for one Pfaffian equation. We show how to eliminate the Pfaffian equation from the mixed system

$$\omega = 0, \qquad F = a, \qquad F: M \to \mathbb{R}^{n-1}, \tag{2.4}$$

outlining the key ideas.

DEFINITION 2.5. A smooth positive function $\rho: M \to \mathbb{R}_+$ is said to be *covering* if it tends to zero along any nonaccumulating sequence of points in M. In other terms, ρ vanishes "at infinity" on M, so that all level hypersurfaces of the covering function are compact subsets of M.

REMARK. This definition applies both to compact and noncompact manifolds, but in the compact case a smooth function is covering if and only if it is everywhere positive, thus automatically bounded away from zero.

Suppose that the manifold M is endowed with a Riemannian volume. Since it is orientable, one may use the duality between functions and n-forms on M. Denote by the asterisk the operator taking an n-form into the function (dividing by the volume form).

Let $F_1, \dots, F_{n-1}$ be the coordinate functions of the map F in (2.4); by introducing them, we explicitly use the Euclidean structure on $\mathbb{R}^{n-1}$.

DEFINITION 2.6. The contact function for the mixed system (2.4) is

$$F_n = *(\omega \wedge dF_1 \wedge \cdots \wedge dF_{n-1}). \tag{2.5}$$

The operator taking the mixed system (ω, F) into the corresponding contact function will be denoted by $\mathfrak{d}: (\omega, F) \mapsto \mathfrak{d}(\omega, F) = F_n$.

Define the two maps by their coordinate functions,

$$F^c = (F_1, \dots, F_{n-1}, F_n), \qquad F^\infty = (F_1, \dots, F_{n-1}, \rho),$$

both taking M to $\mathbb{R}^n$, where F_n is the contact function (2.5), and ρ is the covering function.

THEOREM 2.1 [Kh]. *Suppose that the system* (2.4) *admits regular solutions in the sense of Definition* 2.3. *Then*

$$\mathcal{K}(\omega, F) \leqslant \tfrac{1}{2}\mathcal{K}(\varnothing, F^\infty) + \mathcal{K}(\varnothing, F^c). \tag{2.6}$$

PROOF. Take any regular solution (Γ, L_a) for (2.4), and suppose that the intersection $\Gamma \cap L_a$ consists of d points. Since a is the regular value of the restriction $F|_\Gamma$, any small variation of a may result only in increasing of the number of intersection points. Take a to be a regular value of the map F itself (rather than of its restriction to Γ).

Then the fiber L_a is a one-dimensional smooth manifold, intersecting Γ transversally. By the classification theorem for one-dimensional manifolds, L_a is the union of compact and noncompact connected components, each diffeomorphic to the circle or the line respectively, see Figure 4. These diffeomorphisms provide the

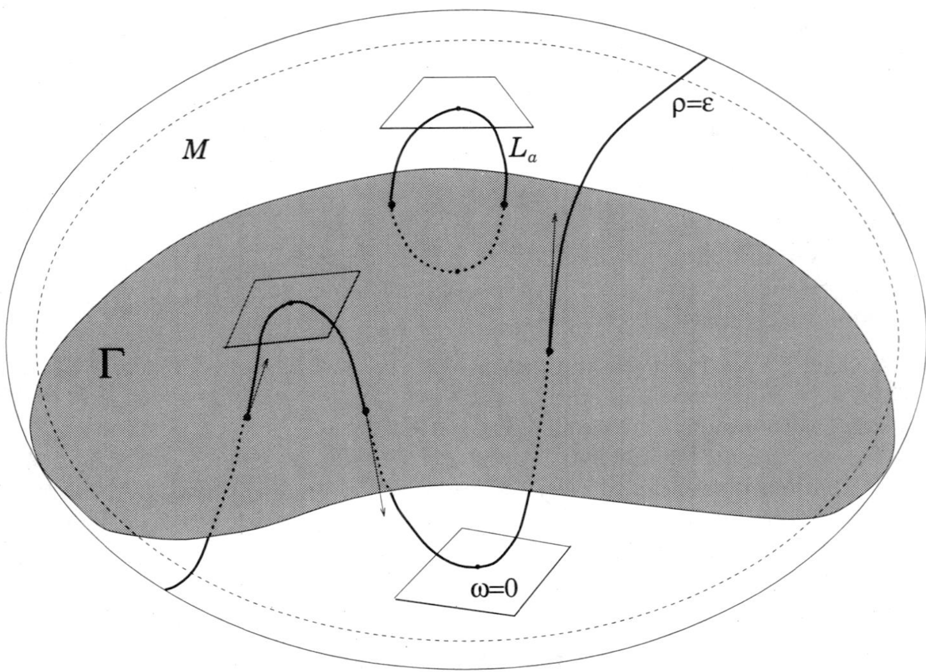

FIGURE 4. Elimination of one Pfaffian equation

components with the parameterizations, therefore each point of intersection with Γ is automatically supplied with the sign of the value taken by the form ω on the velocity vector. This value is always nonzero, since the intersections are transversal, and any reparameterization either preserves all the signs on a given component, or reverses all of them.

The key fact is that on any component the signs alternate: this follows from the topological conditions imposed on separating solutions. Indeed, each component successively enters and leaves the domain D bounded by Γ, and therefore the corresponding signs alternate.

Denote the parameter on the given component by t, $t \in \mathbb{R}$ or $t \in \mathbb{R} \bmod \mathbb{Z}$ respectively, and consider the function $s(t)$ equal to the value taken by the form ω on the velocity vector $\dot{x}(t)$. This function is differentiable, hence between any two points at which it has opposite signs there must be a point $x \in L_a$ at which the function vanishes. But this means that at this point the curve is tangent to the zero space of the linear functional $\omega(x)$, therefore the linear functionals $dF_1(x), \dots, dF_{n-1}(x)$, and $\omega(x)$ are linear dependent, i.e., $F_n(x) = 0$. We call such a point the point of contact.

Thus we proved that between any two points of intersection of L_a with Γ there must be a point of contact. If the component under consideration is compact, then the number of points of contact (geometrically distinct) is no less than the number of the points of intersection of this component with Γ, while on the noncompact component the number of contacts is at most one less then the number of intersections. Therefore the total number of intersections does not exceed the total number of contacts (on all the components) plus the number of different noncompact components of L_a.

It is clear that the first number does not exceed $\mathcal{K}(\varnothing, F^{\mathrm{c}})$. Indeed, zeros of the contact function restricted to the fiber L_a may be not simple, but choosing a regular value δ for the restriction $F^{\mathrm{c}}|_{L_a}$ to be sufficiently small, we may ensure that there will be as many *nondegenerate zeros* for the function $F^{\mathrm{c}} - \delta$ as necessary. In more detail this construction is described in Lemma 3.2.

It remains only to prove that the value $\frac{1}{2}\mathcal{K}(\varnothing, F^{\infty})$ estimates the number of noncompact components. But each noncompact component must intersect the level surface $\rho = \varepsilon$ for any small $\varepsilon > 0$ at least twice. Taking ε to be a regular value of ρ on L_a, we complete the proof of the theorem. □

COROLLARY. *If the manifold M is compact, then*

$$\mathcal{K}(\omega, F) \leqslant \mathcal{K}(\varnothing, F^{\mathrm{c}}). \tag{2.7}$$

PROOF. Indeed, in this case the first term in (2.6) disappears. □

REMARK. The choice of the Riemannian volume form is not essential for the above construction. Indeed, if the volume form vol^n is replaced by a new one $b \cdot \mathrm{vol}^n$, where b is a positive function, then the function F_n will be replaced by $\widetilde{F}_n = b^{-1}F_n$, and the map $\widetilde{F}^{\mathrm{c}} = (F_1, \dots, F_{n-1}, \widetilde{F}_n)$ will have the same zero set.

2.3. The Khovanskiĭ reduction in the general case. Consider now the general case of the mixed system (2.3) with $k > 1$. Suppose that Γ is a separating solution for the Pfaffian system $\Omega = 0$. By definition, this means that there exists a separating solution Γ_k of the Pfaffian equation $\omega_k = 0$ on a separating solution Γ_{k-1} of the Pfaffian system $\Omega' = 0$, where $\Omega' = (\omega_1, \dots, \omega_{k-1})$. Note that if ρ is a covering function on the manifold M, then its restriction to Γ_{n-k} is the covering function for the latter submanifold. Next, one can endow Γ_{k-1} with the Riemannian $(n-k+1)$-volume form $\mathrm{vol}^{n-k}_{\Gamma_{k-1}}$ in such a way that

$$\omega_1 \wedge \dots \wedge \omega_{k-1} \wedge \mathrm{vol}^{n-k+1}_{\Gamma_{k-1}} = \mathrm{vol}^n_M .$$

Since the forms ω_j, $j = 1, \dots, k-1$ are linearly independent in a neighborhood of Γ_{k-1}, this formula defines a volume form near Γ_{k-1}. As was mentioned above, the choice of the Riemannian volume form does not affect the assertion of Theorem 2.1.

Thus one can apply Theorem 2.1 to the mixed system

$$\omega_k = 0, \quad F = a$$

on the manifold Γ_{n-k}. To describe the result, we introduce the following two maps from M to $\mathbb{R}^{n-k+1}$:

$$F^{\mathrm{c}} = (F_1, \dots, F_{n-k}, \rho), \qquad F^{\infty} = (F_1, \dots, F_{n-k}, F_*), \tag{2.8}$$

where ρ is the covering function on the manifold M, and $F_*: M \to \mathbb{R}$ is the smooth function defined by the formula

$$F_* = \mathfrak{d}(\Omega, F) = *(dF_1 \wedge \cdots \wedge dF_{n-k} \wedge \omega_1 \wedge \cdots \wedge \omega_k). \tag{2.9}$$

The above choice of the Riemannian volume on Γ_{k-1} implies that the star operator in the ambient manifold M agrees with the star operator relevant to Γ_{k-1}, therefore the formula (2.9) defines the same function as the formula (2.5): $\omega_k \wedge dF_1 \wedge \cdots \wedge dF_{n-k} = F_* \cdot \mathrm{vol}^{n-k+1}_{\Gamma_{k-1}}$.

THEOREM 2.2. *Let Ω, F, F^{c}, and F^{∞} be as above. Then*

$$\mathcal{K}(\Omega, F) \leqslant \tfrac{1}{2}\mathcal{K}(\Omega', F^{\infty}) + \mathcal{K}(\Omega', F^{\mathrm{c}}). \tag{2.10}$$

COROLLARY. *If the map F is such that any point has a compact preimage (possibly empty), then*

$$\mathcal{K}(\Omega, F) \leqslant \mathcal{K}(\Omega', F^{\mathrm{c}}). \tag{2.10c}$$

PROOF. Straightforward application of Theorem 2.1. □

Iterating the above two statements, we can replace consecutively the Pfaffian equations by the functional ones, obtaining new systems whose Khovanskiĭ numbers estimate from above those of the initial one, according to of the inequality (2.10) and its compact counterpart (2.10c). On each step one can replace a Pfaffian equation either by the contact function, or by the covering function. But once the covering function appears among the functional equations, the fibers (the level sets) $F^{-1}(\cdot)$ become compact, hence on the next steps the Corollary to Theorem 2.2 applies rather than the theorem itself.

Denote by T^{c} and T^{∞} the two operators, transforming the mixed system (Ω, F) into the mixed systems $(\Omega', F^{\mathrm{c}})$ and (Ω', F^{∞}) respectively, where the maps F^{c} and F^{∞} are given by (2.8), (2.9):

$$T^{\mathrm{c}}(\Omega, F) = (\Omega', F, \mathfrak{d}(\Omega, F)), \qquad T^{\infty}(\Omega, F) = (\Omega', F, \rho). \tag{2.11}$$

If we start with the mixed functional–Pfaffian system (Ω, F), with Ω being a k-tuple $(\omega_1, \ldots, w_k)$, and eliminate subsequently the forms ω_k, $\omega_{k-1}, \ldots$, then the following maps from M to $\mathbb{R}^n$ arise:

(1) the map $F_{[0]}$, if on each step the contact function was used,

$$(\varnothing, F_{[0]}) = (\underbrace{T^{\mathrm{c}} \circ \cdots \circ T^{\mathrm{c}}}_{k \text{ times}})(\Omega, F), \tag{2.12}$$

(2) the maps $F_{[j]}$, if on the jth step the covering function was used, while on all other steps the contact ones were, $j = 1, \ldots, k$,

$$(\varnothing, F_{[j]}) = (\underbrace{T^{\mathrm{c}} \circ \cdots \circ T^{\mathrm{c}}}_{k-j \text{ times}} \circ T^{\infty} \circ \underbrace{T^{\mathrm{c}} \circ \cdots \circ T^{\mathrm{c}}}_{j-1 \text{ times}})(\Omega, F). \tag{2.13}$$

Then inductive application of Theorem 2.2 immediately yields the following fundamental result.

THEOREM 2.3. *The Khovanskiĭ number for the mixed system* (2.3) *on a manifold* M *with the covering function* ρ *admits the upper estimate by a linear combination of Khovanskiĭ numbers of some* $k+1$ *auxiliary systems, each of them containing no Pfaffian equations at all*:

$$\mathcal{K}(\Omega, F) \leqslant \mathcal{K}(\varnothing, F_{[0]}) + \frac{1}{2}\sum_{j=1}^{k} \mathcal{K}(\varnothing, F_{[j]}), \tag{2.14}$$

where the maps $F_{[j]}$ *are defined by formulas* (2.11)–(2.13).

2.4. Corollaries. The most important and characteristic feature of the Khovanskiĭ reduction process is its constructive character. This immediately implies several important corollaries, which will be now formulated.

Assume that the manifold M is an open domain in $\mathbb{R}^n$ and admits a polynomial covering function ρ. The principal example is the unit ball $B = \{x \in \mathbb{R}^n\colon \sum_j x_j^2 < 1\}$, for which one may take $\rho(x) = 1 - \sum_j x_j^2$. Then the Riemannian volume form can be chosen to be algebraic, $dx_1 \wedge \cdots \wedge dx_n$.

Assume also that all the forms ω_i, $i = 1, \dots, k$, are polynomial (i.e., with polynomial coefficients), and the map F is at least C^k-smooth. Since the operators T^c, T^∞ introduced above involve only algebraic operations and differentiation of functions, the following holds.

THEOREM 2.4. *If in the system* (2.3) *defined on a semialgebraic subset* $M \subseteq \mathbb{R}^n$ *all Pfaffian forms and the covering function* ρ *are polynomial of degrees* $\leqslant d$, *then all the maps* $F_{[\alpha]}\colon M \to \mathbb{R}^n$, $\alpha = 0, 1, \dots, k$ *constructed in Theorem* 2.3 *are of the form*

$$F_{[\alpha]} = P_\alpha \circ (j^k F), \tag{2.15}$$

where $j^k F$ *is the* k*-jet extension of the map* F *and* P_α *are certain polynomials defined on the jet space* $J^k(\mathbb{R}^n, \mathbb{R}^{n-k})$.

The degrees of all the polynomials P_α *admit an upper estimate by a primitive recursive function of the variables* $n = \dim M$ *and* d.

PROOF. Indeed, the procedure of successive elimination of Pfaffian equations amounts to inductive application of the operator $\mathfrak{d}$ given by the formula (2.9) at most k times and to the substitution the covering function at most once. □

§3. Functional–Pfaffian system for limit cycles

In this section we consider a specified basic system $(\mathcal{T}, \mathcal{S}_a, \mathbf{f}; r)$ obtained from the unspecified basic system (1.1), that is we consider a system (1.1) together with a collection of formal invariants $(c_1, \dots, c_n)$ of all singularities (which determines a point in the λ-space), a collection of hyperbolicity ratios $n_{j_\alpha} : m_{j_\alpha}$ of all resonant saddles and a tuple of sufficiently smooth functions f_j on a sufficiently small open cube $I_r \times B_r$ in the (ε, λ)-space.

Our local goal is to reduce this system to a functional–Pfaffian system having the form described in §2, with the following properties:

- the new system has the form allowing for application of Theorem 2.2;

- the number of *regular solutions* to the functional–Pfaffian system is not less than the number of *isolated* solutions to (1.1), up to a term of order $O(k)$, where k is the number of parameters of the original family.

After application of Theorem 2.2 we will obtain a number of *chain maps* with controlled degrees of the exterior polynomial parts.

3.1. Upper estimate of the number of solutions for the basic system: statement of results. First of all we make the following remark. The algebraic part of the specialization can be identified with a point

$$\mathbf{A} = (c_1, \dots, c_n, n_{j_1}, m_{j_1}, \dots, n_{j_s}, m_{j_s}) \in \mathbb{R}^{n+2s}, \tag{3.0}$$

where $s \leqslant n$ is the number of resonant saddles on the polycycle: the fact that the numbers n_{j_α}, m_{j_α} are natural will become inessential for our constructions.

THEOREM 3.1 (reduction from basic to functional–Pfaffian system). *Consider an unspecified basic system* (1.1) *of a certain type* $\mathcal{T}$ *in codimension* k, *together with an arbitrary specification*

$$\mathcal{S} = (\mathbf{A}, \mathbf{f}; r), \qquad \mathbb{R}^{n+2s} \ni \mathbf{A} \sim \mathcal{S}_a, \quad C^\ell(I_r \times B_r, \mathbb{R}^n) \ni \mathbf{f} \sim \mathcal{S}_f, \quad r > 0.$$

Then one can explicitly construct a functional–Pfaffian system of the form (Ω, F), $\Omega = (\Omega_1, \dots, \Omega_{n+2s})$, $F = (F_1, \dots, F_{n+k+m})$, $m = \sum \mu_j$, *or in a more traditional notation, the mixed system of functional and Pfaffian equations*

$$\Omega = 0, \quad F = a, \tag{3.1}$$

defined in a certain open bounded semialgebraic subset

$$M = M(r) \subset I_r \times B_r \times \mathbb{R}^{2s}$$

(*see Definition* 1.1), *such that the following holds.*

(1) *For any choice of the parameters* $(\varepsilon, \lambda) \in B_r$ *the number of isolated* (x, y)-*solutions of the specified basic system* $(\mathcal{T}, \mathbf{A}, \mathbf{f}; r)$ *can be estimated through the Khovanskiĭ number of the system* (3.1) *on the manifold* $M = M(r)$: *for the value* $\mathfrak{B}$ *defined in* (1.4) *one has the estimate*

$$\mathfrak{B}(\mathcal{T}, \mathbf{A}, \mathbf{f}; r) \leqslant \mathcal{K}_{M(r)}(\Omega, F) + k.$$

(2) *The coefficients of the forms* Ω_k *are polynomial in all their arguments, and also in coordinates of the point* $\mathbf{A} \in \mathbb{R}^{n+2s}$; *the degrees of those polynomials do not exceed a certain upper bound* $d(k)$ *which is of order of* $O(k)$ *and may be computed explicitly; the coefficients of those polynomials are integer numbers with absolute values not exceeding* $d(k)$.

(3) *The covering function* $\rho(\cdot\,; r)$ *for the phase space* $M(r)$ *is polynomial in all its arguments and also in* r, *of the total degree not exceeding the same bound* $d(k) = O(k)$; *the coefficients of this polynomial are integers with the absolute values not exceeding* $d(k)$.

(4) *The coordinate functions of the maps* F_β *are explicitly given as polynomials of the first degree on the* 0*-jet space of functions* $J^0(I_r \times B_r, \mathbb{R}^n)$ *with coefficients* ± 1.

REMARK. The proof of this theorem is completely constructive: we will write explicitly all functional and Pfaffian equations and the covering function of the domain M, see the formulas (3.2). The above formulation stresses the properties of this system, which will be used later.

The construction is carried over in 3.2, 3.3. In 3.4 we describe the result of the application of the Khovanskiĭ reduction to the system (3.1).

REMARK. We would like to stress the fact that in Theorem 3.1 the number of *isolated solutions* to the basic system is estimated through the number of *nondegenerate solutions* (the Khovanskiĭ number) of the system (3.1). All subsequent analysis will be heavily based on nondegeneracy-type arguments and considerations.

3.2. Forms and domains. In this subsection we modify in an appropriate manner the differential equations for the correspondence functions, so that the topological conditions required by the Khovanskiĭ procedure, will be satisfied. There are three major reasons for such a modification:

- the graphs of the standard correspondence maps for resonant saddles, as described in Table 1, do contain singular points (zeros) of the Pfaffian forms occurring in this table;
- even after deleting singularities of the forms, the co-orientation induced by the forms on the graphs, disagree with their co-orientation as boundaries for some open domains;
- the degrees of polynomial forms corresponding to the resonant saddles S_μ, are not bounded from above by a constant depending only on k: indeed, even in a generic 1-parameter family, resonance of any order $(m : n)$ may occur, thus the degree of the polynomial form occurring in the fourth column of Table 1, can be arbitrarily large.

We start with modifying the Pfaffian system in order to achieve a uniform bound for the degrees of the Pfaffian equations, for the saddle resonant case S_μ. The key idea is to introduce the new variables, $z = x^m$ and $w = y^n$: in these new variables the coefficients of the Pfaffian form given in Table 1, have degree $2\mu + 1$ independently of m, n, while the new variables themselves, considered as functions of x, y, satisfy linear Pfaffian equations.

PROPOSITION. *Let* $\Delta(x, \lambda) = \Delta^{(n:m)}_{S_\mu, \mu}(x, \lambda)$ *be the standard correspondence map for the* $(n : m)$*-resonant type* S_μ *(see Table* 1*). Then for each* $\lambda \in \mathbb{R}^{\mu+1}$, *the curve in* $\mathbb{R}^4$, *defined by the equations*

$$y = \Delta(x, \lambda), \qquad z = x^m, \qquad w = y^n, \tag{3.2}$$

is an integral curve for the Pfaffian system consisting of three polynomial Pfaffian equations

$$\omega = y\, P_\mu(w, \lambda)\, dx - \left(P_\mu(w, \lambda) + \frac{n}{m}\right) x\, P_\mu(z, \lambda)\, dy,$$
$$x\, dz = mz\, dx, \qquad y\, dw = nw\, dy.$$

PROOF. An obvious remark. □

In order to satisfy the topological conditions required for the Khovanskiĭ reduction, we have to modify the Pfaffian equations once more. In order to present the results in a compact form, we put them in the following Table 2. For each type of elementary vertex we describe explicitly a submanifold γ that plays the role of the graph of the standard correspondence map, except for the saddle resonant case S_μ. This submanifold is a separating solution (in the domains described in the third column of Table 2) for a Pfaffian system written in the fourth column.

TABLE 2. Separating solutions for Pfaffian systems associated with deformations of elementary singularities

Type	Submanifold γ	Domain M_r, Covering function ρ	Pfaffian system $\Omega = 0$
S_0	$y = \Delta(x, \lambda)$	$0 < x, y < r$, $\lambda \in L_r$, $\rho = xy(r-x)(r-y)\widetilde{\rho}$	$x\,dy - \lambda y\,dx = 0$
S_μ	$y = \Delta(x, \lambda)$, $z = x^m$, $w = y^n$.	$0 < x, y, z, w < r$, $\lambda \in L_r$, $P_\mu(z, \lambda) \neq 0$, $\rho = xyzw(r-x)(r-y)\times$ $(r-z)(r-w)P_\mu^2(z, \lambda)\widetilde{\rho}$	$x\,dz - mz\,dx = 0$, (1) $y\,dw - nw\,dy = 0$, (2) $m\,P_\mu(w, \lambda)\times$ $y\,P_\mu(z, \lambda)\,dx -$ $(mP_\mu(w, \lambda) + n)\times$ $x\,P_\mu^2(z, \lambda)\,dy = 0$ (3)
D_μ^c	$y = \Delta(x, \lambda)$	$\lvert x\rvert, \lvert y\rvert < r$, $x \neq 0$, $\lambda \in L_r$, $\rho = (r^2 - x^2)(r^2 - y^2)x^2\widetilde{\rho}$	$x\,(x\,dy - y\,dx) = 0$
D_μ^h	$y = \Delta(x, \mu)$	$0 < y < r$, $\lvert x\rvert < r$, $\lambda \in L_r$, $Q_\mu(\cdot, \lambda)\rvert_{[x,1]} > 0$, $\rho = y(r-y)(r^2 - x^2)\times$ $Q(x, \lambda)\widetilde{\rho}$	$Q_\mu(x, \lambda)\,dy - y\,dx = 0$

NOTES TO THE TABLE. Here we use the same notation as in Table 1 (and in fact Table 2 continues Table 1). In particular, $n : m$ is the hyperbolicity ratio in the resonant saddle case S_μ.

In the third column of Table 2 the symbol L_r stands for a small r-cube in the $(\mu+1)$-dimensional space of the parameters λ, centered at the localization point $\mathbf{c} = (0, \dots, 0, c) \in \mathbb{R}^{\mu+1}$ corresponding to the unperturbed system:

$$L_r = \left\{\lambda \in \mathbb{R}^{\mu+1} : |\lambda_i| < r,\ i = 0, \dots, \mu - 1,\ |\lambda_\mu - c| < r\right\}.$$

Everywhere in the table the function $\widetilde{\rho} = \widetilde{\rho}(\lambda)$ is the covering function for the set L_r, defined as

$$\widetilde{\rho}(\lambda) = (r^2 - \lambda_1^2)\cdots(r^2 - \lambda_{\mu-1}^2)(r^2 - (c - \lambda_\mu)^2).$$

This is a polynomial of degree $2(\mu+1)$ in all variables λ, r, c. Each covering function ρ is therefore a polynomial (explicitly written in the table).

LEMMA 3.1. *In each of the cases listed in Table 2, the submanifold γ is a separating solution to the system of Pfaffian equations given in the fourth column (with respect to the natural ordering of these forms, if there is more than one such a form) in the corresponding domain M_r.*

The Pfaffian forms constituting the systems have polynomial coefficients of degrees at most $O(\mu)$ in all variables including m, n and the auxiliary variables z, w in the saddle resonant case. The coefficients of these polynomials are integers not exceeding some bound of the order $O(\mu)$.

The domain M_r admits a polynomial covering function $\rho(\cdot)$ of degree at most $O(\mu)$ in all variables including r and c, with integral bounded coefficients. Here and above $O(\mu)$ stands for a certain explicit bound of the specified asymptotics in μ.

REMARK. The last two assertions follow immediately from the explicit formulas given in Table 2 and from the estimates $\deg_{x,\lambda} P_\mu(x, \lambda) = 2\mu+1$, $\deg_{x,\lambda} Q_\mu(x, \lambda) = 2\mu+2$ (see Table 1). In fact we will not need the sharp estimates, for our purposes it is sufficient to point out that the estimates are explicit functions of μ.

PROOF. We prove the statement by studying separate cases.

A. *Case* S_0. The proof for the case S_0 is straightforward: the graph of the correspondence function as the function of the variable x and the parameters λ bounds the hypograph $\{y < \Delta(x, \lambda)\} \cap \{(0, r) \times (0, r) \times L_r\}$ and contains no singular points of the corresponding Pfaffian form from Table 1. The co-orientation condition reduces to the inequality

$$\omega(\partial/\partial y) = x > 0$$

which is valid for the vector field $\partial/\partial y$ transversal to the graph of the correspondence function in the domain $M_r \subset \{x > 0\}$ (see Figure 5, left).

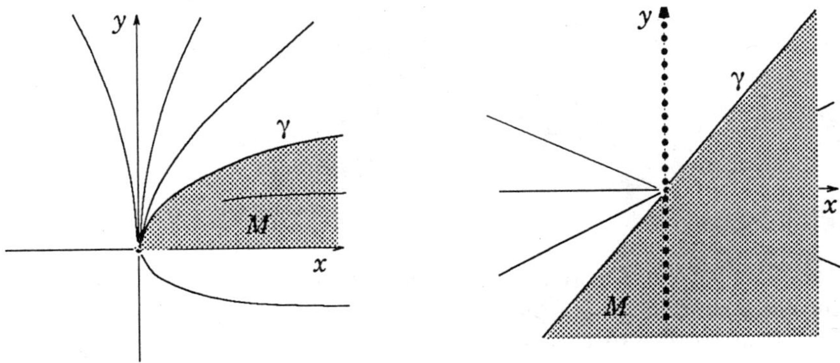

FIGURE 5. Standard Pfaffian systems for the types S_0 (left) and D_μ^c (right): the picture shows the section of the separating solution by a generic plane $\lambda = \text{const}$.

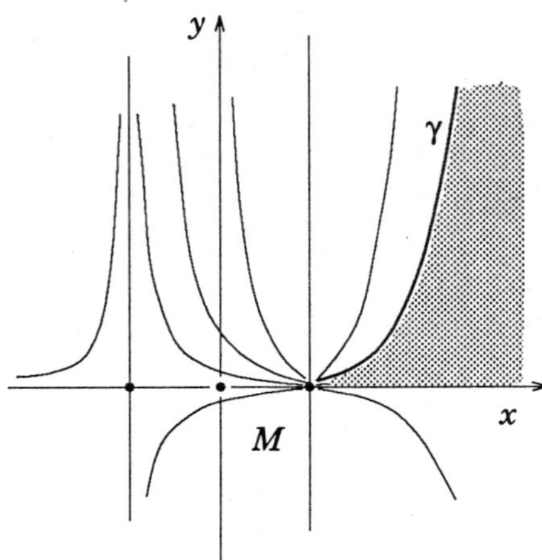

FIGURE 6. Standard Pfaffian system for the type D_μ^h: the picture shows the section of the separating solution by a generic plane $\lambda = \text{const}$.

B. *Case* D_μ^c. The form $\omega = x\,dy - y\,dx$ has a singular point at the origin, while all the graphs $y = C(\lambda)x$ of linear standard correspondence maps pass through it. On the other hand, the graph of the map $(x, \lambda) \mapsto C(\lambda)x$ is the boundary of the hypograph, but the value taken by the form ω on the vector field $\partial/\partial y$, changes the sign (see Figure 5, right).

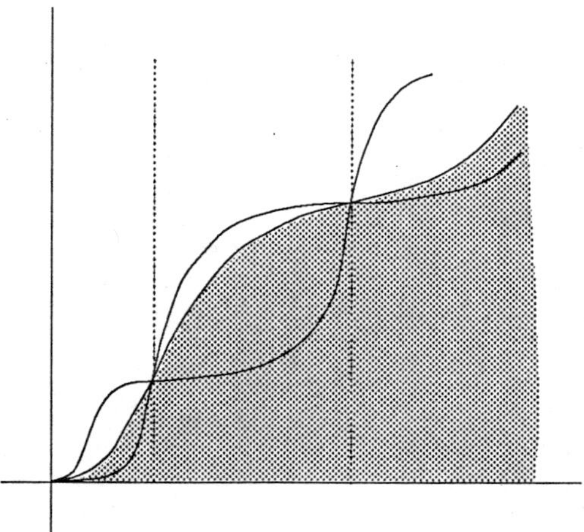

FIGURE 7. Standard Pfaffian system for the type S_μ, $\mu > 0$: the picture shows the section of the separating solution by a generic plane $\lambda = \text{const}$ after the projection on the (x, y)-plane.

If we delete the hyperplane $\{x = 0\}$ from the cube $(-r, r)^2 \times L_r$ to get the ambient manifold M_r, and consider the intersection of γ with it, then the form will be nonsingular, though the co-orientation condition still will not hold. To correct this, we introduce the form $\widetilde{\omega} = x\omega$ so that

$$\widetilde{\omega}(\partial/\partial y) = x^2 > 0 \quad \text{on } \gamma \cap M_r.$$

Thus all the requirements from the definition of separating solution are satisfied.

C. *Case* D^h_μ. The correspondence function for the case D^h_μ is well defined for a value (x, λ) if and only if the polynomial $Q_\mu(\cdot, \lambda)$ is positive on the segment $[x, 1]$ (otherwise the phase curve emanating from $(x, 1)$ will never hit the exit transversal $x = 1$, see Figure 6). This condition defines a semialgebraic domain U in the (x, λ)-space whose intersection with the r-cube L_r we denote by U_r, and the graph γ of the standard correspondence map is the boundary of the hypograph domain over U_r. The vector field $\partial/\partial y$ is transversal to γ and the form $\omega(\partial/\partial y) = Q_\mu(x, \lambda) > 0$ is positive, thus the co-orientation condition is satisfied, and γ is a separating solution to $\omega = 0$ in $M_r = U_r \times (-r, r)$.

D. *Case* S_μ, $\mu > 0$. Consider the curve (3.1), or, more exactly, the graph of the map

$$(x, \lambda) \mapsto (y, z, w), \qquad y = \Delta(x, \lambda), \qquad z = x^m, \qquad w = y^n = \Delta^n(x, \lambda),$$

in the space $\mathbb{R}^{\mu+5}$ with the coordinates (λ, x, y, z, w), $\lambda \in \mathbb{R}^{\mu+1}$. This is a codimension 3 submanifold, and our goal is to describe a Zariski open subset of it as separating solution to a certain Pfaffian system. In Figure 7 there is shown the projection onto (x, y)-plane of the intersection of this manifold with a typical plane $\lambda = \text{const}$.

Consider the chain of submanifolds,

$$\begin{aligned}\Gamma_1 = \{z = x^m\} \subseteq L_r \times (0, r)^4 \cap \{P_\mu(z, \lambda) \neq 0\} = M_r,\\ \Gamma_2 = \Gamma_1 \cap \{w = y^n\} \subset \Gamma_1,\\ \Gamma_3 = \gamma = \Gamma_2 \cap \{y = \Delta(x, \lambda)\} \subset \Gamma_2.\end{aligned}$$

Evidently, the first one is the separating solution to the Pfaffian equation $\omega_1 = 0$, where $\omega_1 = mx\,dz - z\,dx$, since it contains no singular points of ω_1, and $\omega_1(\partial/\partial z) = mx > 0$ in M_r. In the same way Γ_2 is the separating solution to the Pfaffian system $\omega_1 = 0$, $\omega_2 = 0$, where $\omega_2 = y\,dw - nw\,dy$.

On Γ_2 we may choose the coordinates x, y, λ, so that the intersection with the graph $\{y = \Delta(x, \lambda)\}$ is the boundary of the set $\Gamma_2 \cap \{0 < y < \Delta(x, \lambda)\}$, and the vector field ξ tangent to Γ_2 and having y-coordinate identically equal to 1 is transversal to $\Gamma_3 = \gamma$ on Γ_2 and induces a co-orientation of the former.

Denote by $\widetilde{\omega}_3$ the 1-form which occurs in Table 1 (the fourth column for the case S_μ). The form $\omega_3 = m\,P_\mu(z, \lambda) \cdot \widetilde{\omega}_3$ is proportional to the form ω_3 from Table 1, therefore its restriction on Γ_2 vanishes on $T\Gamma_3$, while the value taken by $\widetilde{\omega}_3$ on ξ is equal to

$$x\,(n + mP_\mu(w, \lambda))\,P^2_\mu(z, \lambda),$$

which is positive in M_r if the parameter r is chosen sufficiently small (so that the term in parentheses is of the same sign as the natural $n \geqslant 1$).

Thus the manifold $\Gamma_3 = \gamma$ considered in the domain $P_\mu(z, \lambda) \neq 0$ is a separating solution for the three forms ω_1, ω_2, and ω_3 in the specified order.

The explicit formulas for the covering functions for all types of domains follow easily from their explicit semialgebraic description. The proof of Lemma 3.1 is completed. □

3.3. Principal functional–Pfaffian system. We proceed with writing down the principal functional–Pfaffian system explicitly. Slightly abusing notation, we add the subscript j for objects related to the jth singularity, while letters without this subscript refer to objects related to the entire polycycle. In this notation we omit the reference to the characteristic size, still keeping in mind that all formulas are explicitly polynomial in r.

NOTATIONS. Denote by M_j the domain from Table 2 associated with the jth singular standard map, by γ_j the corresponding manifold (separating solution), and by Ω_j the tuple of Pfaffian forms on it: if the singularity is of the type D_μ or S_0, then Ω_j consists of only one form $\omega_j = A_j\, dx_j + B_j\, dy_j$, while in the case S_μ, $\mu > 0$, there are three forms, of which we denote the third one by $A_j\, dx_j + B_j\, dy_j$ (see Table 2). The covering function for M_j is denoted by $\rho_j \colon M_j \to \mathbb{R}^1_+$.

Construction of the principal system. The phase space for the principal functional–Pfaffian system is the Cartesian product of phase spaces corresponding to all the vertices of the polycycle and the r-cube in the ε-space:

$$\begin{gathered} M = M(r) = M_1 \times \cdots \times M_n \times \widetilde{B}_r, \\ M_j = M_{j,r} \text{ are taken from the second column of Table 2}, \\ \widetilde{B}_r = \{|\varepsilon_i| < r,\ i = 1, \dots, k\}. \end{gathered} \tag{3.2D}$$

Dimension of the phase space is equal to $2n + 2s + k + m$, where:

- k is the number of the parameters ε (the principal integer index);
- $n \leqslant k$ is the number of vertices;
- $s \leqslant n$ is the number of resonant saddles on the polycycle (each such a vertex contributes two additional variables z_j, w_j into the list of independent variables);
- $m = n + \sum \mu_j \leqslant n + k \leqslant 2k$ is the number of additional free parameters $\lambda = (\lambda^1, \dots, \lambda^n)$, $\lambda^j \in \mathbb{R}^{\mu_j+1}$.

The covering function for such a space is the product

$$\rho = \rho_1 \cdots \rho_n \rho_\varepsilon \colon M \to \mathbb{R}^1_+, \tag{3.2C}$$

where the last factor is the covering function for $\widetilde{B}_r$. From Lemma 3.1 it is clear that ρ is a polynomial of degree $O(k)$ in both phase variables and the characteristic size r.

Each form on M_j can be pulled back on M, yielding the form which is independent of all the coordinates except for those related to the jth vertex. Denote by Ω the union of the tuples $\Omega^{(j)}$: thus Ω is itself the tuple of 1-forms on M, containing $n+2s$ of them:

$$\Omega = (\Omega^{(1)}, \dots, \Omega^{(n)}) = (\Omega_1, \dots, \Omega_{n+2s}),$$

$$\text{(3.2P)} \qquad \Omega^{(j)} = \begin{cases} \{\omega_j\} & \text{if } j \text{ is not a resonant saddle,} \\ \{\omega_{j1}, \omega_{j2}, \omega_j\} & \text{otherwise,} \end{cases}$$

where

$$\omega_{j1} = m_j x_j\, dz_j - z_j\, dx_j, \qquad \omega_{j2} = n_j y_j\, dw_j - w_j\, dy_j,$$
$$\omega_j = A_j\, dx_j + B_j\, dy_j.$$

Each γ_j is a separating solution to the Pfaffian equation or system of equations $\Omega^{(j)} = 0$ on M_j, therefore the Cartesian product

$$\Gamma = \gamma_1 \times \dots \times \gamma_n \times \widetilde{B}_r$$

is the separating solution to the Pfaffian system $\Omega = 0$ on M. Indeed, one may consider the chain of submanifolds

$$\Gamma_i = \gamma_1 \times \dots \times \gamma_i \times M_{i+1} \times \dots \times M_n \times \widetilde{B}_r.$$

This chain possesses all the properties required by the definition of a separating solution, see §2: there are no singular points of Pfaffian forms on all the manifolds from this chain, and the topological condition of Γ_{i+1} being the boundary of a domain in Γ_i is trivially satisfied because each γ_{i+1} is the boundary of the corresponding subdomain in M_j. Thus the Pfaffian part of the principal system is constructed.

In this Pfaffian part we have the following information about the polynomials (recall that $\mathbf{A}$ stands for the algebraic part of the specification for the basic system, which is identified by (3.0) with a tuple of real variables):

$$\text{(3.2E}'\text{)} \qquad \begin{gathered} A_j, B_j \in \mathbb{Z}[x, y, \lambda, \mathbf{A}], \qquad \deg A_j, \deg B_j \leqslant d(k), \\ \rho \in \mathbb{Z}[x, y, \lambda, \varepsilon, r], \qquad \deg \rho \leqslant d(k), \\ \text{Coefficients of all polynomials have absolute values } \leqslant d(k), \end{gathered}$$

where $d(k)$ is a certain explicitly computable function of order $O(k)$. This information follows from Lemma 3.1.

Now we proceed with description of the functional part of the principal system. It is given by the map

$$F = (F_1, \dots, F_{n+k+m}) \colon M \to \mathbb{R}^{n+k+m},$$

$$\text{(3.2F)} \qquad F_j = \begin{cases} x_{j+1} - f_j(y_j, \varepsilon), & j = 1, \dots, n \mod (n), \\ \varepsilon_{j-n}, & j = n+1, \dots, n+k, \\ \lambda_{j-n-k}, & j = n+k+1, \dots, n+k+m. \end{cases}$$

The dimension of a generic fiber $F^{-1}(\cdot)$ is equal to the codimension of separating solutions of the Pfaffian system. An essential feature of the above map is the following one: *the coordinate functions of the map F are polynomial combinations of the coordinates on the source space and generic functions f_j* :

$$F_j \in \mathbb{Z}[x, \varepsilon, \lambda, \mathbf{f}], \qquad \deg F_j = 1, \tag{3.2E$''$}$$

and all coefficients of those polynomials are ± 1. A more invariant way of formulating the same property is to say that F is a polynomial map defined on the space of 0-jets of vector-functions

$$\mathbf{f}\colon M \to \mathbb{R}^n,$$

and this makes sense since M is a subset of a Euclidean space.

DEFINITION 3.0. The functional–Pfaffian system with the Pfaffian equations (3.2P), the functional equations (3.2F), defined on the domain (3.2D) considered with the covering function (3.2C), will be called the *principal functional–Pfaffian system*. The information provided by the estimates (3.2E$'$), (3.2E$''$) allows us to say that the principal system is effectively described.

Later on we will refer to the principal system as simply the system (3.2), or even as the system (3.1), which is a symbolic representation of the same set of functional and Pfaffian equations.

From basic to principal system. Our next goal is to prove that the number of *isolated solutions* to the basic system does not exceed the Khovanskiĭ number for the principal mixed functional–Pfaffian system

$$\Omega = 0, \qquad F = a, \tag{3.3}$$

(recall that the latter number is the upper bound for the number of *transversal* intersections of a fiber $F^{-1}(a)$ with separating solutions for all regular values of a). In order to make such a reduction, we use the structure of the (unspecified) basic system (1.1), which allows us to transform it into a family of scalar equations.

DEFINITION 3.1. The Poincaré first return map associated with the specified basic system (1.1) is a family of scalar maps depending on parameters ε and λ and defined as the composition of all the maps Δ_j, f_j in their natural order along the polycycle.

More precisely, let

$$\hat{\Delta}_j = (\Delta_j, \mathrm{id})\colon \mathbb{R}^{m+k+1} \to \mathbb{R}^{m+k+1}, \qquad (x_j, \lambda, \varepsilon) \mapsto (\Delta_j(x, \lambda^j), \lambda, \varepsilon)),$$
$$\hat{f}_j = (f_j, \mathrm{id})\colon \mathbb{R}^{m+k+1} \to \mathbb{R}^{m+k+1}, \qquad (y_j, \lambda, \varepsilon) \mapsto (f_j(y_j, \varepsilon), \lambda, \varepsilon).$$

The first return map $\Delta = \Delta(\cdot, \lambda, \varepsilon)$ for the basic system is defined via the relation

$$\hat{f}_n \circ \hat{\Delta}_n \circ \cdots \hat{f}_1 \circ \hat{\Delta}_1 = (\Delta, \mathrm{id}),$$

and the associated displacement function is

$$D(x_1, \lambda, \varepsilon) = \Delta(x_1, \lambda, \varepsilon) - x_1, \qquad D_{\lambda, \varepsilon} = D(\cdot, \lambda, \varepsilon).$$

REMARK. The definition of the Poincaré return map for the basic system relates to the usual Poincaré return map for a polycycle occurring in a family of vector fields exactly in the same way as the specified basic system (1.1) relates to the system of equations (0.2): in other words, if $\varepsilon \mapsto \lambda(\varepsilon)$ is the reparametrization described in Theorem 1.1, then the standard Poincaré map is induced from the map Δ by the above reparametrization, so that the displacement function can be represented as

$$D_\varepsilon^{\text{standard}}(\cdot) = D(\cdot\,, \lambda(\varepsilon)\,, \varepsilon).$$

Note that the domain of the first return map (and the displacement function $D_{\lambda\,,\varepsilon}$) may be empty for some values of the parameters $\lambda\,, \varepsilon$.

From the chain rule of differentiation one has the following identity:

$$D'_{\lambda\,,\varepsilon}(x_1) = \prod_{j=1}^{n} \Delta'_j(x_j\,, \lambda^j) f'_j(y_j\,, \varepsilon) - 1\,, \tag{3.4}$$

if $(x_1\,, y_1\,, \dots\,, x_n\,, y_n)$ is a solution to the basic system for the specified value of the parameters, and primes denote derivatives in the variables x_j and y_j respectively for Δ_j and f_j. The root of the function $D_{\lambda\,,\varepsilon}$ is *nondegenerate*, if the derivative of $D_{\lambda\,,\varepsilon}$ at this root is nonzero.

DEFINITION 3.2. For any value of the parameters $(\lambda\,, \varepsilon) \in \mathbb{R}^{m+k}$ the *suspension* of a root x_1 of the equation

$$D_{\lambda\,,\varepsilon}(x_1) = \delta\,, \qquad \delta \in \mathbb{R}^1\,, \tag{3.5}$$

is the point $\mathbf{x} = (x\,, y\,, z\,, w\,, \lambda\,, \varepsilon)$ in $\mathbb{R}^{2n+2s+m+k}$ defined by the (recurrent in $i\,, j$) formulas

$$\begin{gathered} y_j = \Delta_j(x_j\,, \lambda^j)\,, \qquad x_{i+1} = f_i(y_i\,, \varepsilon)\,, \\ j = 1\,, \dots\,, n\,, \qquad i = 1\,, \dots\,, n-1\,, \\ z_j = x_j^{m_j}\,,\ w_j = y_j^{n_j}\,, \qquad \text{if } j\text{th vertex is of the type } S_\mu^{(n_j\,:\,m_j)}. \end{gathered} \tag{3.6}$$

The suspension is said to be r-small if the suspended point $(x\,, y\,, z\,, w\,, \lambda\,, \varepsilon)$ belongs to the r-cube $(-r\,, r)^{2n+2s} \times L_r \times \widetilde{B}_r$.

Since we are ultimately interested only in determining the number of limit cycles born close to a polycycle, this means that we need to count only isolated roots of the equation (3.5) with $\delta = 0$, whose suspensions are r-small with $r < r_0$. The bound r_0 should not exceed the size of the neighborhoods of singular points on the polycycle, in which the normalization provided by Theorem 1.1 is possible.

Definition 3.2 immediately implies that

- for $\delta = 0$ the suspension is a solution to the basic system coupled with the equations $z_j = x_j^{m_j}\,,\ w_j = y_j^{n_j}$ for resonant saddles, and
- for any δ the suspension is a solution to the principal system

$$\Omega = 0\,, \qquad F = (\underbrace{0\,, \dots\,, 0}_{n-1 \text{ times}}, \delta\,, \varepsilon\,, \lambda)\,,$$

provided that the suspension belongs to the domain $M(r)$ of the principal system.

Indeed, this follows immediately from the construction of the principal system and the explicit formulas for the functional part (3.2F) of it.

End of the proof of Theorem 3.1. The remaining part of the proof of Theorem 3.1 consists of three steps:

- The number of isolated roots of the displacement function (or, equivalently, the number of isolated solutions of the basic system after suspension (3.6)) does not exceed the upper bound for the number of *nondegenerate* solutions of the equation (3.5) for an arbitrarily small value of δ (Lemma 3.2).
- The suspension of any nondegenerate root of $D_{\lambda,\varepsilon} - \delta$ belonging to the domain $M(r)$ is a *nondegenerate solution* to the principal functional–Pfaffian system (Lemma 3.3).
- The number of r-small roots of (3.5) whose suspension (3.6) does not belong to $M(r)$, does not exceed k.

Two lemmas for the first two steps are proved below. The last estimate is trivial: a root of (3.5) with $\delta = 0$ does not belong to the domain $M(r)$ if there exists j such that either the jth vertex is of the type D^c_μ, and $x_j = 0$, or the vertex is of the type S_μ, and after the suspension (3.6) one has $P_{\mu_j}(x_j^{m_j}, \lambda) = 0$. Because of their polynomial character, these conditions together may exclude from consideration no more than $\sum \mu_j \leqslant k$ discrete points. Thus when passing from the natural domain of the basic system to the reduced domain M, one looses at most k solutions.

LEMMA 3.2. *Let I be an interval (or a circle) and $f : I \to \mathbb{R}^1$ be a C^2-smooth function which has at least p isolated roots on I. Then there exists an arbitrarily small δ such that the function $f - \delta$ has at least p nondegenerate zeros.*

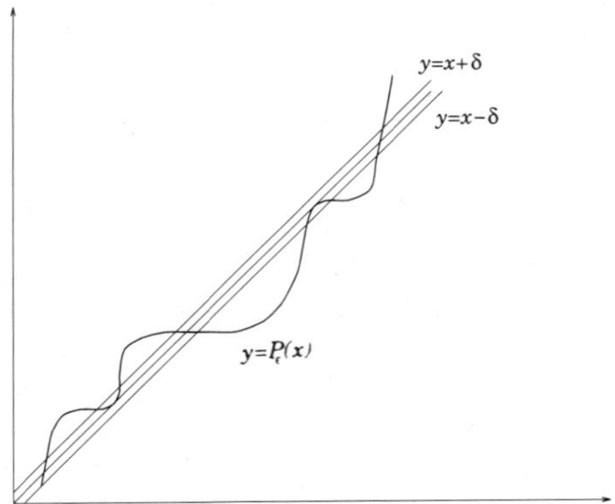

FIGURE 8. From isolated to nondegenerate roots of scalar functions.

PROOF. Cover each of p zeros by an interval that contains no other zeros of f. Three types of such intervals may occur: the function may be either nonnegative on such an interval, or nonpositive, or change its sign when passing through the zero. Denote the numbers of intervals of each type by p_+, p_-, and p_* respectively. Obviously,

$$p_+ + p_- + p_* = p.$$

On any interval of the first kind, the equation $f = \delta$ for any sufficiently small positive δ has at least two roots. The same holds for negative δ on intervals of the second type. On intervals of the third type, independently of the sign of δ there will be at least one root. Thus, denoting by $p(\delta)$ the number of roots for the equation $f = \delta$ and assuming that $\delta > 0$, one has

$$p(\delta) \geqslant 2p_+ + p_*, \qquad p(-\delta) \geqslant 2p_- + p_*.$$

Adding these two inequalities and assuming that both values $\pm\delta$ are regular values for f, we conclude that

$$\max(p(\delta), p(-\delta)) \geqslant \tfrac{1}{2}(p(\delta) + p(-\delta)) \geqslant p.$$

This proves the lemma. □

Now we show that nondegeneracy of a root of the function $D_{\lambda,\varepsilon} - \delta$ implies nondegeneracy of the corresponding solution of the principal system.

LEMMA 3.3. *If a point x_1 is a nondegenerate root of the function $D_{\lambda,\varepsilon} - \delta$ for a certain value of the parameters λ, ε, then its suspension $\mathbf{x} \in \mathbb{R}^{2n+2s+m+k}$ is a nondegenerate solution of the principal system, provided that $\mathbf{x} \in M(r)$.*

PROOF. Obviously, the conditions of the Lemma imply that the point belongs to the intersection of the corresponding fiber $F^{-1}(0, \dots, 0, \delta, \varepsilon, \lambda)$ with the separating solution γ of the Pfaffian system $\Omega = 0$. The only condition to check is that this intersection is *transversal*, that is, the tangent spaces to the fiber and to the integral manifold (at the point) are transversal to each other. The latter condition is equivalent to the (symbolic) inequality

$$dF \wedge \Omega(\mathbf{x}) \neq 0, \tag{3.7}$$

which should be understood as the wedge product (the exterior product) of differentials of all the coordinate functions of the map F and all the forms constituting the tuple Ω. This wedge product is proportional to the volume form, the proportionality coefficient being denoted by $\eth(\Omega, F)$ (see §2). Up to changes of the order of terms (which affects only the sign), we compute this wedge product in (3.9). In this formula we use the notation introduced in 3.3: the (nontrivial) Pfaffian form associated with the jth vertex is denoted by $A_j\,dx_j + B_j\,dy_j$, while f'_j stands for $(f_j)'_{y_j}$.

Let J be a $(2n \times 2n)$-matrix,

$$J = \begin{pmatrix} A_1 & B_1 & & & & & & \\ & -f'_1 & 1 & & & & & \\ & & A_2 & B_2 & & & & \\ & & & -f'_2 & 1 & & & \\ & & & & \ddots & \ddots & & \\ & & & & & & A_n & B_n \\ 1 & & & & & & & -f'_n \end{pmatrix}. \tag{3.8}$$

Then we may compute the wedge product (3.7) explicitly (recall that this product is the form of an upper degree hence proportional to the volume form on the phase space; the proportionality coefficient was denoted by $\mathfrak{d}(\Omega, F)$ in §2). The computation yields

$$\begin{aligned} dF \wedge \Omega &= \mathfrak{d}(\Omega, F)\,(dx \wedge dy \wedge dz \wedge dw \wedge d\varepsilon \wedge d\lambda) \\ &= \bigwedge_{j=1}^{n}(A_j\,dx_j + B_j\,dy_j) \bigwedge_{j=1}^{n}(dx_{j+1} - df_j) \\ &\bigwedge_{i=1}^{s}(x_i\,dz_i - m_i z_i\,dx_i) \bigwedge_{i=1}^{s}(y_i\,dw_i - n_i w_i\,dy_i) \bigwedge_{\alpha=1}^{k} d\varepsilon_\alpha \bigwedge_{\beta=1}^{m} d\lambda_\beta \mod \pm 1. \end{aligned}$$

The wedge product of the term in the second line yields $\det J\,dx \wedge dy \mod \pm 1$, hence all terms proportional to dx_i and dy_i disappear from the terms in the third line. Thus finally the wedge product is equal to

$$dF \wedge \Omega = \det J \cdot \prod_{i=1}^{s} x_i y_i \cdot (\text{volume form}) \mod (\pm 1).$$

The domain $M(r) = \prod_{j=1}^{n} M_j$ was constructed in such a way that the coefficients of the forms B_j are nonvanishing on it, as well as the product $\prod_j x_j y_j$ (see Table 2). Therefore the value $\mathfrak{d}(\Omega, F)$ is nonzero at a point $p \in M$, provided that the matrix J is nondegenerate at this point. To establish the nondegeneracy condition for J, we divide odd lines by the nonvanishing factors $-B_j$, and multiply the even lines by -1. After such a procedure (which preserves degeneracy or nondegeneracy of J), the matrix J turns into the circulant matrix $\widetilde{J}$, whose diagonal entries alternate between the fractions $-A_j/B_j$ and the derivatives f'_j, with the only other nonzero entries $\widetilde{J}_{j,\,j+1\,(\mathrm{mod}\,2n)} = -1$. Such a matrix is nondegenerate if and only if the product of the diagonal terms is different from 1, as this is implied by the following statement.

PROPOSITION. *A circulant matrix $\widetilde{J}$ with the only nonzero entries $\widetilde{J}_{ii} = v_i$, $\widetilde{J}_{i,\,i+1\,(\mathrm{mod}\,2n)} = -1$, $i = 1, \dots, 2n$, is nondegenerate if and only if $\prod v_i \neq 1$.*

PROOF OF THE PROPOSITION. We show that $\operatorname{Ker}\widetilde{J} = 0 \iff \prod v_i \neq 1$. Indeed, the homogeneous system of equations $\widetilde{J}p = 0$, $p = (p_1, \dots, p_{2n})$, admits the explicit solution

$$p_{i+1} = v_i p_i, \quad i = 1, \dots, 2n \text{ cyclically} \iff p_1 = p_1 \cdot \prod_{i=1}^{2n} v_i,$$

therefore any such a solution must be zero if the product of diagonal terms is different from 1. □

Since each of the manifolds γ_j is tangent to the zero space of the form $\omega_j = A_j\,dx + B_j\,dy$, we have

$$\frac{\partial \Delta_j}{\partial x_j}(x_j, \lambda^j) = -\left.\frac{A_j(x_j, y_j, z_j, w_j, \lambda^j)}{B_j(x_j, y_j, z_j, w_j, \lambda^j)}\right|_{\gamma_j}.$$

(in the nonresonant case the coefficients do not depend on z_j, w_j); therefore

$$\det J \neq 0 \iff \prod_{j=1}^{n} \frac{\partial \Delta_j}{\partial x_j}(x_j, \lambda^j) \cdot \frac{\partial f_j}{\partial y_j}(y_j, \varepsilon) \neq 1.$$

But the latter inequality is satisfied by virtue of (3.4) if the point $\mathbf{x}$ is the suspension of a nondegenerate root x_1. We conclude that the matrix J at this point is nondegenerate, and the entire wedge product is nonvanishing. This proves Lemma 3.3 and at the same time Theorem 3.1. □

3.4. Reduction to singularity theory. The system (3.2), whose Khovanskiĭ number majorizes the number of solutions of the basic system (1.1), satisfies the conditions of the Theorem 2.3. The conclusion of the latter theorem claims that the number $\mathcal{K}(\Omega, F)$ is in turn majorized by the combination of Khovanskiĭ numbers for certain $n+2s+1$ systems without Pfaffian equations at all (recall that $n+2s$ is the number of Pfaffian equations in the system, that have to be eliminated). The properties of the principal system listed in Theorem 3.1, yield a complete description of the resulting systems as *chain maps* (the definition is given below).

In what follows we treat the original variables x_j, y_j, the auxiliary variables z_{j_α}, w_{j_α}, and the parameters ε, λ in almost the similar way, as it is suggested by the functional equations (3.2F) of the principal system (3.2). However, the algebraic part $\mathbf{A}$ of the specification plays a different role: the coordinates of the localization points $\mathbf{c}_j$ and the integers n_{j_α}, m_{j_α} determining the hyperbolicity ratios of resonant saddles, would determine the point in the new phase space, around which the resulting chain maps will be considered. Recall that in §1 we introduced the vectors $\mathbf{c}_j$ and $\mathbf{c}$ as

$$\mathbf{c}_j = (0, \dots, 0, c_j) \in \mathbb{R}^{\mu_j+1}, \qquad c_j \in \mathbb{R}^1,$$
$$\mathbf{c} = (\mathbf{c}_1, \dots, \mathbf{c}_n) \in \mathbb{R}^m, \qquad m = n + \sum \mu_j.$$

For our purposes it would be convenient to consider all (new) variables as taking values around the origin in the corresponding phase space. To this end, we make a parallel translation in the λ-space that takes the origin into the point $\mathbf{c}$. Clearly, this translation does not affect the algebraic structure of the principal system (3.2), though it changes the appearance of the equations.

The characteristic size r retains its original meaning.

NOTATIONS. According to what has been said, we introduce the following notations:

$$\mathbf{x} = (x, y, z, w, \varepsilon, \lambda - \mathbf{c}) \in \mathbb{R}^{2n+2s+k+m},$$
$$\mathbf{f} = (f_1, \dots, f_n), \qquad f_j = f_j(y_j, \varepsilon) \iff \mathbf{f} = \mathbf{f}(\mathbf{x}),$$

where $\mathbf{f}$ is now considered as a vector-function of the argument $\mathbf{x}$, although each coordinate function f_j of the vector $\mathbf{f}$ depends in fact only on some of the coordinates of the vector $\mathbf{x}$. By $D^\ell \mathbf{f}$ we denote the collection of all partial derivatives of functions f_j of the order ℓ.

We will also use the same notation $M(r)$ for the domain of the principal system, though in fact it would become a subset of the unit cube $\|\mathbf{x}\| < r$ centered at the origin in the $\mathbf{x}$-space.

Now we can formulate the properties of the systems of equations that appear after elimination of Pfaffian equations from the principal system (3.1) as this was described in §2.

THEOREM 3.2. *For any fixed combinatorial type $\mathcal{T}$ of the principal functional-Pfaffian system* (3.2) *and any choice of the algebraic part* $\mathbf{A}$ *of the specification, the number of nondegenerate solutions of the principal system in the domain* $M(r)$ *for any choice of the characteristic size* $r > 0$ *does not exceed the sum of the Khovanskiĭ numbers for* $n + 2s + 1$ *system of equations in the same domain. Each of these systems has the form*

$$\mathbf{p}\big(\mathbf{x}, \mathbf{f}(\mathbf{x}), D^1\mathbf{f}(\mathbf{x}), \dots, D^{n+2s}\mathbf{f}(\mathbf{x}); \mathbf{A}, r\big) = \mathbf{y}, \qquad \mathbf{x} \in M(r) \subseteq \mathbb{R}^{\mathbf{m}}, \quad \mathbf{y} \in \mathbb{R}^{\mathbf{m}}, \tag{3.10}$$

where

- $\mathbf{m} = 2n + 2s + k + m \leqslant 7k$ *is the total number of variables* (*the dimension of the phase space*);
- $\mathbf{p}$ *is a vector polynomial,* $\mathbf{p} = (p_1, \dots, p_{\mathbf{m}})$, $p_i \in \mathbb{Z}[\mathbf{x}, \dots; \mathbf{A}, r]$; *the degrees and the absolute values of all coefficients of these polynomials are bounded by some explicitly computable* (*in terms of* k *only*) *function which we again denote by* $d(k)$;
- *the domain* $M(r)$ *belongs to the* r*-cube of the space* $\mathbb{R}^{\mathbf{m}}$, *centered at the origin.*

REMARK. Recall that, by definition (see Remark 2 after Definition 2.4), the Khovanskiĭ number for the system (3.10) is equal to the sharp upper bound for the number of solutions for the system of equations (3.10), where $\mathbf{y}$ ranges over the set of all *regular values* of the map whose components are given by the expressions in the left-hand side. This supremum clearly depends on the functional part $\mathbf{f}$ of the specification of the basic system (1.1), and for some bad choice may turn out to be infinite.

PROOF OF THEOREM 3.2. The principal system (3.2) satisfies the conditions of Theorem 2.3 for any choice of the specification $\mathbf{A}$ and any characteristic size $r > 0$. Since the elimination algorithm described in Theorem 2.3 is explicit and involves no more steps than the number of Pfaffian equations in the principal system, and each step involves only polynomial operations and one differentiation (when computing determinants as in (2.9)), the total number of differentiations does not exceed $n + 2s$. The number of polynomial operations to be performed also admits an effective control, and the size of integer coefficients of all polynomials as well, whence comes the first conclusion of the Theorem.

The second conclusion comes from the fact that all systems constructed as in Theorem 2.3, are defined on the same phase space $M(r)$ as the original principal system. □

3.5. Chain maps and related finiteness theorems. Now we proceed with a more invariant description of the geometric object corresponding to the system of equations (3.10).

DEFINITION 3.3. Let $\mathbb{R}^{\mathbf{m}}$ be a Euclidean space with a fixed coordinate system $\mathbf{x} = (X_1, \dots, X_{\mathbf{m}})$, and $U \subseteq \mathbb{R}^{\mathbf{m}}$ a domain of the rectangular form,

$$U = \{\alpha_i < X_i < \beta_i, \ i = 1, \dots, \mathbf{m}\}.$$

Denote by I the index subset $I = \{1, \dots, \mathbf{m}\}$ enumerating the coordinates in $\mathbb{R}^{\mathbf{m}}$, and let for any $j = 1, \dots, \mathbf{n}$, I_j be a nonempty subset of I,

$$\varnothing \neq I_j \subseteq I, \qquad j = 1, \dots, \mathbf{n}.$$

We say that a vector-valued function

$$\mathbf{f}\colon U \mapsto \mathbb{R}^{\mathbf{n}}, \qquad \mathbf{f} = (f_1, \dots, f_{\mathbf{n}}),$$

is a *Cartesian function* of the Cartesian type $\mathcal{I} = (I_1, \dots, I_{\mathbf{n}})$, if for any j the jth component of this function depends only on the coordinates X_i with $i \in I_j$: in other words,

$$\forall i \notin I_j \quad \frac{\partial f_j}{\partial X_i} \equiv 0.$$

For any given Cartesian type $\mathcal{I}$ with $\mathbf{n} = 1$ the set of all C^ℓ-smooth Cartesian maps (scalar functions) of this type constitutes a Banach space with the natural C^ℓ-norm. We denote this space by $\mathbf{C}^\ell_{\mathcal{I}}$, sometimes omitting the explicit reference to the type $\mathcal{I}$ when the latter is clear from context. The space $\mathbf{C}^\ell_{\mathcal{I}}$ will be referred to as the *Cartesian space*. In the same way the *Cartesian spaces of maps* arise. As a consequence, we may talk about *genericity* of Cartesian maps (functions) within the given Cartesian type; the notions of openness and density of subsets are also naturally defined.

DEFINITION 3.4. Let $\mathbf{f}$ be a C^ℓ-smooth Cartesian map of a given Cartesian type $\mathcal{I}$, and $s \geqslant 0$ an nonnegative integer number, $s \leqslant \ell$. A *Cartesian s-jet* of the function $\mathbf{f}$ at a point $\mathbf{x}_0 \in U$ is the equivalence class of all Cartesian functions of the same Cartesian type that differ from $\mathbf{f}$ by a term which is s-flat at $\mathbf{x}_0$:

$$\mathbf{j}^s\mathbf{f}(\mathbf{x}_0) = \{\mathbf{g} \in \mathbf{C}^\ell_{\mathcal{I}}\colon |\mathbf{f} - \mathbf{g}| = o(|\mathbf{x} - \mathbf{x}_0|^s)\}.$$

The space of all s-jets of functions of the given Cartesian type $\mathcal{I}$ at all points $\mathbf{x}_0 \in U$ will be denoted by $\mathbf{J}^s_{\mathcal{I}}(\mathbb{R}^{\mathbf{m}}, \mathbb{R}^{\mathbf{n}})$ or simply by $\mathbf{J}^s$ when the environment is unambiguously defined by the context.

The map

$$\mathbf{x} \mapsto \mathbf{j}^s\mathbf{f}(\mathbf{x})$$

is called the Cartesian s-jet extension of the Cartesian map $\mathbf{f}$.

The space of Cartesian jets of any type and any finite order admits a natural coordinate system, in which the Cartesian jet extension of a map $\mathbf{f} = (f_1, \dots, f_\mathbf{n})$ takes the form

$$\mathbf{x} = (X_1, \dots, X_\mathbf{m}) \mapsto \Bigg(\mathbf{x}, \mathbf{f}(\mathbf{x}), \left\{\tfrac{\partial f_j}{\partial X_i}(\mathbf{x}),\ i \in I_j\right\}, \dots,$$

$$\left.\begin{Bmatrix}\text{all partial derivatives of functions } f_j \text{ of all orders up to } s \\ \text{in the variables on which each } f_j \text{ actually depends}\end{Bmatrix}\right).$$

The Cartesian jet spaces possess almost all properties of the standard jet spaces. In particular, the natural projections

$$\mathbb{R}^\mathbf{m} \supseteq U \xleftarrow{\ \mathrm{pr}_0\ } \mathbf{J}^0 \simeq \mathbb{R}^\mathbf{m} \times \mathbb{R}^\mathbf{n} \xleftarrow{\ \mathrm{pr}_1\ } \cdots \xleftarrow{\ \mathrm{pr}_s\ } \mathbf{J}^s \xleftarrow{\ \mathrm{pr}_{s+1}\ } \cdots$$

are well defined and endow each $\mathbf{J}^s_\mathcal{I}$ with the structure of an affine bundle over $\mathbb{R}^\mathbf{m}$. Thus it makes sense to talk about *polynomial functions* defined on Cartesian bundles.

Definition 3.5. A *chain map* with the *exterior part* $\mathbf{p}$ and the *interior part* $\mathbf{f}$ is a map of the form

$$\mathbb{R}^\mathbf{m} \supseteq U \ni \mathbf{x} \mapsto \mathbf{p}(\mathbf{j}^s_\mathcal{I}\mathbf{f}(\mathbf{x})) \in \mathbb{R}^\mathbf{m},$$

where:

- $\mathbf{f}$ is a Cartesian map from a certain Cartesian space $\mathbf{C}^\ell_\mathcal{I}(\mathbb{R}^\mathbf{m}, \mathbb{R}^\mathbf{n})$, and $\mathbf{j}^s_\mathcal{I}$ is the corresponding s-jet extension of $\mathbf{f}$;
- $\mathbf{p}\colon \mathbf{J}^s_\mathcal{I}(\mathbb{R}^\mathbf{m}, \mathbb{R}^\mathbf{n}) \to \mathbb{R}^\mathbf{m}$ is a vector polynomial (eventually depending polynomially on some additional parameters),
- the composite map is between the spaces of the same dimension: $\dim \mathbf{x} = \dim \mathbf{p} = \mathbf{m}$.

Having introduced the notions of Cartesian functions, maps, jets, etc., we can describe the system (3.10) as a chain map defined on a small cube of some size $r > 0$ with the exterior part $\mathbf{p}$ which is a polynomial with integer coefficients and of a controlled complexity; this polynomial depends on r and some additional variables $\mathbf{A}$ as well, and the interior part $\mathbf{f}$ belongs to some Cartesian space, since the functions f_j depend only on some components of the vector $\mathbf{x} = (x, y, z, w, \varepsilon, \lambda - \mathbf{c})$ (recall that all nonzero coordinates of the vector $\mathbf{c}$ are already included among the variables $\mathbf{A}$). Thus our problem of estimating cyclicity of a polycycle takes the following form: describe the Cartesian maps $\mathbf{f}$ for which the chain map admits an upper estimate for the number of preimages of regular values.

We now can formulate two finiteness results which would imply Theorems I and II from the Introduction.

Theorem I* (Global existential finiteness theorem for chain maps). *Let $\mathcal{T}$ be a fixed Cartesian type, and $\mathbf{p}\colon \mathbf{J}^s(\mathbb{R}^\mathbf{n}, \mathbb{R}^\mathbf{m}) \to \mathbb{R}^\mathbf{m}$ be a vector polynomial of degree d.*

Consider chain maps of the form

$$\mathbf{x} \mapsto \mathbf{G}(\mathbf{x}) = \mathbf{p}(\mathbf{j}^s_\mathcal{I}\mathbf{f}(\mathbf{x})), \qquad \mathbf{x} \in U \subset \mathbb{R}^\mathbf{m}, \tag{3.11}$$

defined in a unit cube of the phase space $\mathbb{R}^\mathbf{m}$.

Then in the space of $\mathbf{C}^\ell_{\mathcal{I}}(U, \mathbb{R}^\mathbf{n})$ *of* ℓ*-smooth Cartesian functions with* $\ell = s + \mathbf{m} + 1$, *there exists an open dense subset* $\mathfrak{F}$ *such that for any* $\mathbf{f} \in \mathfrak{F}$ *the number of nondegenerate preimages of points from* $\mathbb{R}^\mathbf{m}$ *by the chain map* (3.11) *is uniformly bounded:*

$$\sup_{\mathbf{y}\in\mathbb{R}^\mathbf{m}} \#\{\mathbf{x} \in U : \mathbf{G}(\mathbf{x}) = \mathbf{y}, \ \det \mathbf{G}_*(\mathbf{x}) \neq 0\} < +\infty.$$

In general, the supremum depends on $\mathbf{f}$.

The second result gives a constructive version of the previous theorem: for our purposes we need to formulate it for a slightly more general class of chain maps.

Consider chain maps of the form

$$\mathbf{x} \mapsto \mathbf{G}_r(\mathbf{x}) = \mathbf{p}(\mathbf{j}^s_{\mathcal{I}}\mathbf{f}(\mathbf{x}), r), \qquad \mathbf{x} \in U \subset \mathbb{R}^\mathbf{m}, \ r > 0, \tag{3.12}$$

depending polynomially on an additional variable r, so that

$$\mathbf{p} \colon \mathbf{J}^s_{\mathcal{I}}(\mathbb{R}^\mathbf{m}, \mathbb{R}^\mathbf{n}) \times \mathbb{R}^1 \to \mathbb{R}^\mathbf{m}, \quad \mathbf{f} \in \mathbf{C}^\ell_{\mathcal{I}}(U, \mathbb{R}^\mathbf{n}). \tag{3.13}$$

We assume that the polynomial $\mathbf{p}$ and the Cartesian type $\mathcal{I}$ are fixed (and, as before, U denotes a unit cube).

Suppose that the smoothness order ℓ is sufficiently high,

$$\ell > s + \mathbf{m} + 2,$$

and denote the degree of the polynomial $\mathbf{p}$ in *all variables, including* r, by $d = \deg \mathbf{p}$.

THEOREM II* (Local constructive finiteness theorem for chain maps). *There exists a primitive recursive function*

$$(\mathbf{n}, \mathbf{m}, s, d) \mapsto K(\mathbf{n}, \mathbf{m}, s, d),$$

such that:

- *for any polynomial* $\mathbf{p}$ *as in* (3.13) *one may choose a subset* $\mathfrak{F}_\mathbf{p} \subset \mathbf{C}^\ell_{\mathcal{I}}(U, \mathbb{R}^\mathbf{n})$ *in the space of Cartesian functions of the given type, which is open and dense in this space,*
- *for any Cartesian function* $\mathbf{f} \in \mathfrak{F}_\mathbf{p}$ *the number of nondegenerate preimages of an arbitrary point by the chain map* $\mathbf{G}_r$ *in the small ball* U_r *of the same radius* r *admits an upper estimate, as* r *tends to zero:*

$$\forall\, \mathbf{p} \text{ with } \deg \mathbf{p} \leqslant d \quad \exists \mathfrak{F}_\mathbf{p} \subset \mathbf{C}^\ell_{\mathcal{I}} \text{ such that } \forall \mathbf{f} \in \mathfrak{F}_\mathbf{p}$$
$$\limsup_{r\to 0^+} \sup_{\mathbf{y}\in\mathbb{R}^\mathbf{m}} \#\{\mathbf{x} \colon \mathbf{x} \in U_r, \ \mathbf{G}_r(\mathbf{x}) = \mathbf{y}, \ \det(\mathbf{G}_r)_{*,x}(\mathbf{x}) \neq 0\}$$
$$\leqslant K(\mathbf{n}, \mathbf{m}, s, d). \tag{3.14}$$

REMARKS. 1. Nondegenerate preimages of the map $\mathbf{G}_r$ are counted on the ball U_r of radius r *with the same* r.

2. The subspace $\mathfrak{F}_\mathbf{p}$ of admissible functions depends on the choice of the polynomial.

3. The limit estimate (3.14) means that the inequality $\#\{\mathbf{x} \in U_r : \cdots\} \leqslant K$ holds for all positive r smaller than a certain r_0; the value r_0 may depend on the choice of $\mathbf{f}$ within the subspace $\mathfrak{F}_\mathbf{p}$.

4. The value of the upper bound K depends neither on the choice of the polynomial $\mathbf{p}$, nor on the choice of the Cartesian map $\mathbf{f}$, being uniform over all polynomials of the given degree d and all admissible (in the above sense) maps $\mathbf{f}$.

Theorems I* and II* are formulated in such generality which would imply Theorems I and II respectively. But some corollaries can be formulated more easily.

COROLLARY 1 TO THEOREM II*. *Let* $p : \mathbb{R}^{m+n} \to \mathbb{R}^m$ *be a vector polynomial of degree* d. *Then in the space of sufficiently smooth vector-functions* $C^\ell(U, \mathbb{R}^n)$, *where* U *is a unit ball in* $\mathbb{R}^m$, *one may choose an open dense subset* $\mathfrak{F}_p$ *such that for any* $f \in \mathfrak{F}_p$ *the composite map* $x \mapsto G(x) = p(x, f(x))$ *from* U *to* $\mathbb{R}^n$ *admits a uniform upper estimate for the number of small regular preimages of points in the target space* $\mathbb{R}^n$: *for any choice of* $f \in \mathfrak{F}_p$ *one has a limit estimate*

$$\limsup_{r \to 0^+} \sup_{y \in \mathbb{R}^m} \#\{x \in U : |x| < r,\ G(x) = y,\ \det G_*(x) \neq 0\} \leqslant K_1(n, m, d),$$

where K_1 *is some universal algorithmically computable function of three integer arguments* (*the dimensions* n, m *and the degree* d).

PROOF. One should replace the space of Cartesian functions by the usual Banach space of smooth functions, put $s = 0$, and consider polynomials $\mathbf{p}$ independent of r. □

COROLLARY 2 TO THEOREM II*. *For any polynomial* p *as in Corollary* 1 *all generic local* k*-parameter families of maps* $f = f(x, \varepsilon)$ *allow for a uniform estimate of the number of small nondegenerate preimages for the chain map* $G(x, \varepsilon) = p(x, f(x, \varepsilon))$ *by a primitive recursive function*: *for any local family from a certain open dense subset* $\mathfrak{F}'_p$ *one has*

$$\limsup_{r \to 0^+} \sup_{y \in \mathbb{R}^m,\, |\varepsilon| < r} \#\{x \in U : |x| < r,\ G(x, \varepsilon) = y,\ \det G_{*,x}(x, \varepsilon) \neq 0\} \leqslant K_1(n, m + k, d).$$

PROOF. One should consider the extended polynomial map

$$(p, \mathrm{id}) : \mathbb{R}^{m+n} \times \mathbb{R}^k \to \mathbb{R}^n \times \mathbb{R}^k, \qquad (x, \varepsilon) \mapsto (p(x), \varepsilon),$$

and then apply Theorem II* with $s = 0$ to the corresponding chain maps. □

Proof of Theorem II *using Theorem* II*. Theorem 3.2 provides a link between Theorems I and II on one hand and Theorems I* and II*, which are yet to be proved. The chain of reductions from Theorem I to Theorem I* is straightforward: the number of isolated solutions of a specified basic system (1.1) in codimension k with the combinatorial type $\mathcal{T}$ and a given specification $\mathcal{S} = (\mathbf{A}, \mathbf{f}, r)$, which majorizes the number of limit cycles born from the corresponding elementary polycycle, does not exceed by Theorem 3.1, the number of parameters k plus the number of *nondegenerate* solutions of the principal functional–Pfaffian system (3.1). Then Theorem 3.2 reduces the question about the number of nondegenerate solutions of the principal system (3.1), or, in the extended form, (3.2), to the same question about chain maps (3.11). Finally, Theorem I* asserts that the for the most part of functional specifications $\mathbf{f}$ and any choice of the algebraic specifications the latter system admits a uniform bound for the number of preimages for the map

$$\mathbf{x} \mapsto \mathbf{y} = \mathbf{G}(\mathbf{x}), \qquad \mathbf{x}, \mathbf{y} \in \mathbb{R}^{\mathbf{m}}.$$

Taking points $\mathbf{y}$ of the form $(0, \ldots, 0, \varepsilon, \lambda - \mathbf{c})$, we obtain an upper estimate for the number of solutions to the equations (3.10), *uniform* over all values of ε, λ from the small r-cube centered at the point $(\varepsilon, \lambda) = (0, \mathbf{c})$ which corresponds to the unperturbed vector field with the elementary polycycle. In symbolic form,

$$\boxed{\text{Th. I}^*} \overset{\text{Th. 2.4}}{\Longrightarrow} \boxed{\mathcal{K}(\Omega, F) < \infty \text{ for } (3.1)} \overset{\text{Th. 3.1}}{\Longrightarrow} \boxed{\text{Th. 1.2}} \overset{\text{Th. 1.1}}{\Longrightarrow} \boxed{\text{Th. I}}.$$

In almost the same manner Theorem II formulated in the Introduction is an implication if Theorem II*. Indeed, as it was already shown, all integer parameters d, s, n, m (the dimensions of all spaces, the highest order of derivatives and the degrees of all polynomials) occurring in this chain of reductions, can be estimated by some primitive recursive functions of the principal parameter $k = \dim \varepsilon$, the number of parameters:

$$\begin{aligned} &n \leqslant k, \ \mu_j \leqslant k, \ m = n + \textstyle\sum \mu_j \leqslant 2k && \text{in the basic system}, \\ &d = d(k) && \text{in the principal system}, \\ &s \leqslant n, \ \mathbf{m} = 2n + 2s + k + m, \ \mathbf{n} = n && \text{in the chain map.} \end{aligned}$$

Therefore the primitive recursive function $E(\cdot)$ in Theorem I is the composition of primitive recursive functions.

In fact, we have to prove that the estimate (3.14) holds not for a generic Cartesian family $\mathbf{f}$, but rather for a *generic Cartesian family of functions vanishing at the origin.* But note that any smooth Cartesian family satisfying the condition $\mathbf{f}(0, 0) = 0$ (that is, $f_j(0, 0) = 0$ for all $j = 1, \ldots, n$) can be written in the form

$$f_j(y_j, \varepsilon) = y_j f_{j0}(y_j, \varepsilon) + \sum_{\nu=1}^{k} \varepsilon_\nu f_{j\nu}(y_j, \varepsilon), \tag{3.15}$$

with the functions $f_{j\nu}$, $\nu = 0, \ldots, k$, $j = 1, \ldots, n$ belonging to another Cartesian family. This representation can be obtained in a canonical way, using Hadamard's lemma and writing $f_{j\beta}$ through integrals of partial derivatives of f_j:

$$f_{j0}(y_j, \varepsilon) = \int_0^1 \frac{\partial f_j}{\partial y_j}(ty_j, \varepsilon)\, dt,$$

and the same for $f_{j\nu}$. The following proposition is trivial.

PROPOSITION. *In the space of Cartesian functions of a given type, consider the subspace of vector-functions vanishing at the origin, with the natural* C^ℓ*-norm. Then for any open dense subset* $\mathfrak{F}$ *in the space of tuples of Cartesian functions,* $\mathbf{C}^{\ell-1}(\mathbb{R}^{\mathbf{m}}, \mathbb{R}^{\mathbf{n}\cdot\mathbf{m}})$ *the set of vector-functions* $\mathbf{f} = (f_1, \dots, f_{\mathbf{n}})$ *admitting representation* (3.15) *with* $\{f_{j\nu}\} \in \mathfrak{F}$*, constitutes a subset* $\mathfrak{F}'$ *which is* C^ℓ*-open and* $C^{\ell-1}$*-dense.* □

Now if we want to obtain an explicit estimate for the number of small nondegenerate preimages by the chain map as in (3.14) for *generic maps* $\mathbf{f}$ *vanishing at the origin,* we should substitute the representation (3.15) into the chain map (3.11) and consider $\{f_{j\nu}\}$ as the new generic tuple $\widetilde{\mathbf{f}}$, now without the additional restriction $\mathbf{f}(0) = 0$. Since the formulas (3.15) are polynomial in the natural sense, the result of such substitution will be again the chain map of the form

$$\mathbf{x} \mapsto \widetilde{\mathbf{G}}(\mathbf{x}) = \widetilde{\mathbf{p}}(\mathbf{j}^s_{\mathcal{I}}\widetilde{\mathbf{f}}(\mathbf{x}))$$

with the same s, but with the new dimension $\widetilde{\mathbf{n}} = \mathbf{n}\cdot\mathbf{m}$, and the new polynomial $\widetilde{\mathbf{p}}$ of degree $2d = 2\deg\mathbf{p}$. Substituting this data into the primitive recursive function K, we obtain the primitive recursive function

$$E(k) = K(\mathbf{n}(k)\cdot\mathbf{m}(k), \mathbf{m}(k), s, 2d(k)),$$

which would yield the upper estimate in Theorem 1.3, hence in Theorem II:

$$\boxed{\text{Th. II}^*} \overset{\text{Th. 2.4}}{\Longrightarrow} \boxed{\mathcal{K}_{M(r)}(\Omega, F) < \widetilde{E}(k), \ \text{for all sufficiently small } r < r_0} \overset{\text{Th. 3.1}}{\Longrightarrow} \boxed{\text{Th. 1.3}} \overset{\text{Th. 1.1}}{\Longrightarrow} \boxed{\text{Th. II}}$$

(in this chain $\widetilde{E}(k)$ is some intermediate primitive recursive function).

Thus we finished the chain of reductions, proving Theorems I and II modulo Theorems I* and II*. □

The remaining part of the paper is devoted to the proof of Theorems I* and II*; they are of purely analytical nature and their proof is absolutely independent of the first part of the paper. Thus all variables n, m, k and so on, will be used without any connection to their previous meaning.

§4. Stratified sets and contiguity number

For the rest of the paper our main goal is to prove Theorems I* and II* formulated at the end of §3. It is more convenient to prove them in a slightly more abstract form.

If M is a real manifold and $f\colon M \to \mathbb{R}^n$ a smooth mapping of it to the Euclidean space of the same dimension, then for any regular value $a \in \mathbb{R}^n$ the full preimage consists of isolated points in M, and one can define the supremum $\sup_a \#f^{-1}(a)$. If a is not a regular value, it may still have some nondegenerate preimages, and we define

$$\mathcal{G}(f) = \sup_{a\in\mathbb{R}^n} \#\{x \in f^{-1}(a)\colon \det f_*(x) \neq 0\} \leqslant \infty$$

over all values. If this is a finite number, then the map is said to possess the *Gabrielov property*, and $\mathcal{G}(f)$ is the corresponding *Gabrielov number*.

REMARK. Actually the Gabrielov number coincides with the Khovanskiĭ number for the functional–Pfaffian system $(\varnothing, f)$ as it was was introduced in Remark 2 in 2.1. Indeed, if $f^{-1}(a)$ contains N nondegenerate preimages, then for a regular value b sufficiently close to a there will be at least N of them, hence $\mathcal{G}(f) \leqslant \mathcal{K}(\varnothing, f)$. On the other hand, by definition we have $\mathcal{K}(\varnothing, f) \leqslant \mathcal{G}(f)$, since when defining the Khovanskiĭ number, we consider only regular values of f. Thus $\mathcal{G}(f) = \mathcal{K}(\varnothing, f)$.

If the manifold is compact, then any preimage of a regular value is finite, but still the common upper bound $\mathcal{G}(f)$ may be infinite. Nevertheless, it is known from [G], that if M is the open unit cube in $\mathbb{R}^n$ and f a *real analytic* map, then $\mathcal{G}(f) < \infty$. Our objective is to extend this result for the *majority of C^k-smooth functions*, their jet extension maps, and superpositions of the latter with polynomial maps.

In order to prove those results, we have to extend the category of objects to include not only smooth manifolds, but also stratified varieties. In the next section all the relevant definitions and notions are gathered. We failed to find in the existing literature the notion of the contiguity number (introduced below), so we had to re-expose a reasonable part of the theory.

4.1. Stratified varieties: definitions and main properties. Let S be a smooth manifold, which we call the *ambient manifold*.

DEFINITION 4.1. A locally closed subset $M \subseteq S$ is called a *stratified subvariety* (subset, submanifold) of S if it is represented as a locally finite disjoint union of smooth submanifolds of S, called *strata*, of different dimensions in such a way that the closure of each stratum consists of itself and the union of some other strata of *strictly smaller dimensions*, and Whitney Condition (b) is satisfied (see [W2], [Ma1], [Ma2] and the remark below).

REMARK. Recall that the Whitney Condition (b) for two submanifolds $U, V \subset \mathbb{R}^n$, $V \subseteq \overline{U} - U$, means that for any two sequences of points $u_i \in U$, $v_i \in V$ converging to a point $x \in \overline{V} \cap \overline{U}$ the limit direction defined by the segments $[u_i, v_i]$ (if it exists), must belong to the limit position of the tangent subspaces $T_{u_i} U$, provided that the latter also exists. This condition is invariant under diffeomorphisms, therefore it applies to submanifolds of an arbitrary ambient manifold S rather than to subsets of $\mathbb{R}^n$.

We say that the union of submanifolds satisfies Condition (b) if the latter holds for any pair of them (the assertion is void if the closures of the two are disjoint).

Any representation satisfying the conditions from Definition 4.1,

$$M = \bigcup_{\alpha} M_\alpha \tag{4.1}$$

is called the *stratification* of M, the submanifolds being the *strata*. A subset $M \subseteq S$ is *stratifiable*, if it can be partitioned in such a way. If the partition is already fixed, we speak about the given stratification and denote by $\mathcal{M}$ the list of all its strata. Usually the ambient manifold S is known from context, in which case we talk about stratified manifolds (varieties, etc.). We do not assume that the strata are connected: although this additional assumption would simplify some formulations, still some other constructions, e.g., the definition of the pullback stratification, would become more cumbersome. In any case, all our stratified manifolds will always satisfy the following local finiteness assumption: any compact subset of the ambient space intersects only a finite number of connected components of all strata.

Let (4.1) be a stratification of a subset M by the strata M_α having dimensions $n_\alpha = \dim M_\alpha$.

DEFINITION 4.2. Dimension of the stratified set M is the maximum of the dimensions n_α of the strata.

The *smooth points* of the stratified set are those belonging to the strata of the maximal dimension. The complement, i.e., the union of all strata of smaller dimensions is the *skeleton* of the stratification.

REMARK. The dimension of any stratifiable set is independent of the choice of the stratification. On the contrary, smooth points and the skeleton do depend on it. If $\mathcal{M}$ is a certain stratification of the subset $M \subseteq S$, then we denote the set of all smooth points by $\dot{\mathcal{M}}$, and the skeleton by $\operatorname{sk}\mathcal{M}$:

$$\operatorname{sk}\mathcal{M} = \bigcup_\alpha \{M_\alpha\colon \ \dim M_\alpha < n\}, \quad \dot{\mathcal{M}} = \bigcup_\alpha \{M_\alpha\colon \ \dim M_\alpha = n\},$$
$$n = \dim M = \max_\alpha \dim M_\alpha, \qquad M = \dot{\mathcal{M}} \cup \operatorname{sk}\mathcal{M}.$$

Recall definitions of smooth maps and transversality in the category of stratified manifolds.

DEFINITION 4.3. A map $f\colon M \to \mathbb{R}^k$ is *smooth*, if there exists a smooth map $F\colon S \to \mathbb{R}^k$ of the ambient manifold, whose restriction to M coincides with f.

DEFINITION 4.4. If N is a smooth manifold, $g\colon N \to S$ a smooth map and $M \subseteq S$ a stratified subset endowed with a stratification $\mathcal{M}$, then this map is *transversal* to M if it is transversal to each stratum from $\mathcal{M}$:

$$g \pitchfork (M, \mathcal{M}) \iff \forall M_\alpha \in \mathcal{M} \quad g \pitchfork M_\alpha.$$

Let $f\colon M \to \mathbb{R}^n$ be a smooth map of an n-dimensional stratified variety to the Euclidean space of the same dimension. As usual, denote the stratification by $\mathcal{M}$.

DEFINITION 4.5. A point $x \in M$ is the regular point for f if $x \in \dot{\mathcal{M}}$ and $\operatorname{rank}_x f = n$. Otherwise the point is critical.

A point $b \in \mathbb{R}^n$ is a regular value for f if the preimage $f^{-1}(b)$ consists of regular points of f only. Otherwise the point is a critical value.

REMARK. The Sard lemma implies that n-dimensional Lebesque measure of all critical values of a smooth map is zero, because the skeleton is at most $(n-1)$-dimensional, therefore its image also is at most $(n-1)$-dimensional.

The following results are valid in the stratified category and serve as the main tool when constructing stratified varieties.

THEOREM 4.2 [W2]. *If a smooth map $g: N \to S$ is transversal to a stratum $M_\alpha \in \mathcal{M}$ at a point $x \in M_\alpha$, then it is also transversal to all the strata containing M_α in their closure, at all points sufficiently close to x.* □

COROLLARY. *For any stratified subset $(M, \mathcal{M}) \subseteq S$ and any smooth manifold N, maps transversal to $(M, \mathcal{M})$ form a residual subspace in $C^1(N, S)$. If the ambient manifold S is a jet space, $S = J^\ell(N, L)$, then for a generic map $g: N \to L$ its jet extension is transversal to $(M, \mathcal{M})$.* □

THEOREM 4.3 [Ma1, Corollary (8.8)]. *If a C^2-smooth map $g: N \to S$ is transversal to a closed stratified subset $M \subseteq S$, then the full preimage $g^{-1}(M) \subseteq N$ is a stratified subset in the ambient manifold N. The canonical pullback stratification $g^{-1}(\mathcal{M})$ of $g^{-1}(M)$ is formed by the strata $g^{-1}(M_\alpha)$.* □

COROLLARY. *If $L \subseteq S$ is a closed C^2-smooth submanifold of the ambient space, transversal to a closed stratified subset M, then the intersection $L \cap M$ is also a stratified subset.* □

The last notion we need is that of *refinement of a stratification.* Given a stratification $\mathcal{M}$ on M, one may subdivide each stratum $M_\alpha \in \mathcal{M}$ into smaller subsets. If they satisfy Definition 4.1 as well, than what one obtains is the refinement of the initial stratification. The formal definition follows.

DEFINITION 4.6. Let M be a subset endowed with a certain stratification $\mathcal{M}$. Another stratification $\mathcal{M}'$ is called a refinement of $\mathcal{M}$ if any stratum from the list $\mathcal{M}$ is the union of some strata from $\mathcal{M}'$.

In other words, for any two strata $M_\alpha \in \mathcal{M}$, $M'_\beta \in \mathcal{M}'$ either $M'_\beta \subseteq M_\alpha$, or they are disjoint.

4.2. Contiguity number. If M is one-dimensional stratified variety, then it is a locally finite union of points (vertices) and edges (one-dimensional strata). By definition, each edge is either closed (diffeomorphic to the circle), or its closure consists of the edge itself and two endpoints (possibly coinciding); these two points must be vertices of the variety, and we say that the edge *lands* at them. Thus for each vertex the number of edges landing at it is well defined and finite (if an edge lands at the vertex with both endpoints, one has to count this edge twice). One can estimate the topological complexity of a one-dimensional stratified variety by the supremum of these *landing numbers*, taken over all the vertices. The multidimensional generalization of this construction looks as follows.

Let M be a stratified variety of dimension n endowed with a certain stratification $\mathcal{M}$. The set $M^\# = \operatorname{sk}\mathcal{M}$ is again a stratified variety. Assume that $\dim M^\# = n-1$, and let $x \in M^\#$ be a smooth point of $M^\#$ (in other words, x belongs to a certain stratum M_α of dimension exactly $n-1$).

Take any smooth submanifold $L \subseteq S$ of codimension $n-1$ in S and transversal to M_α at x. Such a submanifold has to be transversal to all the strata of $\mathcal{M}$ at all their points sufficiently close enough to x, therefore by Theorem 4.3 the intersection locally is a one-dimensional stratified variety containing x as a vertex.

DEFINITION 4.7. The *local contiguity number* $\nu_x(\mathcal{M})$ of a stratified subset $(\mathcal{M}, M)$ at a point $x \in \operatorname{sk}\mathcal{M}$ belonging to a $(\dim M - 1)$-dimensional stratum, is the landing number at the vertex x of the intersection $M \cap L$ with a generic submanifold L of codimension $\dim M - 1$ in the ambient manifold.

The *global contiguity number* of the stratified variety M is the supremum of all the local contiguity numbers over all points belonging to $(n-1)$-dimensional strata:

$$\nu(\mathcal{M}) = \sup_{\substack{x \in M_\alpha \\ \dim M_\alpha = n-1}} \nu_x(\mathcal{M}).$$

REMARK. The local contiguity number must be finite, at least for algebraic stratified varieties and their full preimages by maps transversal to all strata. If the variety is noncompact, then the above supremum may *a priori* be infinite, but it turns out to be finite in the cases we are interested in.

REMARK. A point $x \in \operatorname{sk} \mathcal{M}$ must belong to an $(n-1)$-dimensional stratum of $\mathcal{M}$, since strata of smaller dimensions do not intersect a generic submanifold L of codimension $\dim M - 1$. By definition, the global contiguity number $\nu(\mathcal{M})$ is zero if there are no $(n-1)$-strata at all, i.e., $\dim \operatorname{sk} \mathcal{M} < \dim M - 1$.

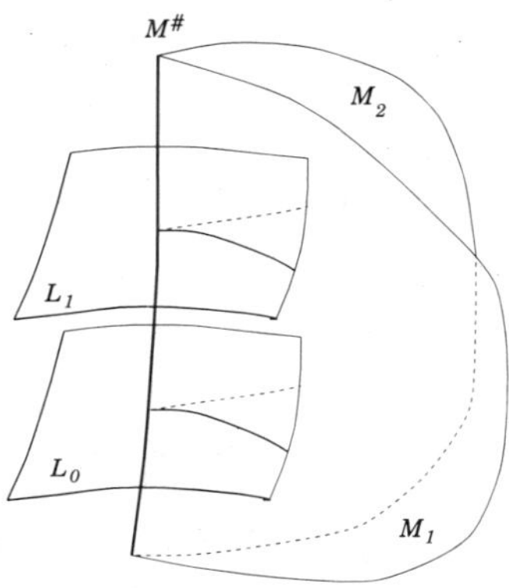

FIGURE 9. Definition of the local contiguity number.

In order to prove that this definition is consistent, that is, the local contiguity number is independent of the choice of the submanifold L as far as the latter remains transversal to the skeleton, we need the following result.

LEMMA 4.1. *If L_0, L_1 are two submanifolds in S of codimension $n-1$, both transversal to the same stratum M_α at two points x_i, $i = 0, 1$, belonging to the same connected component of the stratum M_α, then the germs of stratified one-dimensional intersections $(M \cap L_i, x_i)$, $i = 0, 1$, are homeomorphic.*

COROLLARY. *The local contiguity number is constant along any connected component of any stratum $M_\alpha \in \mathcal{M}$ with $\dim M_\alpha = \dim M - 1$, and can be computed using any submanifold L transversal to the stratum.* □

PROOF OF LEMMA 4.1. This almost evident assertion is a simple corollary to Thom's First Isotopy Lemma [Ma1]. It is sufficient to prove it for any two sufficiently close but distinct points (the general case can be reduced to this one using the transitivity argument).

After this locality is assumed, one can use the local coordinate system in S with the coordinates z, w chosen in such a way that the stratum M_α coincides with the plane $z = 0$. If both points x_i belong to the domain of this coordinate system, then the two manifolds L_i can be thought of as the graphs of two smooth vector-functions $w = \varphi_i(z)$, $i = 0, 1$. Consider the linear homotopy $\varphi_t = t\varphi_1(z) + (1-t)\varphi_0(z)$, $t \in [0, 1]$. This linear homotopy defines a one-parameter family of manifolds L_t (graphs of φ_t) transversal to M_α as well. If $\|z\|$ is sufficiently small, then the graphs of the functions φ_t do not intersect each other for different values of t, so there is a well-defined map from $L_* = \bigcup_{t\in[0,1]} L_t$ onto the segment $I = [w_0, w_1]$, $w_i = \varphi_i(0)$. Because of the Whitney Condition (b), L_* is transversal not only to M_α (by construction), but also to all the strata of greater dimension, adjacent to M_α. This means that the Isotopy lemma is applicable, guaranteeing that the foliation $L_* \cap M \to I$ is topologically locally trivial in a tubular neighborhood of the curve $t \mapsto \varphi_t(0)$. □

REMARK. John Mather explained to us that there exists a more general result implying consistency of the definition of contiguity number.

THEOREM. *If a smooth map $g: N \to S$ is transversal to a compact stratified subset $(M, \mathcal{M})$, then the preimage $g^{-1}(M)$ is stable with respect to C^1-small perturbations of g: for any other map $\widetilde{g}: N \to S$ sufficiently C^1-close to g, the preimage $\widetilde{g}^{-1}(M)$ is a stratified subset homeomorphic to $g^{-1}(M)$.* □

Now we study how the contiguity number of a stratified variety and its transversal inverse image are related to each other.

LEMMA 4.2. *Let $M^n \subseteq S$ be a stratified variety, and $x \in M_\alpha \subseteq \operatorname{sk} \mathcal{M}$ a smooth point on the skeleton, $\dim M_\alpha = n-1$. Then for any map $g: N \to S$ transversal to $(M, \mathcal{M})$, the local contiguity number $\nu_y(M')$ for $M' = g^{-1}(M)$ at any point $y \in g^{-1}(x)$ coincides with the contiguity number of M at x.*

The contiguity number for $M' = g^{-1}(M) \subseteq N$ is to be considered with respect to the canonical pullback stratification $g^{-1}(\mathcal{M})$, see Theorem 4.3.

PROOF. One may construct a submanifold $L' \subseteq N$ transversal to the stratum $M'_\alpha = g^{-1}(M_\alpha)$ at y and having codimension $n-1$ *in* N, in such a way that the restriction $g|_{L'}: L' \to L = g(L') \subseteq S$ would be a local diffeomorphism. Indeed, take any $(n-1)$-tuple of linear independent vectors $v_1, \dots, v_{n-1}$ spanning the subspace complementary to T_xM_α in T_xS. The transversality assumption guarantees that there exist $n-1$ vector $w_1, \dots, w_{n-1}$ in T_yN that are mapped into the former tuple by dg. Then any submanifold L' tangent to $\operatorname{span}(w_i)$ is diffeomorphically mapped by g onto its image L, and the latter automatically is transversal to M_α.

By Lemma 4.1, L and L' can be used to compute the local contiguity numbers of M and $M' = g^{-1}(M)$ respectively. The two germs of stratified sets, $L' \cap M'$ and $L \cap M$, are diffeomorphic one-dimensional stratified varieties, therefore the claim is proved. □

COROLLARY. *The global contiguity numbers for a stratified variety and its transversal inverse image coincide.* □

The contiguity number is in a sense independent of the choice of stratification: if one takes a refinement $\mathcal{M}'$ of the stratification $\mathcal{M}$, then the contiguity number remains almost the same.

LEMMA 4.3. *If a set M is endowed with two stratifications, $\mathcal{M}$ and $\mathcal{M}'$, the latter being the refinement of the former, then for the corresponding two contiguity numbers ν and ν' one has* $\nu' = \max(\nu, 2)$.

PROOF. Take any point $a \in M$ which is smooth on the skeleton $\operatorname{sk}\mathcal{M}'$. Then there are two possibilities: either $a \in \dot{\mathcal{M}}$, or $a \in \operatorname{sk}\mathcal{M}$. In the first case the skeleton $\operatorname{sk}\mathcal{M}'$ is locally a smooth codimension 1 submanifold in the corresponding upper stratum of $\mathcal{M}$, therefore the local contiguity number is equal to 2. Otherwise the local contiguity numbers for both $\mathcal{M}$ and $\mathcal{M}'$ at a coincide. □

4.3. Contiguity number of an algebraic variety. Algebraic varieties provide a primary source of examples of stratified sets. Moreover, the above results allow us to compute different characteristics of stratified varieties obtained as the transversal preimages of algebraic ones.

DEFINITION 4.8. A real algebraic variety is the subset of $\mathbb{R}^n$ defined as the zero locus for a family of real polynomials,

$$V = \{x \in \mathbb{R}^n : P_1(x) = P_2(x) = \cdots = 0\}.$$

When dealing with an algebraic variety, it is sometimes convenient to have its simplest possible representation. We introduce the notion of algebraic complexity, or σ-complexity, of an algebraic variety, in the following way.

DEFINITION 4.9. The σ-complexity of a (real) algebraic variety V is the smallest integer d such that V is the zero locus for a family of polynomials of degree $\leqslant d$:

$$\sigma(V) = \min\{d \in \mathbb{N} : V = \{P_1 = P_2 = \cdots = 0\},\ \deg P_j \leqslant d\}.$$

REMARK. To say that an algebraic variety has the σ-complexity no greater than d means that it is a zero locus for some ideal generated by polynomials of degree less or equal to d. We never deal with the σ-complexities themselves, only with their upper bounds.

The principal fact concerning algebraic varieties is their stratifiability.

THEOREM 4.4 (Whitney stratification theorem, [W1]). *Any algebraic variety can be stratified. There exists a stratification with semialgebraic strata. The skeleton of this stratification is also an algebraic variety.* □

Later on we make a more precise statement about stratification of algebraic subsets (some quantitative estimates). Now we prove that the contiguity number of an algebraic variety can be estimated in terms of the σ-complexity of the latter.

LEMMA 4.4. *For an algebraic variety $V \subset \mathbb{R}^n$ with $\sigma(V) = d$ and an arbitrary stratification $\mathcal{V}$ of it, the contiguity number $\nu(\mathcal{V})$ admits an upper estimate by a (primitive recursive) function of d and n:*

$$\nu(\mathcal{V}) \leqslant \max(2,\ d(2d-1)^{n-1}). \tag{4.2}$$

This estimate is independent of the choice of stratification of the variety.

PROOF. If $d = 1$, that is, the variety is an affine plane, then it is smooth and for any stratification the contiguity number is equal to 2. So assume that $d \geqslant 2$.

Let the dimension of V equal m and x be any smooth point of the skeleton $\operatorname{sk}\mathcal{V}$. For the stratum $V_\alpha \in \mathcal{V}$ containing x we have $\dim V_\alpha = m-1$. Take an affine subspace $L \subset \mathbb{R}^n$ of codimension $m-1$ transversal to V at x. Let $l_1, \dots, l_{m-1}$ be a system of independent linear functions vanishing on L.

The intersection $L \cap V$ is a one-dimensional stratified manifold, and each edge landing at x must intersect sufficiently small ε-sphere S_ε centered at x.

If the initial variety V is defined by polynomials $P_1(x), \dots, P_s(x)$ of degree $\leqslant d$, then the intersection $V \cap L \cap S_\varepsilon$ is the zero locus for the system of polynomials $P_1, \dots, P_s, l_1, \dots, l_{m-1}, Q$, where $Q = \varepsilon - \sum_j x_j^2$. On the other hand, the number of connected components of this intersection is no less then the local contiguity number.

Now we apply the Milnor theorem [Mi1, Theorem 2] estimating the sum of Betti numbers of a real algebraic variety through dimension of the ambient Euclidean space n and the maximal degree of polynomials d:

$$\sum b_j(M) \leqslant d(2d-1)^{n-1}.$$

Applying this estimate to the intersection $V \cap L \cap S_\varepsilon$, we conclude that the number of its connected components is estimated from above by the expression (4.2) uniformly over all points x and independently of the dimension of the variety and the number of polynomial equations (which can be larger than the codimension of the variety). □

REMARK. One can define the contiguity number at a point x that is smooth on the skeleton, in a different way as the number of connected components of the difference $M \setminus \operatorname{sk}\mathcal{M}$ in a sufficiently small ball $B_\varepsilon(x) \subset S$ centered at x (assuming that the ambient manifold is the Euclidean space). Then the finiteness theorem from [Mi2, Appendix A] guarantees that the contiguity number of an algebraic variety is finite, and the references mentioned there allow for its explicit upper estimate.

4.4. Whitney stratification of real algebraic varieties. We describe here a special stratification of an algebraic variety. The main reason for this is to obtain some quantitative characteristics of the stratification provided by Whitney Theorem 4.4; in particular, we find an upper estimate for $\sigma(\operatorname{sk}\mathcal{V})$ in terms of $\sigma(V)$ for an arbitrary algebraic variety V and a suitable stratification of V.

Recall that for a real algebraic variety its algebraic dimension is defined [W1]: if $I_V \subseteq \mathbb{R}[x]$ is the ideal of real polynomials vanishing on V, then the algebraic dimension is the transcendence degree of the field of rational functions on V, that is, the transcendence degree of the quotient $\mathbb{R}[x]/I_V$ over the field of reals. If the variety is l-dimensional and irreducible, then near its generic point it is a smooth l-dimensional submanifold of the ambient Euclidean space [W1], [Mi2]. For a family of functions $\mathbb{R}^n \to \mathbb{R}$ its rank at a point x is the dimension of the linear space spanned by their differentials computed at this point:

$$\operatorname{rank}_x \{p_1(x), \dots, p_s(x)\} = \dim(\mathbb{R} \cdot dp_i(x) + \dots + \mathbb{R} \cdot dp_s(x)).$$

DEFINITION 4.10. The *Whitney stratification* $\mathcal{V}$ of a real algebraic l-dimensional variety $V \subseteq \mathbb{R}^n$ is the stratification with the following characteristic properties. There exists a family of polynomials $p_1, p_2, \dots, p_s \in \mathbb{R}[x_1, \dots, x_n]$ such that:

(1) If V is irreducible or of constant dimension (which means that all its irreducible components have the same dimension l), then all p_i vanish on V, and

$$\max_{x \in V} \operatorname{rank}_x(p_1, \dots, p_s) = n - l,$$
$$\dot{\mathcal{V}} = \{x \in V : \operatorname{rank}_x(p_1, \dots, p_s) = n - l\},$$
$$\operatorname{sk} \mathcal{V} = V - \dot{\mathcal{V}}.$$

(2) If V is reducible, with irreducible components of different dimensions, and $V = V_0 \cup V_1$ is the decomposition of V into the union of irreducible components V_0 of (constant) maximal dimension l, and the part V_1 of dimension $\leqslant l - 1$, then all p_i vanish on V_0, and

$$\max_{x \in V_0} \operatorname{rank}_x(p_1, \dots, p_s) = n - l,$$
$$\dot{\mathcal{V}} = \{x \in V_0 : \operatorname{rank}_x(p_1, \dots, p_s) = n - l\} \setminus V_1,$$
$$\operatorname{sk} \mathcal{V} = V_1 \cup (V_0 - \dot{\mathcal{V}}).$$

We call the tuple $\mathbf{p} = \{p_1, \dots, p_s\}$ the *Whitney family of polynomials* supporting the Whitney stratification $\mathcal{V}$.

LEMMA 4.5. *For any algebraic variety $V \subseteq \mathbb{R}^n$ of dimension l and σ-complexity $\sigma(V) \leqslant d$ there exists a Whitney stratification supported by the Whitney family of polynomials $\mathbf{p} = (p_1, \dots, p_s)$ such that:*

(1) *the degrees of these polynomials admit an upper estimate by a certain primitive recursive function W_0,*

$$\deg p_i \leqslant W_0 = W_0(n, d), \qquad i = 1, \dots, s,$$

(2) *the σ-complexity of the skeleton admits a similar estimate,*

$$\sigma(\operatorname{sk} \mathcal{V}) \leqslant W_1 = W_1(n, d).$$

PROOF. The existence of Whitney stratification is proved in [W1] and [Ma1]. Hence we need only to establish the σ-complexity estimates. To that end, we must reproduce the main part of the proof from [W1], supplying it with constructive estimates from other sources [S], [BN].

1. By Definition 4.9, V is the locus of a polynomial ideal $\mathfrak{p}$ with generators of degree less or equal to d. At the first step we introduce the *ideal of the real variety* V as the ideal I_V of all real polynomials vanishing on V. This procedure is constructive: one may construct a finite family of polynomials generating I_V, and their degrees admit a primitive recursive upper estimate in terms of d and n (the dimension of the ambient space), see [BN].

REMARK. Since we are working over the field of reals, the ideal I_V is *not a radical* of the ideal generated by original polynomials defining V, but rather the *real radical* of the latter: if we denote by $P_1, \dots, P_k, \dots$, $\deg P_i \leqslant d$, the polynomials generating $\mathfrak{p}$, then

$$I_V = \left\{ p \in \mathbb{R}[x] \colon \exists r, m \in \mathbb{N}, q_i \in \mathbb{R}[x] \text{ such that } p^{2r} + \sum\nolimits_{i=1}^{m} q_i^2 \in \mathfrak{p} \right\}.$$

Still, this ideal can also be explicitly computed starting with $\mathfrak{p}$, see [BN].

2. Next, we can effectively construct the irreducible decomposition of V by constructing the primary decomposition of the ideal I_V [S, Proposition 65]: the number and the degrees of real polynomials generating the primary ideals in the representation

$$I_V = Q_1 \cap \cdots \cap Q_s, \qquad Q_i \subseteq \mathbb{R}[x_1, \dots, x_n] \text{ primary ideals},$$

admit a primitive recursive bound in terms of degrees of generators of I_V and n. The zero loci $V_{(i)} \subset \mathbb{R}^n$ of the ideals Q_i are irreducible varieties.

REMARK. The decomposition

$$V = \bigcup_i V_{(i)}$$

is in fact almost independent of the ground field: if we consider the ideal I_V^* of *complex polynomials* vanishing on the *real locus* V and then construct the normal decomposition $I_V^* = \bigcap_i Q_i^*$, $Q_i^* \subseteq \mathbb{C}[x]$, denoting by $V_{(i)}^* \subset \mathbb{C}^n$ the *complex zero loci* for Q_i^*, then by [W1, Lemma 7]

$$V_{(i)} = V_{(i)}^* \cap \{\operatorname{Im} x = 0\}$$

is the same irreducible decomposition of V. Moreover, $V_{(i)}^*$ are minimal complex algebraic varieties in $\mathbb{C}^n$ containing the corresponding parts $V_{(i)}$.

3. Since the intersection of ideals is constructive in the same sense [S, Proposition 2], we may explicitly compute the decomposition

$$V = V_0 \cup V_1, \qquad \dim V_0 = \dim V, \ \dim V_1 < \dim V,$$

into the part of pure dimension $l = \dim V$ and that of the pure dimension $< l$. Hence we can operate with the ideal $I_0 = I_{V_0}$ of real polynomials vanishing on V_0, meaning that a base for this ideal can be constructed and the degrees of all polynomials can be explicitly estimated.

4. Let $\mathbf{p} = \{p_1, \dots, p_s\} \subset \mathbb{R}[x]$ be a base for the ideal $I_0 = I_{V_0}$. The previous arguments show that this tuple may be chosen in such a way that the degrees of all p_i's admit a primitive recursive upper estimate $\deg p_i \leqslant W_0(n, d)$ in terms of the degrees of the original equations describing V: $d \geqslant \sigma(V)$. We claim that the tuple $\mathbf{p}$ is a Whitney family of polynomials.

Indeed, the rank of the tuple $\mathbf{p}$ attains its maximal value $n-l$ on each irreducible component $V_{(i)} \subseteq V_0$ [W1, Lemma 4]. Hence the set

$$Z = \{x \in V_0 \colon \operatorname{rank}_x \mathbf{p} < n - l\}$$

does not contain components $V_{(i)}$ with $\dim V_{(i)} = l$. Thus the irreducibility of each of the latter components implies that $\dim Z < l$.

The condition of rank dropping is algebraic (all minors of order $(n-l) \times (n-l)$ of the Jacobian matrix for the family of functions $p_1, \dots, p_s$ with respect to the variables $x_1, \dots, x_n$ should vanish), and the additional polynomial equations determining Z are of degrees admitting a primitive recursive estimate. Hence $\sigma(Z) \leqslant (W_0(n, d))^{n-l}$, $l \geqslant 0$. By [W1, Theorem 1], the complement $V_0 \setminus Z$ is a union of smooth l-dimensional semialgebraic submanifolds in $\mathbb{R}^n$ and everywhere on U the tuple $\mathbf{p}$ has rank $n - r$.

5. Denote by $\widetilde{V}$ the union $Z \cup V_1$; we have

$$\dim \widetilde{V} < l, \qquad \sigma(\widetilde{V}) \leqslant W_1(n, d),$$

where W_1 is a primitive recursive upper estimate for the degrees of generators of the intersection $I_Z \cap I_1$, $I_1 = I_{V_1}$. This observation allows for inductive iteration of the above described procedure, which finally leads to a partition of V into a disjoint union of semialgebraic smooth submanifolds. It is not yet a stratification, since Condition (b) may fail somewhere, but in the paper [Ma1] it is proved that Condition (b) for two semialgebraic strata fails on a proper algebraic subset of the smaller of the two. This assertion means that the above partition $V = (V_0 \setminus \widetilde{V}) \cup \widetilde{V}$ can be further refined to become a true stratification of V. It is essential to point out that the refinement described in [Ma1] does not affect the upper strata constructed on the first step (operations 1–4 above). Thus we have for the stratification $\mathcal{V}$ obtained after all refinements,

$$\begin{aligned} \dot{\mathcal{V}} &= U = V_0 \setminus (Z \cup V_1), \qquad && \operatorname{rank} \mathbf{p}|_U = n - l, \ \deg \mathbf{p} \leqslant W_0(n, d), \\ \operatorname{sk} \mathcal{V} &= \widetilde{V} = Z \cup V_1, && \sigma(\operatorname{sk} \mathcal{V}) \leqslant W_1(n, d). \end{aligned}$$

The proof of both parts of Lemma 4.5 is complete. □

REMARK. It is this very statement which essentially involves primitive recursive functions: before Lemma 4.5 and after it all expressions occurring in estimates, can be written explicitly (the ones related to the Khovanskiĭ reduction procedure, etc.).

Recent achievements in constructive commutative algebra and constructive real algebraic geometry suggest that the estimates provided by Lemma 4.5 can be made explicit. In [MR] an algorithm for the effective Whitney stratification of algebraic subsets in $\mathbb{C}^n$ is suggested and explicit estimates for the degrees of polynomial formulas defining the strata are given; the authors claim also that their algorithm applies to real algebraic varieties. Some explicit estimates for the degrees of generators of the real radical of an ideal were recently obtained by R. Neuhaus: he proved that these generators can be chose of degrees not exceeding $d^{2^{O(n^2)}}$ [N]. We do not want to enter deeply into this area; later on we formulate precisely the problem from constructive algebraic geometry whose solution would yield an explicit upper estimate for the number $E(k)$ occurring in the assertion of Theorem II from the Introduction.

4.5. Stratification of pairs of algebraic varieties. We show that for any algebraic subvariety $W \subseteq V$ of an algebraic variety V there exists a refinement $\mathcal{V}'$ of the Whitney stratification which provides the *joint stratification* of the pair (V, W).

DEFINITION 4.11. A joint stratification for an algebraic variety V and its algebraic subvariety $W \subseteq V$ is the stratification $\mathcal{V}$ on V such that each stratum of it either belongs entirely to W, or is disjoint with the latter.

In other words, W is represented as the union of some strata from V. Sometimes it is convenient to speak about two stratifications, $\mathcal{V}$ on V and $\mathcal{W}$ on W, such that $\mathcal{W} \subset \mathcal{V}$ (as the lists of strata of the two stratifications).

LEMMA 4.6. *Each pair (V, W) of algebraic varieties, $W \subseteq V$, admits a joint stratification $\mathcal{W}' \subseteq \mathcal{V}'$.*

OUTLINE OF THE PROOF. We show only that upper-dimensional strata may be constructed in the appropriate manner; afterwards one may use induction in the dimension of the larger variety V. If $V = \bigcup V_j$ is the irreducible decomposition of V, then for any V_j the intersection $V_j \cap W$ is either a proper subset in V_j with $\dim V_j \cap W < \dim V_j$, or coincides with V_j. Denote by $\mathcal{V}$ and $\mathcal{W}$ the Whitney stratifications for V and W respectively.

If $\dim W < \dim V = d$, then to obtain upper strata of the stratification $\mathcal{V}'$ it is sufficient to delete from $\dot{\mathcal{V}}$ all points of W: because of the dimension assumption, any stratum thus obtained will be disjoint with W, and to proceed further one has to stratify the pair $(\operatorname{sk} \mathcal{V} \cup W, W)$. The larger set of this pair has dimension $\leqslant \dim V - 1$, and in this case $\mathcal{W}'$ will have no d-dimensional strata at all. The stratification goes then by induction in the dimension of the larger variety.

If $\dim W = \dim V$, then one may take as upper strata of $\mathcal{V}'$ the smooth parts of upper-dimensional irreducible components of V after deleting from them the points from the skeleton of the Whitney stratification $\mathcal{W}$. The upper strata of $\mathcal{W}'$ are those which intersect W. Then any upper stratum of $\mathcal{V}'$ either is disjoint with W, or belongs to W entirely. This follows immediately from irreducibility of the components and the assumption on the dimensions $\dim V = \dim W$. As in the proof of Lemma 4.5, after deleting upper strata the problem is reduced to the case of smaller dimensions. □

§5. Finitely differentiable maps with Gabrielov property

5.1. Stratified and stratifiable maps. In this section we introduce a class of smooth mappings of stratified manifolds into Euclidean spaces of the same dimension, which behave essentially like polynomials.

DEFINITION 5.1. Let $(M, \mathcal{M})$ be a stratified m-dimensional submanifold, and $f: M \to \mathbb{R}^m$ a smooth map into the Euclidean space *of the same dimension*. We say that $\mathcal{M}$ *stratifies* f if the following dichotomy holds: each stratum $M_\alpha \in \mathcal{M}$ entirely consists either of only regular points, or of only critical points of f (in the sense of Definition 4.5).

The map f is *stratifiable* if M admits a stratification (possibly a refinement of the initial one) that stratifies f. If $\mathcal{M}$ is such a stratification, then we say that f is $\mathcal{M}$-stratified.

REMARK. Since any stratum of dimension smaller than $\dim M$ by definition consists of only critical points, the above alternative is nontrivial only for upper-dimensional strata.

DEFINITION 5.2. A *criminant* set for a stratifiable map is the skeleton of a stratification stratifying the map. We denote the criminant set of a map f by $c(f)$.

REMARK. The notion of the criminant set is not uniquely determined by the map itself, but depends also on the choice of the stratification. When speaking about criminant sets, we assume that the stratification is specified by the context.

The following proposition gives a principal example of a stratified map: actually, what we will need is a generalization of the result formulated below. It is similar in spirit to Lemma 2.7 from [Mi2], and coincides with the latter for the case of algebraic varieties of constant dimension.

PROPOSITION. *Any polynomial map*

$$q: \mathbb{R}^N \supseteq V^n \to \mathbb{R}^n, \qquad q = (q_1, \dots, q_n),$$

restricted to an algebraic variety V, *is stratifiable.*

PROOF. This is almost evident. Let $\mathcal{V}$ be the Whitney stratification of V, as it was introduced in 4.4, and $\mathbf{p} = (p_1, \dots, p_s)$ is the Whitney family of polynomials supporting $\mathcal{V}$. Denote by $\mathbf{q} = (p_1, \dots, p_s, q_1, \dots, q_n)$ the tuple of polynomials obtained by adding to $\mathbf{p}$ the coordinate polynomials of the map q.

Consider the algebraic subset

$$W = \{x \in V: \operatorname{rank}_x \mathbf{q} < N\} \subseteq V \subset \mathbb{R}^N.$$

Let $\mathcal{V}'$ be the joint stratification of the pair $(V, W \cup \operatorname{sk} \mathcal{V})$ as it was introduced in Definition 4.11. We claim that $\mathcal{V}'$ stratifies $q|_V$. Indeed, if the rank of the tuple $\mathbf{q}$ drops from its maximal value at some point x, then either $\operatorname{rank}_x \mathbf{p} < N - n$, or $\operatorname{rank}_x \mathbf{p} = N - n$, but the zero space $\bigcap_{i=1}^n \ker dq_i(x)$ is not transversal to the subspace $\bigcap_{j=1}^s \ker dp_j(x)$. In the first case the point is singular by the property of the Whitney family $\mathbf{p}$, i.e., $x \in \operatorname{sk} \mathcal{V}$, while in the second case the point x is smooth, $x \in \dot{\mathcal{V}}$, but the restriction $q|_V$ has a critical point at x in the usual differentiable sense. In both cases we see that W is contained in the criminant set $c(q)$ and contains all smooth points of $c(q)$:

$$W \subseteq c(q), \qquad c(q) \cap \dot{\mathcal{V}} \subseteq W.$$

Thus $W \cup \operatorname{sk} \mathcal{V} = c(q)$, since all points belonging to less-than- n-dimensional strata are critical by definition. □

5.2. Inductive Gabrielov property of stratified maps. We start with an evident generalization of the Gabrielov number construction for the category of stratified sets. Let $(M, \mathcal{M})$ be an n-dimensional stratified set and $f\colon M \to \mathbb{R}^n$ a smooth map of this set into the Euclidean space of the same dimension. Let $y \in \mathbb{R}^n$ be a point in the target space.

DEFINITION 5.3. The *local Gabrielov number* $\mathcal{G}_y(f)$ of a smooth map $f\colon M \to \mathbb{R}^n$ is the number of *regular points* in the full preimage $f^{-1}(y)$:

$$\mathcal{G}_y(f) = \#\{\, x \in \dot{\mathcal{M}}\colon \operatorname{rank}_x f = n = \dim M\,,\ f(x) = y \,\}.$$

For a regular value $y \in \mathbb{R}^n$ one has $\mathcal{G}_y(f) = \# f^{-1}(y)$. The function $y \mapsto \mathcal{G}_y(f)$ is semicontinuous: if $\mathcal{G}_y(f) < \infty$, then

$$\forall y \in \mathbb{R}^n \quad \exists \varepsilon > 0\colon \qquad \|y - \tilde{y}\| < \varepsilon \implies \mathcal{G}_y(f) \leqslant \mathcal{G}_{\tilde{y}}(f).$$

Hence in order to obtain an upper estimate for the Gabrielov number it is sufficient to find an upper estimate for $\mathcal{G}_y(f)$ for all y from some open dense subset in $\mathbb{R}^n$.

The following definition gives a natural generalization of the Gabrielov number for the category of stratified sets.

DEFINITION 5.4. The Gabrielov number $\mathcal{G}(f)$ of a smooth map $f\colon M \to \mathbb{R}^n$ is the upper bound of the number of regular preimages of points,

$$\mathcal{G}(f) = \sup_{y \in \mathbb{R}^n} \mathcal{G}_y(f) \leqslant \infty.$$

The map possesses the *Gabrielov property* if the corresponding Gabrielov number is finite: $\mathcal{G}(f) < \infty$.

Given these definitions, we proceed with the main construction allowing for inductive estimation of the Gabrielov number for a smooth map.

Suppose that M is a subset of an ambient manifold S, $\dim M = n$, and $f\colon M \to \mathbb{R}^n$ is a smooth map which is stratified by a certain stratification $\mathcal{M}$ of M. Let also $\mathbb{R}^{n-1} \subset \mathbb{R}^n$ be a subspace of codimension 1, and $\pi\colon \mathbb{R}^n \to \mathbb{R}^{n-1}$ the orthogonal projection.

The following result claims that, if in addition, the map $\pi \circ f$ restricted to the criminant set $c(f) = \operatorname{sk} \mathcal{M}$ possesses the Gabrielov property, than the initial map f also does. Denote for brevity the restricted map by $f^{\#}$:

$$f^{\#}\colon \operatorname{sk} \mathcal{M} \to \mathbb{R}^{n-1}, \qquad f^{\#} = (\pi \circ f)|_{\operatorname{sk} \mathcal{M}}.$$

LEMMA 5.1. *Let* $(M, \mathcal{M})$ *be a stratified n-dimensional set with the contiguity number* $\nu = \nu(\mathcal{M})$, *and* $f : M \to \mathbb{R}^n$ *a smooth* $\mathcal{M}$*-stratified map with the criminant set* $c(f) = \operatorname{sk}\mathcal{M} \subseteq M$. *Suppose that the map* $\pi \circ f : M \to \mathbb{R}^{n-1}$ *is proper. Then*

$$\forall y \in \mathbb{R}^n \qquad \mathcal{G}_y(f) \leqslant \frac{\nu}{2} \cdot \mathcal{G}_{\pi(y)}(f^{\#}) \leqslant \infty.$$

COROLLARY. *If* f *satisfies the conditions of Lemma* 5.1 *and* $f^{\#}$ *possesses the Gabrielov property, then* f *also does, and* $\mathcal{G}(f) \leqslant \frac{\nu}{2} \cdot \mathcal{G}(f^{\#})$.

PROOF. 0. The idea of the proof is extremely simple: we take the point y and move it along the fiber of the projection π, keeping an eye on the number of preimages in $f^{-1}(y)$. This number is changed only when the point y passes across the image of the skeleton $\operatorname{sk}\mathcal{M}$, since on all upper strata the preimages are regular (the process is shown on Figure 3 in the Introduction, with the notation slightly different from that used above).

At each point of discontinuity the number of preimages can change no more than by the local contiguity number. The number of points at which y meets the image of the skeleton, is basically the cardinality of the corresponding preimage for $f^{\#}$. This gives an estimate for the variation of the number of preimages in $f^{-1}(y)$. At "infinity" this number is zero, because of the properness assumption. Hence we get the estimate for its maximal value. More accurate arguments are as follows.

1. Without loss of generality we may assume that the map f has the full rank n on all upper strata of $\mathcal{M}$. Indeed, if there is a stratum on which the rank drops down from its maximal value, then this occurs everywhere on the stratum, hence by deleting the stratum we do not change the number of *regular* preimages of any point $y \in \mathbb{R}^n$. If after such a procedure there will be no n-dimensional strata at all, then the assertion of the lemma trivially holds. Note that the criminant set $c(f)$ is always at most $(n-1)$-dimensional.

2. Next, it is sufficient to prove the estimate for a generic point $y \in \mathbb{R}^n$, because of the semicontinuity of the function $y \mapsto \mathcal{G}_y(f)$. Thus without loss of generality we may assume that y is such that $y' \in \mathbb{R}^{n-1}$ is the regular value both for the map $f^{\#}$ and for the restriction $\pi \circ f|_{\dot{\mathcal{M}}}$. We prove the uniform estimate of the required type for all points on the line $\pi^{-1}(y')$ passing through the point y (therefore for the point itself).

By the transversality arguments, *the inverse image*

$$L = (\pi \circ f)^{-1}(y') = f^{-1}(\pi^{-1}(y')) \subseteq M$$

will be a compact stratified one-dimensional manifold with the contiguity number not exceeding $\nu = \nu(\mathcal{M})$.

Indeed, let $F : S \to \mathbb{R}^n$ be the map whose restriction to M is f. We have chosen the point y' in such a way that the set $Z = (\pi \circ F)^{-1}(y') \subset S$ is transversal to M. Then one can easily see that $L = Z \cap M$. By Corollary to Theorem 4.3, L is a one-dimensional stratified manifold itself. Its contiguity number $\nu(L)$ does not exceed $\nu(\mathcal{M})$ by virtue of Corollary to Lemma 4.1, because $\operatorname{codim} Z = \dim M - 1$. Moreover, L is compact since $\pi \circ f$ is proper. Denote the induced stratification of L by $\mathcal{L}$.

3. Thus we have the map from the stratified subset L onto the real line $\mathbb{R}^1 \simeq \pi^{-1}(y')$, which is induced by the original map $f\colon M \supseteq f^{-1}(\mathbb{R}^1) = L \xrightarrow{f} \mathbb{R}^1$. Since upper strata (edges) of $\mathcal{L}$ come from upper strata of $\mathcal{M}$, $\dot{\mathcal{L}} = \dot{\mathcal{M}} \cap Z$, this map is submersive on any upper stratum $L_\alpha \in \mathcal{L}$.

4. The number of regular preimages of a point on $\mathbb{R}^1$ does not exceed the number of edges of $\mathcal{L}$, because the submersivity of the map means that only one preimage on each edge can occur. But there can be neither edges landing "at infinity" (noncompact ones), nor edges diffeomorphic to the circle (the latter possibility is ruled out by the submersivity, since a circle, being compact, cannot be mapped onto the real line without critical points). Therefore each edge must land at two vertices of sk $\mathcal{L}$ (in fact, an edge cannot land on the same vertex by both endpoints by virtue of the same submersivity arguments). The number of vertices is equal to $\mathcal{G}_{y'}(f^\#)$ by definition of the latter number (recall that y' was chosen to be a regular value, therefore all preimages are regular points for $f^\#$), and no more than ν edges can land at any single vertex. Thus we can estimate from above the number of edges by $\frac{\nu}{2} \cdot \mathcal{G}_{y'}(f^\#)$. □

5.3. Nice maps. Now we proceed with the main construction and define a class of *nice* maps allowing for iterated application of Lemma 5.1. This class is similar to the class of maps arising in the Thom–Boardmann theory [AVG], and a sufficient condition for a smooth map to be nice is given in the next section, where we prove in particular that a generic chain map is nice.

Let M be a stratified set and $f\colon M \to \mathbb{R}^n$ a smooth map from it to the Euclidean space of the same dimension $n \geqslant 0$. Recall that a *complete flag* in the Euclidean space $\mathbb{R}^n$ is a decreasing chain of subspaces

$$\mathbb{R}^n \supset \mathbb{R}^{n-1} \supset \cdots \supset \mathbb{R}^1 \supset \mathbb{R}^0 = \{0\}.$$

We consider a tuple of orthogonal projections associated with this flag,

$$\mathbb{R}^n \xrightarrow{\pi_n} \mathbb{R}^{n-1} \xrightarrow{\pi_{n-1}} \cdots \xrightarrow{\pi_2} \mathbb{R}^1 \xrightarrow{\pi_1} \mathbb{R}^0 = \{0\}. \tag{5.1}$$

DEFINITION 5.5. A map f of an n-dimensional stratified variety $(M, \mathcal{M})$ to the Euclidean space of the same dimension is *nice* with respect to the flag (5.1), if there exists a decreasing sequence of subsets

$$M = M^n \supseteq M^{n-1} \supseteq \cdots \supseteq M^1 \supseteq M^0,$$

and a sequence of smooth maps $f_j\colon M^j \to \mathbb{R}^j$, $j = 0, \dots, n$, $f_n = f$, such that:

- the diagram

$$\begin{array}{ccccccccc} M^n & \xleftarrow{i_n} & M^{n-1} & \xleftarrow{i_{n-1}} & \cdots & \xleftarrow{i_2} & M^1 & \xleftarrow{i_1} & M^0 \\ {\scriptstyle f_n}\downarrow & & {\scriptstyle f_{n-1}}\downarrow & & & & \downarrow{\scriptstyle f_1} & & \downarrow{\scriptstyle f_0} \\ \mathbb{R}^n & \xrightarrow{\pi_n} & \mathbb{R}^{n-1} & \xrightarrow{\pi_{n-1}} & \cdots & \xrightarrow{\pi_2} & \mathbb{R}^1 & \xrightarrow{\pi_1} & \mathbb{R}^0 \end{array} \tag{5.2}$$

  where i_j are the natural embeddings, is commutative,

- each map $\pi_j \circ f_j$ is proper,
- each subset M^j is a stratified variety that admits a stratification $\mathcal{M}^j$, and each f_j is $\mathcal{M}^j$-stratified map,
- the set M^{j-1} is the criminant set for f_j, i.e.,

$$M^{j-1} = \operatorname{sk} \mathcal{M}^j, \qquad j = 1, \dots, n. \tag{5.3}$$

We say that the map f is *nice* (without referring to any flag), if there exists a complete flag such that f is nice with respect to this flag in the above sense.

REMARK. We *do not require* in the definition of a nice map that the stratification $\mathcal{M}^{j-1}$ on M^{j-1} coincides with the stratification of the latter set as the skeleton of the stratification $\mathcal{M}^j$: the condition (5.3) above means only the point set equality.

Denote by $\nu_n, \dots, \nu_1$ the contiguity numbers for the stratifications $\mathcal{M}^j$ and fix *any* flag in $\mathbb{R}^n$.

THEOREM 5.1. *A nice map possesses the Gabrielov property:* $\mathcal{G}(f) < \infty$. *Moreover, if* f *is nice with respect to an arbitrary flag* (5.1) *and* $\nu_n, \dots, \nu_1$ *are the corresponding contiguity numbers for the stratifications* $\mathcal{M}^n, \dots, \mathcal{M}^1$ *from the diagram* (5.2), *then*

$$\mathcal{G}(f) \leqslant 2^{-n} \nu_n \nu_{n-1} \cdots \nu_2 \nu_1 \cdot \# M_0.$$

PROOF. Each square of the commutative diagram (5.2) satisfies the conditions of Lemma 5.1, if we set $M = M^j$, $\mathcal{M} = \mathcal{M}^j$, $f = f_j$, $\pi = \pi_j$, $f^\# = f_{j-1} = (\pi_j \circ f_j)|_{\operatorname{sk} \mathcal{M}^j} = \pi_j \circ f_j \circ i_j$. Henceforth we have a series of inequalities,

$$\mathcal{G}(f_j) \leqslant \frac{\nu_j}{2} \mathcal{G}(f_{j-1}), \qquad j = n, \dots, 1, \tag{5.4}$$

valid by virtue of Corollary to Lemma 5.1. But since the map f_0 is proper, the last zero-dimensional set M^0 must be finite, and by definition $\mathcal{G}(f_0) = \# M^0$. □

REMARK. In Definition 5.5 it is implicitly contained the assumption that $\dim M^j = j$ for all $j = 0, 1, \dots, n$, because the definition of stratifiability assumes that the dimension of the source and the target spaces are equal. But this is not a very serious restriction: if $f = f_n$ has at least one regular preimage and is stratifiable, then the skeleton must be exactly $(n-1)$-dimensional. Indeed, the image $f(M)$ in this case must have a nonvoid interior, hence its boundary must be the image of at least $(n-1)$-dimensional criminant set.

5.4. Local finiteness. The conclusion of Theorem 5.1 is not an effective estimate, since one cannot in general estimate the last factor $\# M_0$ in the product. But it is possible to find a constructive estimate for the Gabrielov number of a nice map restricted to a sufficiently small neighborhood of a point of the source space M.

DEFINITION 5.6. Let $M \subseteq \mathbb{R}^N$ be a stratified n-dimensional manifold, $f \colon M \to \mathbb{R}^n$ a smooth map of it into the Euclidean space of the same dimension, and $a \in M$ a point (eventually, singular). We define the *local Gabrielov number* of f at a as

$$\mathcal{G}^{\text{loc}}(f, a) = \limsup_{r \to 0^+} \mathcal{G}(f|_{M \cap B_r(a)}),$$

where $B_r(a)$ is a small open ball of radius r centered at the point a.

REMARK. Between the two Gabrielov-type numbers (Definitions 5.3 and 5.6) there exists a natural inequality $\mathcal{G}^{\text{loc}}(f, a) \geqslant \mathcal{G}_{f(a)}(f)$, but the inequality may be strict, as the following example shows.

Let $M \subseteq \mathbb{R}^N$ be an one-dimensional stratified subvariety with the vertex at the origin, $\mathbb{R}^1$ a line passing through the origin, and f a projection $\mathbb{R}^N \to \mathbb{R}^1$ restricted to M such that $f^{-1}(0) = 0$. Denote by $\nu = \nu_0(\mathcal{M})$ the landing number at the origin.

Then $\mathcal{G}^{\text{loc}}(f, 0) \leqslant \nu$, and generically $\mathcal{G}^{\text{loc}}(f, 0) > \mathcal{G}_0(f) = 0$ (see Figure 10).

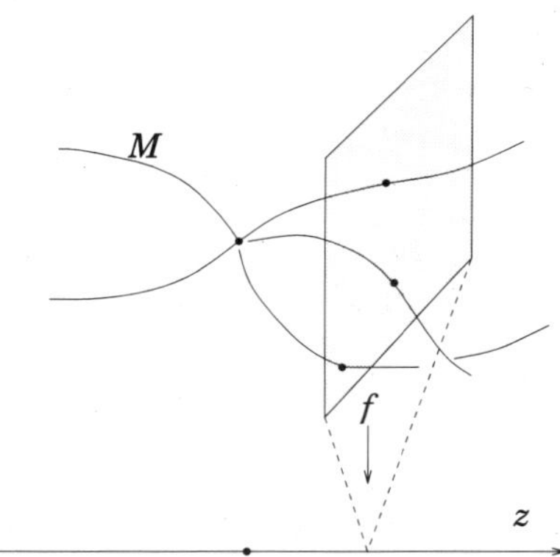

FIGURE 10. Confluent preimages: here $\nu_0(\mathcal{M}) = 5$, $\mathcal{G}^{\text{loc}}(f, 0) = 3$, $\mathcal{G}_0(f) = 0$.

Now we proceed with the definition of a locally nice map. Let M, f and a be the same as in Definition 5.6. Consider the Euclidean space $\mathbb{R}^{N+1} = \mathbb{R}^N \times \mathbb{R}^1$, the additional coordinate being denoted by z, the set

$$\widetilde{M_a} = \{(x, z) \in \mathbb{R}^{N+1} : x \in M, \ |x - a|^2 \leqslant z^2, \ |z| \leqslant 1\},$$

and the *suspended* map

$$\widetilde{f} : \widetilde{M_a} \to \mathbb{R}^{n+1}, \qquad (x, z) \mapsto (f(x), z). \tag{5.5}$$

The target space $\mathbb{R}^{n+1}$ is thus supplied with the Cartesian structure, the last coordinate (denoted also by z) being distinguished.

DEFINITION 5.7. A *special flag* in $\mathbb{R}^{n+1}$ is a complete flag (as in (5.1)) starting from the entire space $\mathbb{R}^{n+1}$ and ending with the origin $\mathbb{R}^0 = \{0\}$ such that all the subspaces constituting this flag, are parallel to the z-axis, and all the projections $\pi_{n+1}, \ldots, \pi_2$ (except for the last one π_1) preserve the z-coordinate.

DEFINITION 5.8. A map $f : M \to \mathbb{R}^n$ is *locally nice* at a point $a \in M$ if the suspended map $\widetilde{f}|_{\widetilde{M_a}}$ is nice with respect to some special flag.

If a map is locally nice, then all the objects relevant to Definition 5.5 do exist: the sequence of stratified varieties $\widetilde{M}^{n+1} = \widetilde{M_a}, \widetilde{M}^n, \ldots, \widetilde{M}^1, \widetilde{M}^0$, the sequence of maps $\widetilde{f}_{n+1} = \widetilde{f}|_{\widetilde{M}^{n+1}}, \widetilde{f}_n, \ldots, \widetilde{f}_0$, the associated contiguity numbers $\widetilde{\nu}_{n+1}, \ldots, \widetilde{\nu}_1$, etc.

THEOREM 5.2. *If a map* $f\colon M \to \mathbb{R}^n$ *is locally nice at a point* $a \in M$, *then*

$$\mathcal{G}^{\mathrm{loc}}(f, a) \leqslant 2^{-n} \prod_{j=1}^{n+1} \widetilde{\nu}_j.$$

PROOF. 1. One has to estimate the number of regular preimages of a point $(y, z) \in \mathbb{R}^n \times \mathbb{R}^1$ by the map $\widetilde{f}$ uniformly over all $y \in \mathbb{R}^n$ and all *sufficiently small* values of $z \in (\mathbb{R}^1, 0)$. Indeed, $\widetilde{f}^{-1}(y, z) = f^{-1}(y) \cap B_z(a)$, and a regular preimage $x \in B_z(r)$ for a point y by the map f is the first component of a regular preimage of any point (y, z') by $\widetilde{f}$ for all $z' > z$.

Take any value of y and construct the sequence of points,

$$b_{n+1} = (y, z),\ b_n = \pi_{n+1}(b_{n+1}),\ \dots,\ b_1 = \pi_2(b_2),\ b_0 = \pi_1(b_1) = z.$$

Applying Lemma 5.1 n times, one gets the following chain of inequalities,

$$\mathcal{G}_y(f) = \mathcal{G}_{b_{n+1}}(\widetilde{f}_{n+1}) \leqslant \frac{\widetilde{\nu}_{n+1}}{2} \mathcal{G}_{b_n}(\widetilde{f}_n) \leqslant \cdots \leqslant 2^{-n} \prod_{j=2}^{n+1} \widetilde{\nu}_j \cdot \mathcal{G}_{b_0}(\widetilde{f}_1).$$

2. The last Gabrielov-type number $\mathcal{G}_{b_0}(\widetilde{f}_1)$ is the number of regular preimages of the small value $b_0 = z \in (\mathbb{R}^1, 0)$ by the map $\widetilde{f}_1$ which is the restriction of the map $(x, z) \mapsto z$ to the one-dimensional stratified variety $\widetilde{M}^1 \subseteq \mathbb{R}^{N+1}$. From the construction of the variety $\widetilde{M}_a = \widetilde{M}^{n+1}$ it is clear that $\widetilde{M}^{n+1}$ belongs to the cone $\{|x - a|^2 \leqslant z^2\}$, therefore all the other varieties $\widetilde{M}^j$ do belong to the same cone.

Thus the inverse image $\widetilde{f}_1^{-1}(0) \subseteq \widetilde{M}^1$ consists of only one point $(a, 0)$. The set $\widetilde{M}^0$ is discrete and zero-dimensional. Hence $\widetilde{M}^0 \setminus \{0\}$ consists of a finite number of points, and their z-coordinates are positive and greater than some z_0. Therefore for all sufficiently small values of $z \in (0, z_0)$ the preimage $\widetilde{f}_1^{-1}(z)$ will consist of only regular points, and their number will not exceed the number of edges landing at the origin, since on each edge only one regular preimage may occur. By definition, the number of edges of $\widetilde{M}^1$ landing at the origin, does not exceed the contiguity number $\widetilde{\nu}_1$, see Figure 10. Thus $\mathcal{G}_{b_0}(\widetilde{f}_1) \leqslant \widetilde{\nu}_1$, and the required estimate is established. □

5.5. Appendix: nice maps depending on parameters. Assume that in Definition 5.6 the map f additionally depends on some parameters $\lambda \in B$, where B is a closed ball in a finite-dimensional space: $B = \{\lambda \in \mathbb{R}^k : |\lambda| \leqslant \rho\}$. Define the suspension of f as the map

$$F\colon \widetilde{M}_a \times B \to \mathbb{R}^{n+k+1}, \qquad (x, \lambda, z) \mapsto (f(x), \lambda, z). \tag{5.6}$$

Assume that some flag in $\mathbb{R}^{n+k+1}$ is chosen in such a way that all subspaces of dimensions $\geqslant k+1$ are parallel to the (λ, z)-subspace, and all projections $\pi_{n+k+1}, \dots, \pi_{k+2}$ preserve λ and z; the choice of the other subspaces and projections is inessential.

THEOREM 5.3. *Assume that the suspended map (5.6) is nice with respect to the flag with the above described properties, and* $\nu^*_{n+k+1}, \dots, \nu^*_{k+1}, \dots, \nu^*_1$ *are the corresponding contiguity numbers.*

Then for any $\lambda \in B$ *the map* $f_\lambda = f(\cdot, \lambda)$ *admits an upper estimate of the form*

$$\mathcal{G}^{\mathrm{loc}}(f_\lambda, a) \leqslant 2^{-n} \prod_{j=k+1}^{n+k+1} \nu^*_j, \tag{5.7}$$

which is uniform over all values of $\lambda \in B$.

PROOF. 1. The first step of the proof coincides with the first step of the proof of Theorem 5.2. As a result, we estimate the local Gabrielov number for the suspended map F by

$$2^{-n} \prod_{j=k+2}^{n+k+1} \nu^*_j \cdot \mathcal{G}_{b_k}(f^*_{k+1}),$$

where f^*_{k+1} is the projection $(x, \lambda, r) \mapsto (\lambda, r)$ restricted to some $k+1$-dimensional stratified manifold $M^{k+1,*}$. From the construction of this manifold it follows that $M^{k+1,*}$ belongs to the set $\{|x-a|^2 \leqslant z^2, \ |\lambda| \leqslant \rho\}$.

2. By the semicontinuity of the function $y \mapsto \mathcal{G}_y(f)$ it is sufficient to prove the estimate (5.7) only for generic values of $\lambda \in B$. But the generic section $\lambda = \lambda_0 = \text{const}$ intersects $M^{k+1,*}$ at a one-dimensional stratified manifold whose contiguity number at the point $(a, \lambda_0, 0)$ does not exceed the contiguity number ν^*_{k+1}.

3. Now the same arguments as those used at the second stage of the proof of Theorem 5.2, prove the required estimate (5.7). □

§6. Universal criminant sets

The objective of this section is to prove that the requirements imposed on a nice map (Definition 5.5) can be formulated in terms of transversality conditions. More precisely, we show that a smooth chain map of the form $P \circ f \colon \mathbb{R}^n \supseteq U \to \mathbb{R}^n$ is nice when the exterior term is polynomial and the jet extension of the inner term is transversal to certain algebraic stratified subsets (depending on P) of some Cartesian jet spaces.

6.1. Cartesian transversality theorem. Recall that in 3.4 we introduced the notion of *Cartesian maps* and the associated *Cartesian jet spaces*; for a given Cartesian type $\mathcal{I}$, that is, the list of variables on which the components of a map are allowed to depend, the space of s-jets was denoted by $\mathbf{J}^s$ (the combinatorial type, hence also all dimensions, being fixed once and forever).

To show that for any polynomial p the set of Cartesian maps f such that the chain map $p \circ \mathbf{j}^s f$ is nice, is open and dense, we need the strong form of Thom's transversality theorem for Cartesian functions.

CARTESIAN TRANSVERSALITY THEOREM. *If* $S \subseteq \mathbf{J}^s$ *is a submanifold in the Cartesian jet space, then maps whose Cartesian* s*-jet extensions are transversal to* S, *constitute a residual subset in the corresponding Cartesian space* $\mathbf{C}^{s+1}$

Moreover, the above assertion holds if S *is a stratified submanifold of the Cartesian space.*

PROOF (O. Shelkovnikov). The only nontrivial assertion is that the subset of maps transversal to S is dense. Consider the universal deformation which adds to each component all the monomials of degree $\leqslant s$ in the variables on which this component actually depends, with the coefficients at these monomials being considered as parameters of the deformation, cf. [GG]. Such a deformation, being submersive, is obviously transversal to anything in the Cartesian jet space. But this immediately implies that for almost all values of the parameters the perturbed map is transversal to S (see [GG], Corollary 4.7). The case of stratified subsets is treated in the same way, since the transversality to each stratum can be obtained on a residual subset of the Cartesian space. □

6.2. Stratifiability of chain maps. Our goal for the next two subsections will be to characterize stratifiability of (chain) maps as it occurs in the definition of nice maps, in terms of transversality. We start with the following principal technical lemma.

Let V be an algebraic subvariety in $\mathbf{J}^s$ of a certain codimension l, endowed with the Whitney stratification $\mathcal{V}$, and $P\colon \mathbf{J}^s \to \mathbb{R}^{n-l}$ a polynomial map with the coordinate functions $P_1, \dots, P_{n-l}$. Then for any $\mathbf{C}^s$-smooth Cartesian map f one may consider the chain map $P \circ (\mathbf{j}^s f)\colon \mathbb{R}^n \to \mathbb{R}^{n-l}$ and its restriction φ to the full preimage $V(f) = (\mathbf{j}^s f)^{-1}(V)$. Assume that f is C^{s+1}-differentiable and its Cartesian jet extension is transversal to V. Then the preimage $V(f)$ is the stratified subset in $\mathbb{R}^n$ of codimension l (and dimension $n-l$), and the chain map restricted to $V(f)$ takes the latter into the Euclidean space of the same dimension $n-l$:

$$\varphi\colon V(f) \to \mathbb{R}^{n-l}, \qquad P\colon \mathbf{J}^s \to \mathbb{R}^{n-l},$$
$$V(f) = (\mathbf{j}^s f)^{-1}(V), \qquad \varphi = P \circ (\mathbf{j}^s f)|_{V(f)}.$$

We are interested in conditions guaranteeing that φ is stratifiable. The following result is the main application of all the theory exposed in §§4–5.

PRINCIPAL LEMMA 6.1. *The full preimage* $V_* = \mathrm{pr}_{s+1}^{-1}(V) \subseteq \mathbf{J}^{s+1}$ *can be stratified by a certain stratification* $\mathcal{V}_*$ *in such a way that for any* C^{s+2}*-smooth Cartesian map (of the same type) whose* $(s+1)$*-jet extension* $\mathbf{j}^{s+1} f$ *is transversal to* $\mathcal{V}_*$, *the restriction* φ *constructed above, is stratifiable. The stratification which stratifies* φ *is given by the natural pullback* $(\mathbf{j}^{s+1} f)^{-1}(\mathcal{V}_*)$.

Moreover, both the σ*-complexity of the skeleton* $\operatorname{sk}\mathcal{V}_*$ *and the contiguity number for the stratification* $\mathcal{V}_*$ *are estimated from above by a certain primitive recursive functions in the variables* n, s, *the degree* D *of the polynomial* P, *and* $d = \sigma(V)$.

Before proving the lemma we make the following evident assertion.

PROPOSITION. *Let* $(M, 0)$ *be a germ of a smooth codimension* l *submanifold in* $(\mathbb{R}^n, 0)$ *and* $g_1, \dots, g_\varkappa$, $\varkappa < \infty$, *be any finite system of germs vanishing on* M *such that the rank of this functional system is equal to* l *for any point* $x \in M$.

Then for any smooth germ $G\colon (\mathbb{R}^n, 0) \to \mathbb{R}^{n-l}$, $G = (G_1, \dots, G_{n-l})$ *the restriction* $G|_M$ *has the full rank if and only if the rank of the system of functions* $\{g_i, G_j : i = 1, \dots, k, \ j = 1, \dots, n-l\}$ *is equal to* n.

PROOF OF THE PROPOSITION. This fact is implied by the following evident assertion from linear algebra: if a subspace $T \subseteq \mathbb{R}^n$ is the zero locus for a system of linear functionals $\gamma_1, \dots, \gamma_\varkappa$ of the total rank l equal to the codimension of T, and $\Gamma_1, \dots \Gamma_{n-l}$ are the components of the *linear* map $\Gamma: \mathbb{R}^n \to \mathbb{R}^{n-l}$, then the restriction of Γ to T has the full rank $n-l$ if and only if the rank of the system of linear functionals $\gamma_1, \dots, \gamma_\varkappa, \Gamma_1, \dots, \Gamma_{n-l}$ is equal to n. Indeed, the latter condition means that T is transversal to $\operatorname{Ker}\Gamma$. □

PROOF OF LEMMA 6.1. We start with the surgery of V according to Definition 4.10: let $\mathcal{V}$ be the Whitney stratification of V, and $\mathbf{p} = \{p_1, \dots, p_\varkappa\}$ the corresponding Whitney family of polynomials (now we denote by $\varkappa$ their number). Let $V = V_0 \cup V_1$ be the decomposition of V into the upper-dimensional part V_0 (the union of $(n-l)$-dimensional irreducible components) and the union V_1 of irreducible components of lower dimensions.

If f is a smooth map whose s-jet extension is transversal to V, then it must be also transversal to both V_0 and V_1. Denote by $V_i(f)$ the preimage $(\mathbf{j}^s f)^{-1}(V_i)$, $i = 0, 1$. The transversality implies that both $V_i(f)$ are stratified submanifolds in $\mathbb{R}^n$, and $\operatorname{codim} V_1(f) > \operatorname{codim} V_0(f) = l$. Thus all regular points of $V(f)$ lie on $V_0(f)$. Define at least C^2-smooth functions $g_i: \mathbb{R}^n \to \mathbb{R}$ by the formula $g_i = p_i \circ (\mathbf{j}^s f)$, $i = 1, \dots, \varkappa$. Then $V_0(f) = \{x \in \mathbb{R}^n: g_i(x) = 0,\ i = 1, \dots, \varkappa\}$. Because of the transversality of $\mathbf{j}^s f$ to V_0, one has

$$\operatorname{rank}\{dg_1(x), \dots, dg_\varkappa(x)\} = \operatorname{rank}\{dp_1(u), \dots, dp_\varkappa(u)\},$$
$$\forall x \in V_0(f),\quad u = \mathbf{j}^s f(x) \in V.$$

Since by construction the rank of the system of polynomials p_i is equal to l at all smooth points of V_0, the family of functions g_i has rank l at all smooth points of $V_0(f)$ and at the points of the skeleton the latter rank drops down. In other words, the condition $\operatorname{rank}\{dg_i(x)\} = l$ holds for all smooth points of $V_0(f)$ and only for them.

Denote by $G_j(x)$ the smooth functions $P_j \circ (\mathbf{j}^s f)$, $j = 1, \dots, n-l$, and construct the Jacobian matrix $J(x)$ for the family of functions $\{g_i, G_j\}$: this is $(\varkappa + n - l) \times n$-matrix formed by first-order partial derivatives of the functions g_i, G_j. By virtue of the above proposition, the rank of the matrix $J(x)$ drops down from its maximal value n at a point x in the following two cases: either the point is singular for $V_0(f)$, or the point is smooth but the rank of the restriction of G to $V_0(f)$ drops down from its maximal value $n-l$.

Thus we have proved that *critical points of the map* φ (in the sense of Definition 4.5) *are either points of* $V_1(f)$, *or points of* $V_0(f)$ *at which the Jacobian matrix has the rank less than* $n = l + (n-l)$:

$$c(\varphi) = V_1(f) \cup \{x \in V_0(f): \operatorname{rank} J(x) < n\}.$$

The rest of the proof consists in showing that the initial stratification $(\mathbf{j}^s f)^{-1}(\mathcal{V})$ on $V(f)$ admits a refinement that stratifies φ. To do this, we show that the critical set $c(\varphi)$ is the preimage of a certain algebraic subset Σ of V_* under the jet extension map $\mathbf{j}^{s+1} f$, and that this subset is determined by the variety V and the polynomials P_j.

To construct Σ and prove its algebraicity, we make the following evident remark. If $p\colon \mathbf{J}^s \to \mathbb{R}$ is a real polynomial and f a C^{s+1}-smooth Cartesian map, then the differential of the chain map $g = p \circ (\mathbf{j}^s f)$ is a 1-form with coefficients that are polynomial in the derivatives of f up to the order $s+1$. More formally,

$$dg(x) = \sum_{j=1}^{n} (b_j \circ \mathbf{j}^{s+1} f)(x)\, dx_j\,,$$

where $b_j\colon \mathbf{J}^{s+1} \to \mathbb{R}$, $j = 1, \dots, n$ are some real polynomials of degrees $\leqslant \deg p - 1$. Indeed, this identity is just the chain rule of differentiation.

Obviously, all the entries of the Jacobian matrix $J(x)$ are of the same form, being compositions of some polynomials with the jet extension map $\mathbf{j}^{s+1} f$. Therefore, any $(n \times n)$-minor of $J(x)$ has the form $\widetilde{b}_\beta \circ \mathbf{j}^{s+1} f$, $\beta \in \mathfrak{B}$, where $\mathfrak{B}$ is the list of all such minors, and $\widetilde{b}_\beta$ is a polynomial. Hence

$$\operatorname{rank} J(x) < n \iff \forall \beta \in \mathfrak{B} \quad (\widetilde{b}_\beta \circ \mathbf{j}^{s+1} f)(x) = 0.$$

Thus we conclude that the set of critical points of the restriction of φ to $V_0(f)$ is the preimage $(\mathbf{j}^{s+1} f)^{-1}(\Sigma_0)$ of the set $\Sigma_0 \subseteq \operatorname{pr}_{s+1}^{-1}(V_0)$ defined as the zero locus for all polynomials $\widetilde{b}_\beta$:

$$\Sigma_0 = \left\{ u \in \mathbf{J}^{s+1} \colon \widetilde{b}_\beta(u) = 0\,,\ \operatorname{pr}_{s+1} u \in V_0\,,\ \beta \in \mathfrak{B} \right\}.$$

Consider the algebraic subset $\Sigma \subseteq V_* = \operatorname{pr}_{s+1}^{-1}(V)$ defined as $\Sigma_0 \cup \operatorname{pr}_{s+1}^{-1}(V_1)$. Being the union of two algebraic subsets, Σ is also algebraic, and by the construction we have $c(\varphi) = (\mathbf{j}^{s+1} f)^{-1}(\Sigma)$.

The last step in the proof is to construct the stratification $\mathcal{V}_*$ on $V_* = \operatorname{pr}_{s+1}^{-1}(V)$ with the required properties. We do this by applying Lemma 4.6 to the pair (V_*, Σ), $\Sigma \subseteq V_*$ in $\mathbf{J}^{s+1}$. There exist such a stratification, and the complexity of its skeleton and the contiguity numbers admit the required estimates because of the explicit character of the construction process. □

If the variety $V \in \mathbf{J}^s$ is stratified not by the Whitney stratification, but by some refinement $\mathcal{V}'$, and this refinement is algebraic (that is, the skeleton $\operatorname{sk} \mathcal{V}'$ is a (proper) algebraic subvariety of V), then the same statement about stratifiability of the restricted map φ holds. Indeed, in this case we take V_1 to be the skeleton of such a stratification and put Σ being the union $\Sigma_0 \cup \operatorname{pr}_{s+1}^{-1}(V_1)$. This proves the following generalization of Lemma 6.1.

COROLLARY. *The assertion of Lemma* 6.1 *remains true if the Whitney stratification* $\mathcal{V}$ *of the variety* V *is replaced by a refinement* $\mathcal{V}'$ *of* $\mathcal{V}$ *with an algebraic skeleton.* □

To make references to the process described in Lemma 6.1, we introduce some terminology.

DEFINITION 6.3. The stratification $\mathcal{V}_*$ of the set $V_* = \operatorname{pr}_{s+1}^{-1}(V)$, which was constructed in Lemma 6.1, is called the *P-uplifting* of the original stratification $\mathcal{V}$, or simply uplifting, if the polynomial is known from context:

$$\mathcal{V}_* = \upuparrows_s(\mathcal{V}, P).$$

6.3. Construction of universal critical sets: the global case. Lemma 6.1 allows for an explicit description of nice maps introduced in §5. This is almost evident. Fix a certain type $\mathcal{I}$ of Cartesian maps and a polynomial map $P\colon \mathbf{J}^s_{\mathcal{I}}(\mathbb{R}^n, \mathbb{R}^m) \to \mathbb{R}^n$. We are going to prove that for a generic smooth Cartesian map of the prescribed type, the chain map

$$P \circ (\mathbf{j}^s f)\colon \mathbb{R}^n \to \mathbb{R}^n \tag{6.1}$$

restricted to the unit ball in $\mathbb{R}^n$, is nice. This will prove the (global) Theorem I* from §3.

Let B be the closed unit ball in $\mathbb{R}^n$ defined by the algebraic condition $\sum_j x_j^2 \leqslant 1$, and S its boundary, the unit sphere. By induction we define a sequence of algebraic subsets $\Sigma^j \subseteq \mathbf{J}^{s+j} = \mathbf{J}^{s+j}_{\mathcal{I}}(\mathbb{R}^n, \mathbb{R}^m)$, $j = 0, \dots, n$, depending only on P such that as soon $\mathbf{j}^{s+j} f$ is transversal to all of them, the restriction of $P \circ (\mathbf{j}^s f)$ to the unit ball is the nice map. Therefore, by Theorem 5.1 it admits a uniform estimate for the number of nondegenerate preimages.

The subsets Σ^j are defined by induction, the first step being slightly different from the remaining ones. Denote by $\pi = \mathrm{pr}_s \circ \cdots \mathrm{pr}_1 \circ \mathrm{pr}_0 \colon \mathbf{J}^s \to \mathbb{R}^n$ the natural projection (taking each Cartesian s-jet into the source point of it).

The induction base. The set $\Sigma^0 \subseteq \mathbf{J}^s$ is the entire Cartesian jet space $\mathbf{J}^s$, endowed not with the trivial stratification, but rather with the stratification $\mathcal{U}_0 = \{\pi^{-1}(U_-), \pi^{-1}(U_+), \pi^{-1}(U_0)\}$, where

$$U_\pm = \{x \in \mathbb{R}^n \colon \pm(|x|^2 - 1) > 1\}, \qquad U_0 = \{x \in \mathbb{R}^n \colon |x|^2 = 1\}. \tag{6.2}$$

The skeleton of such a stratification is U_0.

The sequence of polynomial maps. Define the sequence of polynomial maps $P^{(j)}\colon \mathbf{J}^{s+j} \to \mathbb{R}^{n-j}$, $j = 0, \dots, n$ from the commutative diagram

$$\begin{array}{ccccccccccc}
\mathbb{R}^n & \xleftarrow{\pi} & \mathbf{J}^s & \xleftarrow{\mathrm{pr}_{s+1}} & \mathbf{J}^{s+1} & \xleftarrow{\mathrm{pr}_{s+2}} & \cdots & \xleftarrow{\mathrm{pr}_{s+j}} & \mathbf{J}^{s+j} & \xleftarrow{\mathrm{pr}_{s+j+1}} & \cdots \\
 & & \downarrow{\scriptstyle P=P^{(0)}} & & \downarrow{\scriptstyle P^{(1)}} & & & & \downarrow{\scriptstyle P^{(j)}} & & \\
 & & \mathbb{R}^n & \xrightarrow{\pi_n} & \mathbb{R}^{n-1} & \xrightarrow{\pi_{n-1}} & \cdots & \xrightarrow{\pi_{n-j+1}} & \mathbb{R}^{n-j} & \xrightarrow{\pi_{n-j}} & \cdots,
\end{array}$$

where the bottom line corresponds to the flag of subspaces. In other words, the polynomial map $P^{(j)}$ is the map obtained from the initial map P by dropping the last j coordinate functions of the former, and pulling the result back to the space $\mathbf{J}^{s+j}$.

The first induction step. To construct the next set Σ^1, we apply the Corollary to Lemma 6.1 rather than the lemma itself, to the map $P \circ (\mathbf{j}^s f)$ restricted to the space $\mathbb{R}^n$. We consider the latter space as being stratified by three strata (6.2). Thus an algebraic subvariety $\Sigma^1 \subseteq \mathbf{J}^{s+1}$ appears. Note that by construction, the stratification $\{U_\pm, U_0\}$ on $\mathbb{R}^n$ is the pullback of the stratification $\mathcal{U}_0$ by the map $\mathbf{j}^s f$:

$$\Sigma^1 = \operatorname{sk} \mathcal{U}_1, \qquad \mathcal{U}_1 = \pitchfork_s(\mathcal{U}_0, P^{(0)}).$$

Further steps of induction. The algebraic variety Σ^j, $j \geqslant 1$, being constructed, we apply Lemma 6.1 (rather than the Corollary, since the Whitney stratification is considered from now on) to the restriction of the map $P^{(j)} \circ (\mathbf{j}^s f)$ to the preimage $(\mathbf{j}^{s+j} f)^{-1}(\Sigma^j)$, and obtain an algebraic subset $\Sigma^{j+1} \subseteq \mathbf{J}^{s+j+1}$:

$$\Sigma^{j+1} = \operatorname{sk} \mathcal{U}_{j+1}, \qquad \mathcal{U}_{j+1} = \pitchfork_{s+j}(\mathcal{U}_j, P^{(j)}).$$

LEMMA 6.2. *The above sequence of sets $\Sigma^j \subseteq \mathbf{J}^{s+j}$, $j = 0, 1, \dots, n$ has the following properties*:

- *all the sets from the sequence are algebraic*;
- *codimensions of these sets are strictly increasing*;
- *σ-complexities and the contiguity numbers of the varieties Σ^j admit an upper estimate by a primitive recursive function of the parameters n, m, $d = \deg P$, s (recall that the contiguity number is almost independent of the choice of stratification, see Lemma* 4.3);
- *if $\mathbf{j}^{s+j} f \pitchfork \mathcal{U}_j$ for $j = 0, \dots, k+1$, then the first k squares of the commutative diagram* (5.2) *may be constructed subject to the stratifiability condition from Definition* 5.5, *if one puts*

$$M^j = (\mathbf{j}^{s+n-j} f)^{-1}(\Sigma^{n-j}), \qquad f_j = P^{(j)} \circ \mathbf{j}^{s+j} f.$$

PROOF. The inductive application of Lemma 6.1. □

6.4. Construction of the universal critical sets: the local case. The same construction applies also to the local case: the suspended map $\widetilde{f}$ has the same chain form, being polynomial (identical!) in the last component. The only difference is that the initial stratification $\mathcal{U}_0$ is defined as follows: take the cone $C_a = \{|x - a|^2 - z^2 = 0\}$ and the hyperplane $L = \{z = 1\}$ and consider the minimal stratification of $\mathbb{R}^{n+1} = \{(x, z)\}$ for which the union $C_a \cup L$ is the skeleton. This stratification defines the natural cylindrical pullback stratification on the space $\mathbf{J}^s_{\mathcal{I}}(\mathbb{R}^{n+1}, \mathbb{R}^{m+1})$. Denote this stratification by $\mathcal{U}_0$ and proceed further using the same construction.

All of the rest goes without any change, leading to the sequence of algebraic subsets $\widetilde{\Sigma}^1, \dots, \widetilde{\Sigma}^{n+2}$. The number of terms in this sequence is $n+2$, since the dimension of the phase space has been increased by 1.

The sets $\widetilde{\Sigma}^j$ have the same properties as the sets Σ^j from Lemma 6.2: they are algebraic, of increasing codimensions, and of controlled σ-complexity. For a generic Cartesian map f they determine the sets M^j and the maps f_j which fill the first squares of the diagram (5.2) for the suspended map (5.5).

6.5. Genericity of nice maps. If P is a vector polynomial and a C^{s+n+1}-smooth Cartesian map f has all its Cartesian $(s+j)$-jet extensions transversal to the corresponding varieties Σ^j, then the diagram (5.2) exists for the map $P \circ (\mathbf{j}^s f)$. Indeed, one may take the coordinate flag (the chain of subspaces of $\mathbb{R}^n$ spanned by several first coordinate vectors). Then the construction from 6.3, 6.4 implies that the sets M^j defined as $(\mathbf{j}^{s+n-j} f)^{-1}(\Sigma^{n-j})$ satisfy the conditions imposed in Definition 5.5.

LEMMA 6.3. *If f is a C^{s+n+1}-smooth Cartesian map defined in a neighborhood of the unit ball in $\mathbb{R}^n$, and*

$$\forall j = 0, \dots, n+1 \qquad \mathbf{j}^{s+j} f \pitchfork \Sigma^j ,$$

then the chain map (6.1) *is nice.*

COROLLARY. *If the above jet extensions are transversal to the varieties $\widetilde{\Sigma}^j$, then the chain map is locally nice.*

PROOF. If f is such a map, then one can extend it to the entire space $\mathbb{R}^n$ preserving the transversality, thus the construction of the two preceding subsections applies, and we conclude that the sequences of stratified varieties M^j and the maps f_j can be constructed.

In the global case we need then to restrict the maps f_j to the closed unit ball $B = \{|x| \leqslant 1\}$, replacing the sets M_j by their intersections with this ball. To make sure that these intersections remain stratified sets, we note that the universal sets $\Sigma^j \subseteq \mathbf{J}^{s+j}$ remain stratified when the strata outside the cylinders C_j in $\mathbf{J}^{j+s}$ over the ball B are deleted. This follows from their algebraicity, since each stratum of the stratifications $\mathcal{U}_j$ either belongs to such a cylinder or does not intersect it. Evidently, $M^j \cap B = (\mathbf{j}^{s+n-j} f)^{-1}(\Sigma^{n-j}) \cap C_{s+n-j}$ (cf. the last assertion from Lemma 6.2). Thus $M^j \cap B$ are stratified compact sets, and all conditions from Definition 5.5 are satisfied.

In the local case we do the same with the suspended map, replacing the ball B with the set $\{(x, z) \in \mathbb{R}^{n+1} : 0 < z \leqslant 1, \ |x|^2 \leqslant z^2\}$, which is also compact.

Therefore all the conditions from Definition 5.5 are satisfied, and the chain map is (locally) nice. □

6.6. Proofs of Theorems I* and II*: chains of reductions. The chain of reductions proving Theorems I* and II* looks as follows:

- A generic C^{s+n}-smooth (resp., C^{n+s+1}-smooth, if the local case is considered) Cartesian map f is transversal to all the algebraic subsets Σ^j (resp., $\widetilde{\Sigma}^j$) [Cartesian transversality theorem, §6.1], therefore
- the chain map $P \circ \mathbf{j}^s f$ is (locally) nice [Lemma 6.3 and the Corollary], therefore
- the chain map $P \circ \mathbf{j}^s f$ is Gabrielov-finite (possesses the Gabrielov property), and any local Gabrielov number admits an upper estimate by the product of the contiguity numbers [Theorems 5.1 and 5.2], which in turn are computable, therefore
- the number of (small) nondegenerate preimages of points by the chain map $P \circ \mathbf{j}^s f$ is finite (admits a primitive recursive upper bound) [Theorems I* and II*]. □

The implications

$$\begin{aligned} \text{Theorem I}^* &\implies \text{Existential finite cyclicity (Theorem I)} \\ \text{Theorem II}^* &\implies \text{Constructive finite cyclicity (Theorem II)} \end{aligned}$$

was already discussed in §3. Thus both Theorem I and II are completely proved.

6.7. Appendix: simple arguments proving the existence of $E(k)$ for any finite k. The assertion about algorithmic computability of the number $E(k)$ in Theorem II is heavily based on the theory of effective computations in the ring of real polynomials, in particular, on the estimate of the degrees of generators of the real radical of an ideal.

In fact, the constructions leading to the demonstration of Theorem II* prove the existence of the finite upper bound $E(k)$ without referring to these deep and rather technical results.

Indeed, Theorem 3.2 reduces the question about the number of isolated solutions for a (specified) basic system (1.1) in any finite codimension k to estimating the local Gabrielov number for a *finite number* of chain maps of the form $P_\alpha \circ \mathbf{j}^s f$, where P_α are polynomials *with integer coefficients* but depending on additional parameters $\mathbf{A}$ (the algebraic part of the specification). For each of these polynomials one can construct the corresponding chains of universal criminant sets Σ_α^j as this was described in the proof of Lemma 6.1 (see 6.3, 6.4). When performing these operations, we do not care any more about the effective bounds for the complexity of all operations, in particular, we use only the existential part of Lemma 4.5, which follows from results of Whitney [W1] without any reference to effectiveness of the computations.

Denote by l the number of additional parameters $\lambda = \mathbf{A}$. The above arguments applied to each chain map $P_\alpha \circ \mathbf{j}^s f$ with a generic (Cartesian) f, allow for the construction of the first n squares of the diagram (5.2), starting from the original manifold M_α^{n+l+1} and ending with the manifold $M_\alpha^{l+1} \subseteq \mathbb{R}^{n+l+1}$, which is projected by the corresponding map $f_{k+1,\alpha}$ onto the (x, λ)-space.

At the final stage of the demonstration we use Theorem 5.3 instead of Theorem 5.2 and obtain the estimate for the local Gabrielov number for any chain map associated with any polynomial P_α. This estimate is now only existential, but because of the finiteness of the number of different polynomials P_α, the maximum over all α is finite, and this maximum will finally determine the number $E(k)$ in the same way as before.

In addition to making the proof of Theorems I and II self-contained, the above arguments in fact suggest a way for explicit computation of the values $E(k)$ for small k and for writing down explicitly the genericity conditions for the Cartesian maps f. Indeed, since in this case all chain maps have exterior parts with integer coefficients (and the list of all such polynomials may be explicitly composed), all further computations can be made using symbolic computation programs, like `Maple` or `Mathematica`; these computations would yield explicit polynomial representation of the universal sets Σ_α^j in the ring $\mathbb{Q}[x]$ of polynomials with rational coefficients, hence giving in principle all relevant information about the contiguity numbers, Gabrielov-type indices, etc.

References

[AVG] V. I. Arnold, A. N. Varchenko, and S. M. Gusein-Zadeh, *Singularities of differentiable mappings*, vol. 1, "Nauka", Moscow, 1982; English transl., Birkhäuser, Basel, 1986.

[B] N. N. Bautin, *On the number of limit cycles appearing with variations of coefficients from an equilibrium state of the type of a focus or a center*, Mat. Sb. **30** (1952), 181–196; English Transl., Amer. Math. Soc. Transl. 1954, 1954; Reprinted in: Stability and Dynamical Systems, Amer. Math. Soc. Transl. Series 1, vol. 5, Amer. Math. Soc., Providence, RI, 1962, pp. 396–413.

[BN] E. Becker and R. Neuhaus, *Computation of real radicals of polynomial ideals*, Computational Algebraic Geometry (Nice, 1992), Progress in Math., vol. 109, Birkhäuser, Boston, 1993, pp. 1–20.

[FP] J. P. Françoise and C. C. Pugh, *Keeping track of limit cycles*, J. Differential Equations **65** (1986), 139–157.

[G] A. M. Gabrielov, *On projections of semianalytic sets*, Functional Anal. Appl. **2** (1968), 282–291.

[GG] M. Golubitsky and V. Guillemin, *Stable mappings and their singularities*, Springer-Verlag, New-York, Heidelberg, Berlin, 1973.

[H] G. Hermann, *Die Frage der endlich vielen Schritte in der Theorie der Polynomialideale*, Math. Ann. **95** (1926), 736–788.

[I] Yu. S. Ilyashenko, *Dulac memoir "Sur les cycles limites" and related problems of the local theory of differential equations*, Uspekhi Mat. Nauk **40** (1985), no. 6, 41–78; English transl. in Russian Math. Surveys **40** (1985).

[IY1] Yu. S. Ilyashenko and S. Yu. Yakovenko, *Finite-differentiable normal forms for local families of diffeomorphisms and vector fields*, Russian Math. Surveys **46** (1991), no. 1, 1–43.

[IY2] ———, *Nonlinear Stokes phenomena in smooth classification problems*, Nonlinear Stokes Phenomena (Yu. S. Ilyashenko, ed.), Adv. Soviet Math., vol. 14, Amer. Math. Soc., Providence, RI, 1992.

[IY3] ———, *Finite cyclicity of elementary polycycles*, C. R. Acad. Sci. Paris Ser. I **316** (1993), 1081–1086.

[Kh] A. G. Khovanskiĭ, *Fewnomials*, Amer. Math. Soc., Providence, RI, 1991.

[Man] Yu. I. Manin, *A course in mathematical logic*, Springer-Verlag, Berlin, Heidelberg, New York, 1977.

[Ma1] J. Mather, *Stratifications and Mappings*, Dynamical Systems (M. Peixoto, ed.), Academic Press, New York and London, 1973, pp. 195–232.

[Ma2] ———, *Notes on topological stability*, Mimeographed notes, Harvard Univ., 1970.

[Mi1] J. Milnor, *On the Betti number of real varieties*, Proc. Amer. Math. Soc. **15** (1964), 275–280.

[Mi2] ———, *Singular points of complex hypersurfaces*, Annals of Mathematical Studies, vol. 61, Princeton Univ. Press, Princeton, NJ, 1968.

[Mou] A. Mourtada, *Cyclicité finie des polycycles hyperboliques de champs de vecteurs du plan. Algorithme de finitude*, Annales Inst. Fourier Grenoble, **41**, fasc. 3 (1991), 719–753.

[MR] I. Mostowski and E. Rannou, *Complexity of computation of the canonical Whitney stratification of algebraic sets in $\mathbb{C}^n$*, Applied Algebra, Algebraic Algorithms and Error-Correcting Codes (H. F. Mattson, T. Mora, and T. R. N. Rao, eds.), Proc. 9th Int. Symp. AAECC-9 (New Orleans, LA, October 1991) Lecture Notes Computer Science, vol. 539, Springer-Verlag, Berlin, Heidelberg, and New York, 1991, pp. 281–291.

[N] R. Neuhaus, Personal communication based on a forthcoming paper (1994).

[R] R. Roussarie, *A note on finite cyclicity and Hilbert's 16th problem*, Dynamical Systems, Proc. of the Chilean Symp. (Valparaiso, 1986) (R. Bamon, R. Labarca, and J. Palis, eds.), Lecture Notes Math., vol. 1331, Springer-Verlag, Berlin, 1988, pp. 161-188.

[S] A. Seidenberg, *Constructions in algebra*, Trans. Amer. Math. Soc. **197** (1974), 273–313.

[W1] H. Whitney, *Elementary structure of real algebraic varieties*, Ann. of Math. **66** (1957), no. 3, 545–556.

[W2] ———, *Local properties of analytic varieties*, Differential and Combinatorial Topology (S. Cairns, ed.), Princeton Univ. Press, Princeton, NJ, 1965, pp. 205–244.

Translated by S. YAKOVENKO

YU. I.: MOSCOW STATE UNIVERSITY AND MOSCOW MATHEMATICAL INSTITUTE, MOSCOW, RUSSIA

E-mail address: yuilyashenko@glas.apc.org

S. YA.: THE WEIZMANN INSTITUTE OF SCIENCE, REHOVOT, ISRAEL

E-mail address: yakov@wisdom.weizmann.ac.il

Amer. Math. Soc. Transl.
(2) Vol. **165**, 1995

Desingularization in Families of Analytic Differential Equations

S. TRIFONOV

Introduction

Parameter and individual desingularization. The procedure of resolution of singularities, or desingularization (called also *blowing-up*, or *σ-process*) plays an important role in the theory of analytic vector and line fields on two-dimensional manifolds, resolving degenerate singular points (equilibria) into simpler points. Recall that a singular point of a planar vector field is called *elementary* if at least one of its eigenvalues is nonzero. A singular point of a line field is elementary if near this point the line field is spanned by an analytic vector field with an elementary singular point. The basic result from this subject, the Bendixson–Seidenberg theorem, claims that by a finite number of blowing-ups any singularity of an analytic line field can be split into a finite number of elementary singularities.

Therefore it is important to generalize the desingularization procedure for families of line fields, in order to reduce them to families having only elementary singularities. The paper describes one such construction (another approach was recently suggested by R. Roussarie and Z. Denkowska [15]). *We will consider only families of differential equations on two-dimensional manifolds.*

We prove that by a finite composition of blowing-ups and *base transformations* (reparametrizations) one may resolve all singularities of all differential equations from a given family into elementary ones. Moreover, the blow-ups depend nicely on the parameters of the family. This result was first announced in [20].

Unfortunately, the suggested desingularization procedure cannot rule out (and even generates itself) the case of *singular perturbations*, that is, families of vector fields unfolding nonisolated singularities. An example of such family is the line field associated with the parameter family of differential equations

$$\begin{cases} \dot{x} = f_0(x, y)\,h(x, y) + \varepsilon f_1(x, y), \\ \dot{y} = g_0(x, y)\,h(x, y) + \varepsilon g_1(x, y). \end{cases}$$

Here $(x, y) \in \mathbb{C}^2$, $\varepsilon \in \mathbb{C}$, all functions f_0, f_1, g_0, g_1, h are holomorphic on

1991 *Mathematics Subject Classification*. Primary 34C05, 34C20.

$\mathbb{C}^2$ and relatively prime, and $\{h(x, y) = 0\}$ is a nonvoid one-dimensional set. This effect is discussed in §6.

The structure of the paper. The paper is organized as follows.

§1. Definitions and formulation of the main result of the paper, the theorem on parameter desingularization.

§2. Description of the structure of the singular locus for families of analytic line fields. Examples.

§3. Main tools used in the construction: rectification of the singular locus (Lemma S).

§4. More detailed investigation of the blowing up for an individual line field. We estimate the number of blowing-ups necessary to resolve all degenerate singularities of a given analytic line field.

§5. Proof of the main theorem.

§6. Applications of the parameter blowing up to investigation of properties of analytic families. Here we discuss the above mentioned phenomenon of singular perturbations.

The paper is supplied with three appendices containing necessary technical results.

A. Boundedness of multiplicities of essential singularities in analytic families. This result is essentially related to the Gabrielov–Teissier theorem on finiteness properties of analytic maps.

B. Proof of Lemma S in the complex ground field case. The main argument used here is the Hironaka theorem on desingularization of analytic sets.

C. Proof of Lemma S for real analytic families and demonstration of the Main Theorem in the real case.

Acknowledgments. I want to thank A. Shcherbakov for helping me to prove the lemma on rectification of analytic sets (Lemma R), see §3. My sincere gratitude goes to Yu. Ilyashenko who suggested this problem, for numerous discussions and his permanent attention to my work. I also thank S. Yakovenko who read the manuscript and helped to reduce it to the final form. I am grateful to Z. Denkowska and P. Milman for many valuable remarks.

§1. Definitions and formulation of the main result

Conventions. Throughout this paper we use without further remarks the following basic conventions.

1. *Analyticity.* All manifolds and mappings occurring in the main body of the paper are assumed to be complex analytic. The only exceptions are the parts of the paper treating the real analytic case (final remarks after each section and Appendix C). Each time when we consider real rather than complex case, we use the following terminology. A *strict real analog* of a definition (assertion, example) is the statement obtained by formally substituting the ground field $\mathbb{R}$ for $\mathbb{C}$ and real analyticity for complex analyticity.

2. *Countable compactness.* All manifolds are assumed countably compact, that is, they can be represented as countable unions of compact subsets. Sometimes we consider a nonconnected manifold as the collection of its connected components, $N = \{N_i\}$.

3. *Genericity.* We say that a local property holds for a *generic* point of a manifold if the exceptional set of points for which this property fails, is a proper analytic subset of the ambient manifold.

1.1. Background. In order to formulate the main result of the paper, we need some basic definitions and constructions from algebraic geometry, differential equations and differential topology.

1.1.1. *Line fields.*

DEFINITION. Let M be a manifold. A *line prefield with singularities* is the family of vector fields v_α defined in some open domains U_α, such that the union $\mathcal{U} = \bigcup_\alpha U_\alpha$ is an atlas of charts on the manifold, and for any two charts U_α, U_β with nonempty intersection there exist two functions $f_\alpha, f_\beta \not\equiv 0$ such that on the intersection $U_\alpha \cap U_\beta$ one has $f_\alpha v_\alpha = f_\beta v_\beta$. A *line field* is the maximal family of vector fields with the above property.

In this definition maximality means that the family contains any vector field which agrees (in the natural sense) with other vector fields from the family. We always consider maximal families of vector fields. Any of those vector fields is said to span the line field. In [15] a similar definition was introduced to define a *local vector field*.

DEFINITION. A point $p \in M$ is a *singular point* (singularity) of the line field, if it is singular for any vector field v_α spanning the line field.

Let $H\colon \widetilde{M} \to M$ be an analytic map which is biholomorphic at a generic point of a manifold $\widetilde{M}$. Then for any vector field v_α spanning a line field ℓ on M, the *pullback* $H^* v_\alpha = H_*^{-1} v_\alpha$ is defined on an open dense subset U of $\widetilde{M}$, and one can easily see that pullbacks of different vector fields spanning the same ℓ span the same line field $H^*\ell$ on U. In some important cases this line field may be extended on the whole of $\widetilde{M}$; this extension will be also denoted by $H^*\ell$ and referred to as the *pullback of the line field*.

PROPOSITION 1.1. *Let $H\colon \widetilde{M} \to M$ be an analytic map between two manifolds of the same dimension which is a local diffeomorphism outside a proper analytic subset $D \subset M$, $\dim D < \dim \widetilde{M} = \dim M$. Then the pullback $H^*\ell$ is well defined for any line field ℓ on M.*

PROOF. The assertion of the proposition is local. Take any point $\widetilde{a} \in \widetilde{M}$, let $a = H(\widetilde{a})$ and assume that the line field ℓ on M near a is spanned by an analytic vector field v. Introducing local coordinates, we may associate analytic vector-functions $v(x)$ with v and $H(\widetilde{x})$ with H. Now take the vector field on $\widetilde{M}$ defined by the formula

$$\widetilde{v}(\widetilde{x}) = (\det H_*(\widetilde{x})) \cdot H_*^{-1}(\widetilde{x}) v(H(\widetilde{x})),$$

where $H_*(\widetilde{x})$ is the Jacobian matrix for $H(\widetilde{x})$. This vector field is proportional to the pullback of v wherever this pullback is defined (i.e., outside D). On the other hand the matrix $(\det H_*(\widetilde{x})) \cdot H_*^{-1}(\widetilde{x})$ is analytic (being a conjugate of an analytic matrix-valued function), hence defined everywhere. Thus the vector field $\widetilde{v}$ spans a (uniquely defined for the trivial reasons) line field $H^*\ell$, all the construction being independent of the choice of v. $\square$

1.1.2. *Line fields on two-dimensional surfaces. Blow-ups.* Consider now the two-dimensional case. One may easily see that in this case all singularities are isolated. Indeed, in any local coordinates (x, y) on M a line field is spanned by some vector field v of the form

$$f(x, y)\,\partial/\partial x + g(x, y)\,\partial/\partial y,$$

so that the singular locus of v is the set of common zeros for f and g. If these two functions are relatively prime, then the singularities are isolated by virtue of uniqueness theorem for analytic functions. Otherwise, if f and g admit a common analytic factor, then the vector field

$$h^{-1}(x, y)\,f(x, y)\,\partial/\partial x + h^{-1}(x, y)\,g(x, y)\,\partial/\partial y$$

spans the same line field and for an appropriate h has all singularities isolated.

For studying singular points of line fields on surfaces there exists an efficient tool, known as the *blow-up* (also σ-process, resolution of singularities, etc.). Since we will need several versions of essentially the same construction, we provide these terms with slightly different meanings: the usual blow-up will be referred to as a *simple blow-up*, while the composition of several blow-ups is called a *resolution*. The entire process of constructing an appropriate resolution will be termed as *desingularization*; later we introduce *blow-ups of sections*.

Let p be a point which is the origin in an appropriate coordinate system (x, y), $x, y \in \mathbb{C}$. The blow-up of M at p is a new manifold $\widetilde{M}$ that is obtained by replacing the point p by a complex projective line $\Sigma = \mathbb{C}P^1$. In the local coordinates we choose two charts on $\widetilde{M}$ as

$$(1) \qquad (x, u), \quad u = \frac{y}{x} \qquad \text{and} \qquad (y, v), \quad v = \frac{x}{y}.$$

Thus, instead of a neighborhood of the origin, we obtain a neighborhood of two lines $\{x = 0\}$ and $\{y = 0\}$ which together constitute a neighborhood of the projective line, since $u = 1/v$. The construction defines a map $\widetilde{M} \to M$, which is one-to-one outside the point p and squeezes the entire projective line into p.

The procedure of blowing-up is invariant: if we choose another local coordinates (x, y), then the two manifolds obtained as above are biholomorphic. There exists also a generalization for the multidimensional case, but we do not need its formal description (see, for example, §4 of the introductory paper to this volume).

DEFINITION. A *simple blow-up* of a surface M is a map $\sigma\colon \widetilde{M} \to M$ which is biholomorphic except for the preimage $\Sigma \subset \widetilde{M}$ of just one point $p \in M$; the exceptional set Σ is biholomorphic to the projective line $\mathbb{C}P^1$ and the germ $\sigma\colon (\widetilde{M}, \Sigma) \to (M, p)$ is left-right equivalent to the standard blow-up described above.

The last condition means that in a small neighborhood of Σ two local charts can be introduced, (x, u) and (y, v), with the transition functions between them $y = xu$, $v = 1/u$, such that the map σ has in these charts the form

$$(2) \qquad (x, u) \mapsto (x, ux), \qquad (y, v) \mapsto (yv, y).$$

A *resolution* of the surface M is an analytic map $\theta : \widetilde{M} \to M$ which is the composition of a finite number of simple blow-ups.

After blowing up an isolated singularity of the vector field, we obtain a new vector field (the pullback of the original field by the squeezing map), for which the whole projective line is usually, though not always, singular. But this field is analytic, hence the line field spanned by it has only isolated singularities on the projective line. It would be natural to assume that the new singular points are in a sense simpler than the original singularity so that perhaps after a number of successive blowing-ups the singularity would be resolved into some simplest singularities. This is in fact true, and the role of the simplest singularities is played by elementary points.

DEFINITION. A singular point p of a vector field on a surface is said to be *elementary* if the linearization of the field at this point has at least one nonzero eigenvalue. A singularity of a line field is elementary, if this field is spanned by a vector field with an elementary singular point.

BENDIXSON–SEIDENBERG THEOREM. *An analytic line field on a surface after a finite number of appropriate blowing-ups yields a line field having only elementary singularities.*

REMARK. This result was first formulated by Bendixson who never gave a formal proof. In the complex case the proof belongs to Seidenberg [13]. The real smooth case was considered by Dumortier [7], who proved the same assertion under the natural assumption that all singular points satisfy the so called Łojasiewicz condition.

1.1.3. *Fibrations.* It is natural to consider parameter depending families of surfaces (and k-dimensional manifolds in general) together with line fields on them as *fibrations* over the parameter space: according to the terminology introduced below, a fibration is a foliation which has an additional structure of an analytic manifold on the space of fibers. It is essential to point out that *we allow for parameter spaces consisting of several connected components*. Below we introduce the corresponding language and show how families of manifolds can be reparametrized.

DEFINITIONS. 1. A *holomorphic fibration* is the triple (M, B, π), where B is an n-dimensional (not necessary connected) manifold (the parameter space), called also the *base*, the manifold M is the *total space* of the fibration, and $\pi : M \to B$ is an analytic map of the constant rank $n = \dim B$ at any point of M. The preimages $M_\varepsilon = \pi^{-1}(\varepsilon)$ are called *fibers*, and each fiber is a k-dimensional nonsingular manifold, $k = \dim M - \dim B$. Note that we do not require the fibers be globally biholomorphically equivalent.

2. The *restriction* of a fibration (M, B, π) to an open subset of the total space $W \subset M$ is the fibration $(W, \pi(W), \pi|_W)$, and the restriction to an open subset of the base $U \subset B$ is the family $(\pi^{-1}(U), U, \pi|_{\pi^{-1}(U)})$.

3. A *connected component* of a fibration is its restriction to a connected component of the total space.

4. By virtue of the theorem on rank, near each point $p \in M$ one may choose a local coordinate system (x, ε), $x \in (\mathbb{C}^k, 0)$, $\varepsilon \in (\mathbb{C}^n, 0)$, in such a way that the map π becomes the standard projection, $\pi : (x, \varepsilon) \mapsto \varepsilon$. Such a local coordinate

system is called an *adapted* coordinate system, and only adapted local coordinates will be considered.

5. A *section* of the fibration (M, B, π) is a holomorphic map $S: B \to M$ such that $\pi \circ S = \mathrm{id}_B$. We identify the section S with its image $S(B) \subset M$.

Any transformation of the base of a fibration yields a new fibration.

PROPOSITION 1.2. *Let (M, B, π) be a fibration and $\rho: \widetilde{B} \to B$ an analytic map. Then there is defined another fibration $(\widetilde{M}, \widetilde{B}, \widetilde{\pi})$ (the pullback of the original fibration) and an analytic map $H: \widetilde{M} \to M$ such that the diagram*

$$\begin{array}{ccc} \widetilde{M} & \xrightarrow{H} & M \\ {\scriptstyle\widetilde{\pi}}\downarrow & & \downarrow{\scriptstyle\pi} \\ \widetilde{B} & \xrightarrow{\rho} & B \end{array}$$

is commutative and the restriction of H to any fiber $\widetilde{M}_{\widetilde{\varepsilon}}$ is a biholomorphic map $\widetilde{M}_{\widetilde{\varepsilon}} \to M_{\varepsilon}$.

Moreover,

(1) *if for some $\Omega \subset M$ the restriction $\pi|_{\Omega}$ was proper, then the restriction of $\widetilde{\pi}$ to $H^{-1}(\Omega)$ is also proper, and*
(2) *if ρ has the form $\rho = \varphi \circ \pi$, where $\varphi: \widetilde{B} \to M$ is a section of the fibration, then the set $H^{-1}(\varphi(B))$ contains at least one section of the pullback $\widetilde{\pi}$.*

PROOF. Consider the Cartesian product $E = M \times \widetilde{B}$ and its standard projections:

$$P_M : E \to M, \quad P_{\widetilde{B}} : E \to \widetilde{B}.$$

Then the new total space $\widetilde{M}$, the projection $\widetilde{\pi}$ and the mapping H may be represented as

$$\widetilde{M} = \{(p, \widetilde{\varepsilon}) \mid \pi(p) = \rho(\widetilde{\varepsilon})\}, \quad H = P_M|_{\widetilde{M}}, \quad \widetilde{\pi} = P_{\widetilde{B}}|_{\widetilde{M}}$$

(the assumption on the rank of π guarantees that $\widetilde{M}$ is nonsingular).

(1) Let $\widetilde{K} \subset \widetilde{B}$ be compact. Then, since ρ is continuous and $\pi|_{\Omega}$ is proper, the sets $K = \rho(\widetilde{K})$ and $\pi^{-1}(K) \cap \Omega$ are compact. Hence the set $\widetilde{\pi}^{-1}(\widetilde{K}) \cap h^{-1}(\Omega) \subset E$ is compact as a closed subset of compact $(\pi^{-1}(K) \cap \Omega) \times \widetilde{K}$.
(2) The required section $S: \widetilde{B} \to \widetilde{M} \subset E$ is defined as $\widetilde{\varepsilon} \mapsto (\varphi(\widetilde{\varepsilon}), \widetilde{\varepsilon})$. $\square$

REMARK. Sometimes we refer to the pullback fibration $(\widetilde{M}, \widetilde{B}, \widetilde{\pi})$ as the *induced* fibration $(\rho^* M, \rho^* B, \rho^* \pi)$.

1.2. Families of line fields and their singularities. Here we give definitions related to families of line fields depending on parameters. These definitions are illustrated by examples in 2.1.

DEFINITION. A *family of line fields* on a family of surfaces (represented by the fibration (M, B, π), $\dim M - \dim B = k = 2$) is a line field on M tangent to all fibers M_ε at each point.

The singular locus of a family of line fields has codimension 2 in the total space M. Indeed, near each point in adapted local coordinates the line field is spanned by a vector field

$$f(x, y, \varepsilon)\,\partial/\partial x + g(x, y, \varepsilon)\,\partial/\partial y + 0 \cdot \partial/\partial \varepsilon.$$

As before, one may put f and g being relatively prime, hence the singular locus Ω is the set $\{f = g = 0\}$ of codimension 2.

Consider now the intersection $\Omega_\varepsilon = \Omega \cap M_\varepsilon$ with an arbitrary fiber. Three different cases may occur:

(1) $\dim \Omega_\varepsilon = 0$: all singularities are isolated on the fiber;
(2) $\dim \Omega_\varepsilon = 1$: the singular locus of the restriction is a curve;
(3) $\dim \Omega_\varepsilon = 2$: the entire connected component of the fiber M_ε consists of singular points. We call such fibers *critical*.

In the second case the restriction of the line field to the fiber M_ε may be extended to a line field with isolated singularities.

DEFINITION. Let α be a family of line fields on a family of surfaces (M, B, π). A point $p \in M_\varepsilon \subset M$ is an *essential singularity* of the family, if it lies on a non-critical fiber M_ε and is a singular point of the restriction $\alpha_\varepsilon = \alpha|_{M_\varepsilon}$ (recall that this means that the point p remains singular after the maximal extension of the restricted line field α_ε).

The set of essential singularities will be denoted as Ω_{ess}. Note that all essential singularities are isolated on the corresponding fibers.

DEFINITION. An essential singularity p is *elementary* if it is an elementary singularity for the restriction α_ε, $\varepsilon = \pi(p)$.

DEFINITION. A (complex analytic) *proper line family* is a line family without critical fibers and such that the projection π restricted to closure of the singular locus $\overline{\Omega_{\text{ess}}}$ is proper.

NOTE. As this will be shown in §2, for a (complex analytic) proper line family the set $\overline{\Omega_{\text{ess}}}$ is analytic (Lemma E). Below, in 2.1, together with examples of families of vector fields which are not proper, we give also examples of families of proper line fields, for which the restriction of the projection π restricted to the whole singular locus Ω is not a proper map.

Speaking loosely, the properness condition means that essential singularities occurring in the family cannot escape from the domain of consideration across its boundary, when the parameters of the family change.

DEFINITION. An *elementary* line family is a line family satisfying the following two conditions:

(1) all essential singularities are elementary, and
(2) the set of essential singularities in any connected component of the fibration is contained in a finite union of sections.

Obviously, in the real case the strict real analogs of the notions introduced before make sense, at least formally. However, for the future constructions we need the following modifications concerning the definition of proper families of line fields in the real case. For a real family of line fields (M, B, π, α) there exists at least one complexification, $({}^{\mathbb{C}}M, {}^{\mathbb{C}}B, {}^{\mathbb{C}}\pi, {}^{\mathbb{C}}\alpha)$, a (complex analytic) family of line fields which extends the original family. This means that

$$ {}^{\mathbb{C}}M \supset M, \quad {}^{\mathbb{C}}B \supset B, \quad {}^{\mathbb{C}}\pi|_M = \pi, \quad {}^{\mathbb{C}}\alpha \cap TB = \alpha. $$

DEFINITION. A real family of line fields will be called $\mathbb{R}$-*proper*, if some complexification of this family is a (complex analytic) proper family of line fields.

Apparently this definition is somewhat more restrictive than the strict real analog.

1.3. Constructions in parameter blowing-up. Formulation of the main result. Consider an arbitrary family of line fields. In general, some singularities occurring in this family, are nonelementary. If they depend nicely on parameters, that is, belong to an analytic section S of the fibration (M, B, π), then without loss of generality we may assume that in an adapted coordinates this section is constant, $S \subseteq \{x = y = 0\}$ (this can be achieved by an analytic biholomorphism preserving the fibers of the fibration). In the latter case it is easy to blow up all points of this section, in short, to *blow up the section*, by the same simple blow-up (1), (2). In the original coordinates we obtain the family of blowing-up transformations analytically depending on the parameters. Details and the precise definition are given in 3.2. The problem becomes more complicated if the singular locus is *not a section*; in this case we perform a reparametrization of the family as it was described in 1.1.3.

DEFINITION. An *unfolding* of an analytic manifold B is a surjective analytic mapping $\rho : \widetilde{B} \to B$ of another analytic manifold $\widetilde{B}$ of the same dimension such that ρ is biholomorphic at a generic point of $\widetilde{B}$, and for any compact $K \subseteq B$ there exists a compact $\widetilde{K} \subseteq \widetilde{B}$ such that $\rho(\widetilde{K}) = K$.

REMARK. The manifold $\widetilde{B}$ is not assumed to be connected or compact. Moreover, it may consist of an infinite number of connected components (however, we always make the countable compactness assumption, see Convention 2 above). This situation naturally occurs in the procedure of desingularization in families.

EXAMPLES. 1. If $\{U_\alpha\}$ is an open countable covering of B, then the disjoint union $\widetilde{B} = \coprod_\alpha U_\alpha$ together with the map $\rho = \coprod_\alpha \rho_\alpha$, $\rho_\alpha \colon U_\alpha \to B$, is an unfolding.

2. Any covering map with a finite number of leaves is an unfolding.

3. Let $B = \mathbb{C}^2$ and ρ be the standard blow-up of the origin, given by the formulas (1), (2) in which x is replaced by ε_1 and y by ε_2. Then ρ is an unfolding, and this example may be generalized for the multidimensional case.

According to 1.1.3, any unfolding of the base of a fibration (M, B, π) yields another fibration, induced from the original one, and the pullback of the family of line fields (the induced family) is also well defined by virtue of Proposition 1.1. Let (M, B, π) be a fibration with two-dimensional fibers, interpreted as the parameter family of surfaces.

PRINCIPAL DEFINITION. A *resolution* of the family (M, B, π) is another fibration (= a family of analytic surfaces) $(\widetilde{M}, \widetilde{B}, \widetilde{\pi})$ together with two maps $H: \widetilde{M} \to M$, $\rho: \widetilde{B} \to B$ such that:

(1) the diagram

$$\begin{array}{ccc} \widetilde{M} & \xrightarrow{H} & M \\ {\scriptstyle\widetilde{\pi}}\downarrow & & \downarrow{\scriptstyle\pi} \\ \widetilde{B} & \xrightarrow{\rho} & B \end{array}$$

is commutative, and

(2) the restriction of the mapping H to each two-dimensional fiber $\widetilde{M}_{\widetilde{\varepsilon}} \subset \widetilde{M}$ is a resolution of the surface M_ε in the sense of 1.1.2.

A resolution of a family of line fields ℓ defined on the family of surfaces (M, B, π) is the pullback $H^*\ell$ of the original family ℓ onto $\widetilde{M}$.

EXAMPLE. Unfoldings of the base B, blow-ups of section, and their superpositions are natural examples of resolutions of families.

MAIN THEOREM. *For any proper family of line fields there exists an elementary resolution.*

In other words, the assertion of the theorem means that for any family (M, B, π) of surfaces and any proper family of line fields on it, there exists another proper family of line fields with only elementary essential singularities defined on another family of surfaces $(\widetilde{M}, \widetilde{B}, \widetilde{\pi})$. The latter family of surfaces is a resolution of the original family in the above sense.

REMARKS. 1. The assertion of the Main Theorem is constructive: any elementary resolution may be constructed as a finite superposition of unfoldings of bases and blow-ups of sections.

2. The Main Theorem has a strict real analog for $\mathbb{R}$-proper families, see the remark at the end of 2.2.

NOTE. As this will be explained in §6, the resolution of a proper analytic family of line fields having only essential singularities, in general, will have nonisolated singularities corresponding to singular perturbations of differential equation. However, the resulting family will be again proper.

§2. Singularities of line fields

2.1. Examples. We start with several examples illustrating definitions introduced in §1.

2.1.1. *Critical fibers.* Consider the family of line fields spanned by the following family of vector fields depending on two parameters:

$$v(p, \varepsilon) = \varepsilon_1 v_1(p) + \varepsilon_2 v_2(p),$$

where $(\varepsilon_1, \varepsilon_2) \in \mathbb{C}^2 = B$, $p \in \mathbb{C}$, and v_1, v_2 are two polynomial vector fields on $\mathbb{C}^2$ with isolated singularities and functionally independent coordinate functions.

For the family of line fields spanned by v, the fiber $(\varepsilon_1, \varepsilon_2) = (0, 0)$ is critical. But if we consider the restriction of the family to any line in the base B that goes through the origin, then all singularities of this restriction are essential. Indeed, if one puts $\varepsilon_i = \delta e_i$, $|e_i| = 1$, $\delta \in \mathbb{C}$, then the restricted line field is spanned by the vector field $e_1 v_1 + e_2 v_2$ which has only isolated singularities on each fiber (in fact, the family is independent of the parameter δ). Now it is also clear that two different restrictions may possess different essential singularities on the fiber M_ε for $\varepsilon = (\varepsilon_1, \varepsilon_2) = (0, 0)$.

Consider the family of polynomial vector fields of degrees $\leqslant n$ on the plane (n is arbitrary),

$$\frac{dx}{dy} = \frac{P_n(x, y)}{Q_n(x, y)}, \qquad \deg P, \deg Q \leqslant n.$$

This family has the coefficients of the polynomials P and Q as the natural parameters, so B is just $\mathbb{C}^{(n+2)(n+1)}$, and the origin in this space corresponds to the critical fiber of the family. But if we consider the projectivization $\mathbb{C}P^{(n+2)(n+1)-1}$, then the new family would have no critical fibers at all.

2.1.2. *Runaway singular points.* Look again at the same polynomial family as before. If we consider it as defined on a family of surfaces with the fiber $\mathbb{C}^2$, then the restriction of the projection π to the singular locus would not be proper, since with parameters changing, common zeros of P and Q may disappear at infinity. But if we extend the line field to a family of surfaces with the projective plane $\mathbb{C}P^2$ as a fiber (the *Poincaré compactification*), then the obtained family would be proper.

2.1.3. *The essential singular locus may be nowhere dense in the variety of singular points of a family.* Indeed, for the family spanned by the vector field $(xy+\varepsilon)\,\partial/\partial x + y^2\,\partial/\partial y$ the critical locus Ω is the entire x-axis, while the only essential singularity is at the origin:

$$\Omega = \{(x, 0, 0) \mid x \in \mathbb{C}\}, \qquad \Omega_{\text{ess}} = \{(0, 0, 0)\}.$$

2.1.4. *The essential singular locus may be not an analytic set.* For the family spanned by

$$(\varepsilon x - x^2)\,\partial/\partial x + (\varepsilon + xy)\,\partial/\partial y$$

one has

$$\Omega = \{x = \varepsilon, y = -1\} \cup \{x = 0, \varepsilon = 0\},$$
$$\Omega_{\text{ess}} = \{x = \varepsilon, y = -1, \varepsilon \neq 0\} \cup \{(0, 0, 0)\},$$

and Ω_{ess} is only semianalytic and not closed.

2.1.5. The last two examples show that for a proper family, the restriction of the projection π to the whole critical locus Ω may be not a proper map.

2.2. Discretization of the singular locus.

LEMMA E (on the structure of essential critical locus). *Let α be a complex analytic family of line fields. If the restriction of the projection π to the set $\overline{\Omega_{\text{ess}}}$ is proper, then the set $\overline{\Omega_{\text{ess}}}$ itself is analytic.*

In other words, for a complex analytic proper family of line fields, the closure of the essential singular locus $\overline{\Omega_{\text{ess}}}$ is an analytic set. The proof of Lemma E and Lemma D (below) are given at the end of this subsection.

DEFINITION. Let $\Omega \subset M$ be an analytic subset of the total space of a fibration (M, B, π). We say that this set is *π-isolated* at a point $q \in \Omega$, if q is an isolated point of the intersection $\Omega \cap M_\varepsilon$, $\varepsilon = \pi(q)$, on the fiber M_ε.

The set of points at which Ω is π-isolated, is denoted by Ω_{is}. Sometimes we say that a point of Ω is π-isolated, if Ω is π-isolated at this point.

PROPOSITION 2.1. *The complement $\Omega \setminus \Omega_{\text{is}}$ is an analytic set. Furthermore, if the restriction $\pi|_\Omega$ is proper, then the image $\pi(\Omega \setminus \Omega_{\text{is}})$ is also analytic, and of dimension strictly less than* $\dim \Omega$.

We postpone the proof of this proposition until Appendix B.

DEFINITION. Let α be a family of line fields on a family of surfaces (M, B, π). A finite collection of analytic sets Ω_i is called *discretization* of the essential singular locus of the family α if at each essential singularity $q \in \Omega_{\text{ess}}$ at least one of the sets Ω_i is π-isolated,

$$\Omega_{\text{ess}} \subseteq \bigcup_i (\Omega_i)_{\text{is}},$$

and all the restrictions $\pi|_{\Omega_i}$ are proper.

EXAMPLES. For the family from 2.1.3, the discretization may be chosen as a single set $\Omega_0 = \{(0, 0, 0)\}$. As for the family described in 2.1.4, the discretization may be chosen as

$$\Omega_0 = \{x = \varepsilon, y = -1\}, \quad \Omega_1 = \{(0, 0, 0)\}.$$

LEMMA D (on discretization). *A proper family of line fields admits discretization of its essential singular locus.*

The proofs of Lemmas D and E are based on the following technical result from analytic geometry. As before, let α be a proper family of line fields on the fibration (M, B, π), and Ω the singular locus of this family.

LEMMA 2.2. *For any analytic subset $N \subseteq B$ one may find an analytic subset $\Omega_N \subseteq M$ such that $\dim \Omega_N \leqslant \dim N$ and*

$$\Omega_{\text{ess}} \cap \pi^{-1}(N) \subseteq \Omega_N \subseteq \Omega \cap \pi^{-1}(N).$$

REMARK. If N is a nonsingular submanifold of B, then Ω_N is the singular locus for the family of line fields restricted to $(\pi^{-1}(N), N, \pi)$.

PROOF. First note that the line field on a complex manifold M can be completely described by an irreducible analytic set in the projective tangent bundle of M. This set must intersect the generic fiber of the bundle at a single point.

Let (PTM, M, χ), $\chi\colon PTM \to M$ be the projectivization of the tangent bundle of M, and denote by $S \subset PTM$ the set defining the family of line fields. Evidently, for any nonsingular point $p \in M \setminus \Omega$ the set $\chi^{-1}(p) \cap S$ consists of a single element.

Without loss of generality we can assume N to be an irreducible subset. Consider the set $M_N = \pi^{-1}(N) \subseteq M$. Obviously, it is a nonempty irreducible analytic set, and $\dim M_N = \dim N + 2$. So $S_N = (S \cap \chi^{-1}(M_N)) \setminus \chi^{-1}(\Omega)$ is an analytic irreducible set in the manifold $PTM \setminus \chi^{-1}(\Omega)$ (since the family has no critical fibers, the set S_N is nonempty). By definition of the set S, $\dim M_N = \dim S_N$. By general properties of analytic sets, the closure $\overline{S}_N$ of the set $S_N \subset S \subset PTW$ is an irreducible analytic subset of the whole PTM, and

$$\dim \overline{S}_N = \dim M_N = \dim N + 2, \qquad \dim(\overline{S}_N \cap \chi^{-1}(\Omega)) \leqslant \dim \overline{S}_N - 1.$$

Now put $\Omega_N = \chi(\overline{S}_N \setminus (\overline{S}_N)_{\mathrm{is}})$. The properties of the set S imply that the sets of points at which $\overline{S}_N$ and $\overline{S}_N \cap \chi^{-1}(\Omega)$ are not π-isolated, coincide. Since the mapping χ is proper, by Proposition 2.1, the set $\Omega_N \subset M$ is analytic, and

$$\dim \Omega_N \leqslant \dim S_N \cap \chi^{-1}(\Omega) - 1 \leqslant \dim N.$$

Finally, we have to prove that any point $p \in \Omega_{\mathrm{ess}} \cap \pi^{-1}(N)$ belongs to Ω_N. Perform the same construction but with the set N replaced by a single point $\varepsilon = \pi(p) \in B$. Then the following sets will be constructed,

$$S_\varepsilon = (S \cap \chi^{-1}(M_\varepsilon)) \setminus \chi^{-1}(\Omega), \qquad \Omega_\varepsilon = \chi(\overline{S}_\varepsilon \setminus (\overline{S}_\varepsilon)_{\mathrm{is}}).$$

It is easy to see that the set $\overline{S}_\varepsilon$ determines the line field on the fiber M_ε, so the set Ω_ε is exactly the union of all the essential singularities on the fiber M_ε: $p \in \Omega_{\mathrm{ess}} \cap M_\varepsilon = \Omega_\varepsilon$. So we have to prove that $\Omega_\varepsilon \subseteq \Omega_N$. But since $\varepsilon \in N$ and the fiber M_ε is noncritical, $S_\varepsilon \subseteq S_N$, $\overline{S}_\varepsilon \subseteq \overline{S}_N$, $\overline{S}_\varepsilon \setminus (\overline{S}_\varepsilon)_{\mathrm{is}} \subseteq \overline{S}_N \setminus (\overline{S}_N)_{\mathrm{is}}$, so $\Omega_\varepsilon \subseteq \Omega_N$. □

Proof of Lemma E. We prove the analyticity of the essential critical locus after taking closure by induction in the dimension of the base B.

If $\dim B = 0$, then the family in fact consists of a discrete collection of line fields, all singularities are essential and the assertion is trivially true.

Now consider a family of line fields with a k-dimensional base in a precompact neighborhood of a point in the total space. Let as usual Ω be the singular locus, $\dim \Omega = k$. By Lemma 2.2, the set $\Omega' = \Omega \setminus \Omega_{\mathrm{is}}$ of points of Ω nonisolated on fibers, is analytic. Evidently, a generic point of Ω is essentially singular, hence the analytic set $\overline{\Omega_{\mathrm{is}}}$ is the closure of the set $\Omega_{\mathrm{ess}} \setminus \Omega'$.

In many cases the latter set coincides with the closure of Ω_{ess}, which makes the subsequent obsolete. In the general case, however, the following arguments apply. Since $\Omega' \subset \overline{\Omega_{\mathrm{ess}}}$, the restriction $\pi|_{\Omega'}$ is proper, and the set $N = \pi(\Omega') \subset B$ is analytic.

Take a partition of N into strata homogeneous in dimension:

$$N = \bigcup_i N_i \subset B, \qquad \dim N_i > \dim N_{i+1}.$$

Consider the restrictions of the original family to each stratum N_i. All of them will be analytic families with the dimensions of a base less than k. Naturally, the corresponding essential singular loci are exactly $\Omega_{\mathrm{ess}} \cap \pi^{-1}(N_i)$. By the induction assumption, the closure of each one is analytic in $\pi^{-1}(N_i)$. But each stratum N_i is a difference of two analytic sets, hence for any i the set $\overline{\Omega_{\mathrm{ess}} \cap \pi^{-1}(N_i)} \subset M$ is analytic. □

Proof of Lemma D. Fix the first set of discretization as the singular locus of the family $\Omega_1 = \overline{\Omega_{\rm ess}}$. By Lemma E, this set is analytic. Since it is contained in the singular locus Ω, its dimension does not exceed $\dim M - 2 = \dim B$. Since the family is proper, the mapping $\pi|_{\Omega_1}$ is proper.

Define the next sets by the following inductive procedure: if the set $\Omega_i \subseteq \Omega_1$ is nonempty, put

$$N_{i+1} = \pi(\Omega_i \setminus (\Omega_i)_{\rm is}) \subset B, \quad \Omega_{i+1} = \Omega_{N_{i+1}}.$$

The set N_{i+1} is analytic by Proposition 2.1, and the set Ω_{i+1} is analytic by Lemma 2.2. Since $\Omega_{i+1} \subset \Omega$, the mapping $\pi|_{\Omega_{i+1}}$ is proper. Proposition 2.1 and Lemma 2.2 guarantee also that

$$\dim \Omega_{i+1} \leqslant \dim N_{i+1} < \dim \Omega_i, \qquad \Omega_{\rm ess} \subseteq \bigcup_{j=1}^{i} (\Omega_j)_{\rm is} \cup \Omega_{i+1}.$$

Thus the sequence $\{\Omega_i\}$ consists of analytic sets of strictly decreasing dimensions, hence it must terminate after at most $n = \dim B$ steps. By construction, the sets realize the discretization of the essential singular locus of the family α. □

2.3. Discretization in the real case. All examples from 2.1 may be considered as real, with the complexifications given by the same formulas over $\mathbb{C}$. The definition of the real $\mathbb{R}$- proper family in 1.2 is rather natural and simple. In particular, all polynomial families and also all polynomial deformations of singular points of analytic vector fields on the plane, are $\mathbb{R}$- proper.

Definition. A *discretization for a real analytic family of line fields* is a strict real analog of the discretization for a complex analytic family of line fields (see the definition at the beginning of 2.2) and 1.1.1.

This definition immediately implies the following reformulation of Lemma D for the real case.

Lemma $^{\mathbb{R}}$D (on discretization in the real domain). *An $\mathbb{R}$-proper real analytic family of line fields admits a discretization of its essential singular locus.* □

§3. Special resolutions

As was mentioned in 1.3, there are two special kinds of resolutions, namely unfoldings of the base of a fibration and blowing-up of an analytic section. We show that by making an appropriate unfolding, one may always place all essential singularities of a proper family of line fields onto a finite number of analytic sections (eventually, intersecting). In this case we speak about *rectification* of the singular locus, slightly abusing the term, since usually there are many sections, and only one of them may be really rectified to become a constant section.

3.1. Unfoldings of the base. Rectification of essential singular locus. Unfoldings of the base of a fibration correspond to a reparametrization of the family of line fields. Clearly, such a procedure does not change the essential nature of singular points. For example, the preimage of the essential singular locus of the original family will be the essential singular locus of the induced family as the latter was defined in 1.1.3. One can also see that the pullback of a proper family will be itself a proper family of line fields. However, such unfoldings may simplify dependence of singular points on parameters.

LEMMA R (rectification of analytic subsets). *Let* $\{\Omega_i\}$ *be a finite collection of analytic subsets in the total space of an analytic fibration* (M, B, π). *Assume that the restriction of the projection* π *to each set* Ω_i *is proper and their dimensions do not exceed that of the base* B. *Then there exists an unfolding* $\rho : \widetilde{B} \to B$ *of the base of the fibration such that in each connected component of the induced fibration* $(\widetilde{M}, \widetilde{B}, \widetilde{\pi} = \rho^*\pi)$ *the closure of each set* $H^{-1}((\Omega_i)_{\text{is}})$ *belongs to a finite union of analytic sections.*

This lemma is proved in Appendix B. The proof is based on the Hironaka theorem on desingularization of analytic sets. In fact both the formulation and the proof do not use the fact that the fibers are two-dimensional: the assertion is true for an arbitrary dimension of fibers of an analytic fibration.

Lemma D and Lemma R together imply the following assertion.

LEMMA S (rectification of essential singular locus). *Any proper analytic family of line fields on a fibration with two-dimensional fibers admits an unfolding of the base, after which all essential singularities of the induced family in any connected component of the induced fibration, belong to the union of a finite number of analytic sections of the latter.*

3.2. Blowing-up of analytic sections. Blowing-up of an analytic section $S: B \to M$ is a resolution $(H, \rho): (\widetilde{M}, \widetilde{B}) \to (M, B)$ of the fibration which preserves the base of the fibration, $\rho = \mathrm{id}_B$, and whose restriction $H|_{\widetilde{M}_\varepsilon}$ to each fiber $\widetilde{M}_\varepsilon = \widetilde{\pi}^{-1}(\varepsilon)$ is a resolution of this fiber (in the sense of the definition given in 1.1.2). To make the construction explicit, we assume that the section in an adapted system of local coordinates has the form $\{x = y = 0\}$. Then one may perform the simple blow-up of each fiber using the map (1) for all $\varepsilon \in B$ simultaneously (within the given local chart). The resulting map H is analytic and proper on the whole set $\widetilde{M}$, and defines a simple blow-up of each fiber. Speaking formally, we constructed a resolution

$$\begin{array}{ccc} \widetilde{M} & \xrightarrow{H} & M \\ \widetilde{\pi}\downarrow & & \downarrow\pi \\ B & \xrightarrow{\mathrm{id}} & B \end{array}$$

such that for any $\varepsilon \in B$ the restriction $H_\varepsilon : \widetilde{M}_\varepsilon \to M_\varepsilon$ is a simple blow-up of the point $S(\varepsilon)$. Topologically the result would be (either locally with respect to the parameter ε or globally under the assumption that B is simply connected) as if we delete the section S and instead paste in the cylinder

$$\Sigma = \mathbb{C}P^1 \times B = \bigcup_{\varepsilon \in B} \mathbb{C}P^1_\varepsilon.$$

PROPOSITION 3.1. *After a simple blow-up a nonsingular point yields an elementary singularity which is a nondegenerate saddle.* □

Thus one can blow up sections without paying attention, whether these sections entirely consist of singularities or not: if the family of line fields had no singularity at a certain point $p = S(\varepsilon)$, then the pullback family would have hyperbolic saddles on Σ after the maximal extension, see §4.

Suppose that instead of just one section we have several sections with a nonvoid intersection. Consider the simplest case of two sections S_1 and S_2. Then after a simple blow-up of the first section the preimage $H^{-1}(S_2)$ is a well-defined analytic subset in $\widetilde{M}$, but it may happen that this preimage is not a section anymore. To restore the structure of a section, we need to perform an additional unfolding of the base.

Example. Consider the two-parameter family of surfaces in the adapted coordinates,

$$M = \{(x, y, \varepsilon_1, \varepsilon_2) \in \mathbb{C}^4\}, \qquad B = \{(\varepsilon_1, \varepsilon_2) \in \mathbb{C}^2\}$$

and two sections

$$S_1 = \{x = 0, y = 0\}, \qquad S_2 = \{x = \varepsilon_1, y = \varepsilon_2\}.$$

Then after blowing-up of the section S_1 the preimage of S_2 becomes an irreducible set that is not an analytic section of the new fibration with the same base B: the intersection of the fiber over $(\varepsilon_1, \varepsilon_2) = (0, 0)$ and the preimage of S_2 is the entire projective line $\mathbb{C}P^1_{(0,0)}$.

To correct this situation, we need to make an additional blow-up of the parameter space B at the origin $(0, 0)$ in the same way as it was done with the fiber: we delete the origin and paste in the projective line so that there appears a new base $\widetilde{B}$ and the projection $\rho\colon \widetilde{B} \to B$. In the pullback fibration over $\widetilde{B}$ preimages of all points of S_2 except those that were already blown up on the previous step, would belong to analytic sections.

This situation is quite generic, but it turns out that similar correction may be always done by implementing an auxiliary unfolding of the base that blows up the exceptional subset over which $H^{-1}(S_2)$ is not a section: this is guaranteed by Lemma R. What we need is to guarantee that the process of subsequent simple blow-ups of sections and unfoldings of the bases would not run into an infinite loop.

First we show that Lemma R applied to an analytic set that intersects a generic fiber of the fibration (M, B, π) by exactly one point, yields a unique section carrying the preimage of this set.

Proposition 3.2. *Assume that an analytic subset $\Omega \subset M$ is of the uniform dimension $n = \dim B$ and for any $\varepsilon \in B$ the intersection $\Omega_{\text{is}} \cap M_\varepsilon$ is either empty or consists of a single point. Let $(H, \rho)\colon (\widetilde{M}, \widetilde{B}) \to (M, B)$ be a resolution of the fibration (M, B, π) such that $\Omega' = H^{-1}(\Omega_{\text{is}})$ belongs to the union of analytic sections. Then in each connected component of $\widetilde{M}$ the set Ω' in fact belongs to exactly one analytic section.*

Proof. Without loss of generality we may consider the case of a connected fibration $(\widetilde{M}, \widetilde{B}, \pi)$. Let $\Omega'' \subset \widetilde{M}$ be the set of points whose H-images are π-isolated on Ω, $\Omega'' = H^{-1}(\Omega) \setminus H^{-1}(\Omega \setminus \Omega_{\text{is}})$. Obviously, this set intersects a generic fiber of π at a single point. The closure $\Omega' = \overline{\Omega''}$ is an analytic set, which is irreducible (by virtue of connectedness of the base and the above unique-point-on-a-fiber condition) and coincides with $\overline{H^{-1}(\Omega_{\text{is}})}$. So if Ω' belongs to the union of analytic sections, then in fact it belongs to only one of them. $\square$

COROLLARY 3.3. *Let α be a proper family of line fields on the fibration (M, B, π). Then there exists a resolution (H, ρ) such that each essential singularity of α will be blown up at least once,*

$$\forall p \in \Omega_{\text{ess}} \quad \dim H^{-1}(p) > 0.$$

PROOF. 1. Applying Lemma S, one can resolve the essential singular locus of the family to belong to a finite union of analytic sections of the corresponding fibration in each connected component of the latter. Fix one of the components and denote the number of sections by ℓ.

2. Make a simple blow-up of one of these sections, say, S_ℓ. Then the remaining sections $S_1, \dots, S_{\ell-1}$ will be transformed into some analytic subsets $S'_1, \dots, S'_{\ell-1}$ which may be not analytic sections anymore, but still intersect a generic fiber at a single point.

3. Again, applying Lemma S, one can make yet another unfolding of the base so that the sets $(S'_j)_{\text{is}}$ would be placed on a finite number of analytic sections $\widetilde{S}_1, \dots, \widetilde{S}_N$, $N < \infty$.

4. Now Proposition 3.2 allows us to choose exactly $\ell - 1$ among the sections $\widetilde{S}_j$ which would carry preimages of all unblown essential singularities of the original family of line fields. Thus we reduced the number of sections by 1 and blew up all essential singular points from S_ℓ. Iterating this construction, we achieve the desired result. $\square$

REMARK. The strict real analog of Lemma R will be proved in Appendix C. All other results from this section hold for both the real and the complex case.

§4. Effective bounds for the Bendixson–Seidenberg theory

4.1. Sketch of the proof of Bendixson–Seidenberg theorem after Van den Essen. In this section we give an explicit estimate on the number of simple blow-ups in the nonparameter case that are necessary to split an isolated singularity into elementary ones. The subject is treated in more details by Kleban, see paper [17] in this volume. The exposition is based on the approach by Van den Essen [8, Appendix 1], [16].

DEFINITION. The *multiplicity* of an isolated singularity of the vector field $f(x, y)\,\partial/\partial x + g(x, y)\,\partial/\partial y$ at the origin is the multiplicity of the germ $(\mathbb{C}^2, 0) \to (\mathbb{C}^2, 0)$, $(x, y) \mapsto (f(x, y), g(x, y))$.

The multiplicity of an isolated singularity of a *line field* is the minimal multiplicity of a vector field spanning this line field.

REMARK. The multiplicity of an analytic germ $F = (f, g)\colon (\mathbb{C}^2, 0) \to (\mathbb{C}^2, 0)$ with an isolated preimage of zero at the origin is by definition the (differential topologic) degree of the map

$$\frac{F}{\|F\|}\colon \mathbb{S}^3_r \to \mathbb{S}^3_1, \qquad (x, y) \mapsto \|F(x, y)\|^{-1} \cdot (f(x, y), g(x, y)),$$

$$\|F(x, y)\| = \sqrt{|f(x, y)|^2 + |g(x, y)|^2},$$

where $\mathbb{S}_r^3$ is a sufficiently small sphere $|x|^2 + |y|^2 = r^2$. In Appendix A we give an equivalent definition which is more convenient for our purposes.

The multiplicity of an isolated (over $\mathbb{C}$) singularity is always positive and finite.

DEFINITION. A singularity is *cuspidal* if it is spanned by a vector field with nonzero nilpotent linear part: in an appropriate coordinate system such field has the form

$$v(x, y) = x\,\partial/\partial y + (\text{nonlinear terms}).$$

LEMMA 4.1. 1. *If an isolated singularity of a line field is not elementary or cuspidal, then after one simple blow-up the sum of multiplicities of all singularities on the pasted in projective line is strictly less than the multiplicity of the original singularity.*

2. *After blowing up an elementary singularity or a nonsingular point, only elementary singularities appear. In particular, in the nonsingular case hyperbolic saddles appear.*

LEMMA 4.2. *A cuspidal singularity becomes elementary after a finite number of iterations of simple blow-ups.*

These two results together imply the Bendixson–Seidenberg theorem: one iterates simple blow-ups until all points become either elementary or cuspidal, in no more than μ steps, where μ is the multiplicity of the original singularity. Then applying Lemma 4.2, one resolves all cuspidal points into elementary ones.

4.2. Number of simple blow-ups. In order to estimate the number of simple blow-ups necessary to resolve a degenerate isolated singularity into elementary ones, it is sufficient to do this only for cuspidal points.

LEMMA 4.3 [17]. *A cuspidal point of multiplicity μ is resolved into elementary singularities after no more than $\left[\frac{\mu}{2}\right] + 2$ simple blow-ups.*

COROLLARY 4.4. *A singularity of multiplicity μ can be completely resolved after no more than $\mu + 2$ simple blow-ups.*

REMARK. This effective estimate is valid in the complex case; the real case requires preliminary complexification. However, the number of simple blow-ups can be less than the multiplicity of the complexified singularity, since some of the intermediate points may be not real.

4.3. Uniform bound for multiplicity of all singularities. The above result gives an upper estimate for the number of simple blow-ups necessary to resolve a singular point of multiplicity μ.

LEMMA 4.5. *For an arbitrary analytic family of line fields restricted to any precompact subset of the total space, there exists a common upper bound for multiplicities of all essential singularities.*

The proof of this assertion is given in Appendix A. The assertion of this lemma is the same for both real and complex families, and the real version is an automatic corollary to the complex one.

§5. Proof of the Main Theorem

5.1. The proof. Consider an arbitrary proper family of line fields α on the family of surfaces (M, B, π).

1. By Lemma 4.5 there exists a finite open covering U_α of the base B such that in any domain of this covering the multiplicity of all essential singularities is bounded by a certain constant μ_α. This covering may be considered as an unfolding of the base, so that in any connected component the multiplicity is bounded by μ_α. By Corollary 4.4, each singularity in the component U_α would be completely resolved after no more than $\nu = \mu_\alpha + 2$ simple blow-ups.

2. If all singularities are in fact elementary, then in order to satisfy the second condition from the definition of an elementary family, we apply Lemma S from 3.2. After an appropriate unfolding of the base, all singularities (already elementary) can be placed on analytic sections and the final goal is achieved.

3. If there are nonelementary singularities, then using Corollary 3.3, one may construct a resolution of the family such that each essential singularity would be blown up at least once. This reduces the degeneracy degree of all essential singularities, and after iterating this step no more than ν times we obtain a family of vector fields with all essential singularities being elementary. Then the procedure of the step 2 applies. □

5.2. Remarks and examples. The most simple case of a family of surfaces is the Cartesian product $M = F^2 \times B$ with some two-dimensional fiber F and the canonical projection $\pi: M \to B$. After implementing the construction described in the Main Theorem, this structure can be lost.

EXAMPLE. Consider the trivial family of surfaces with the fiber $\mathbb{C}^2$, that is, the Cartesian product

$$(\mathbb{C}^2 \times \mathbb{C}, \mathbb{C}, \pi), \qquad \pi: \mathbb{C}^2 \times \mathbb{C} \to \mathbb{C},$$

and let S_i, $i = 1, 2$, be two sections,

$$S_1 = \{(x, y, \varepsilon) \mid x = y = 0\}, \quad S_2 = \{(x, y, \varepsilon) \mid x = y = \varepsilon\}.$$

It is not difficult to verify that after the blowing up of one of these sections the other remains an analytic section. Now we may also blow-up this second section. As a result, we obtain the family with two projective lines pasted into each fiber. For $\varepsilon \neq 0$ these projective lines are disjoint, while for $\varepsilon = 0$ they intersect each other. Thus the fiber M_0' of the resulting family (M', B, π') over the point $\varepsilon = 0$ is not holomorphically equivalent to other fibers, hence the latter family is not trivial.

This example also shows that in general the result of desingularization depends on the order in which separate analytic sections are blown up. If we reverse the order in which the two sections S_1, S_2 are blown up, then the fiber M_0'' of the resolution (M'', B, π'') will be *not biholomorphically equivalent* to M_0'.

REMARK. The proof of the Main Theorem in the real case remains exactly the same, since both Lemma S and Lemma 4.5 have strict real analogs.

§6. Abundance of singular perturbations

The desingularization scheme established above allows us to construct elementary resolutions for any family, in particular, for the universal family of polynomial line fields of degree $\leqslant n$ (see 2.1.1, 2.1.2) over the projective space of coefficients of the polynomials.

Still the structure of *all* singular locus may be quite complicated even if all essential singularities from this locus are elementary. To describe phenomena occurring after desingularization, we introduce the notion of a *singular perturbation* of a family of line fields.

6.1. Singular perturbations. Let α be a family of line fields on the fibration (M, B, π), Ω the corresponding singular locus and Ω_{ess} the set of essential singularities.

DEFINITION. A family α is said to exhibit a *regular perturbation* of a singularity $q \in \Omega \subset M$, if the sets Ω and Ω_{ess} locally coincide near that point:

$$(\Omega, q) = (\Omega_{\text{ess}}, q).$$

Otherwise we say that the family α is a *singular perturbation* of the equilibrium q. The corresponding value $\varepsilon = \pi(q)$ of the parameters will be referred to as the *singular parameter value*.

An analytic family exhibits singular perturbation at points that form a set of uniform codimension 1 on fibers. Generically most of them *are not* essential singularities.

Below follow three examples of line families, the first two defined on the Cartesian fibration $\mathbb{C}^2 \times \mathbb{C} \to \mathbb{C}$, $\pi\colon (x, y, \varepsilon) \mapsto \varepsilon$.

EXAMPLES. 1. (See the example in 2.1.3) The family $(xy + \varepsilon)\,\partial/\partial x + y^2\,\partial/\partial y$ exhibits the singular perturbation of points on the line $\{y = \varepsilon = 0\}$. In this case $\Omega_{\text{ess}} = \{(0, 0, 0)\}$.

2. The family $x\,\partial/\partial x + \varepsilon y\,\partial/\partial y$ exhibits a singular perturbation of points on the line $\{x = \varepsilon = 0\}$. The essential singular locus for this family is $\Omega_{\text{ess}} = \{(0, 0, \varepsilon) \mid \varepsilon \neq 0\}$.

3. The family $x\,\partial/\partial x + (1 - \varepsilon y^2)\,\partial/\partial y$ extended on the fibration $\mathbb{C}P^2 \times \mathbb{C} \to \mathbb{C}$ exhibits the singular perturbation of one of the points at infinity: in the projective coordinate system $(x, y) \mapsto (x : y : 1)$ this point is $(0 : 1 : 0)$. Indeed, in the new chart $(x, z = 1/y)$ the family is spanned by $z\,\partial/\partial x - (z - \varepsilon)\,\partial/\partial z$ and the essential set is $\{x = 0, z = \varepsilon\}$.

The last example shows that after perturbation of a polynomial vector field by terms of higher degree, singular perturbations occur at infinity. This phenomena is generic in the polynomial category. In the next section we discuss this phenomenon in more details.

6.2. Singular perturbations in the polynomial category. In this section we consider the universal polynomial family α_n of line fields on $\mathbb{C}P^2$ of degree $\leqslant n$ over the projective space of coefficients $B_n = \mathbb{C}P^{(n+1)(n+2)}$. Each such field is spanned by a vector field $P(x, y)\,\partial/\partial x + Q(x, y)\,\partial/\partial y$ and extended onto the projective plane $\mathbb{C}P^2$.

Recall that for a generic pair of polynomials P, Q the corresponding vector field (without parameters) extends as a line field with isolated singularities onto the projective plane in such a way that the infinite line $\mathbb{C}P^1 \subset \mathbb{C}P^2$ is invariant (tangent to the line field). This case is referred to as *nondiacritical*. For exceptional values of coefficients the vector field extends as a line field transversal to the infinite line at a generic point of the latter. This is the *diacritical* case. The necessary and sufficient condition for that case is formulated in terms of principal homogeneous parts P_n and Q_n of the polynomials P, Q: the diacritical case occurs when $y\,P_n(x\,,\,y) - x\,Q_n(x\,,\,y) \equiv 0$.

The same terminology applies to singular points: if after a simple blow-up the pasted in projective line is transversal (resp., tangent) to the blown-up field, then the case is called diacritical (resp., nondiacritical), and a similar algebraic condition allows us to distinguish between the two cases.

Proposition 6.1. *For the universal polynomial family* α_n *the singular parameter values constitute an analytic subset of codimension* $(n+1)$ *in* B_n.

Proof. Singular parameter values correspond to one of the two cases,

(1) vector field with nonisolated singularities, or
(2) diacritical infinite line.

The second case imposes $n+2$ independent linear restrictions to coefficients of the polynomials P and Q, which after factorization determine a subset of B_n of codimension $n+1$. One can easily see that pairs of polynomials with a common factor constitute an algebraic subset of codimension $> n+1$ in B_n. □

Thus we see that with n growing, both the dimension and the codimension of the set of singular parameter values in B_n grows to infinity. This remark implies that the universal family always exhibits singular perturbations, while *generic* k-parameter families of polynomials of degree $\leqslant n$ for sufficiently large n do not do that (k is assumed fixed).

6.3. Singular perturbations arising after blowing-ups. Blowing-up of analytic sections is the main tool of desingularization. We show that, unlike the other tool, namely, unfoldings of bases, blow-ups can create singular perturbations, even if they did not occur in the original family.

To illustrate this claim, consider a family of line fields whose isolated singularity belongs to the constant section $\{x = y = 0\}$ of the trivial fibration over the n-dimensional parameter $\varepsilon = (\varepsilon_1\,,\,\dots\,,\,\varepsilon_n)$. This family in the adapted coordinates is spanned by the vector field

$$\begin{cases} \dot{x} = f_n(x\,,\,y\,;\,\varepsilon) + o((|x| + |y|)^n)\,, \\ \dot{y} = g_n(x\,,\,y\,;\,\varepsilon) + o((|x| + |y|)^n)\,, \end{cases}$$

where f_n and g_n are homogeneous polynomials of degree n depending on the parameter, which do not vanish identically at the same time. Let $r_n = y\,f_n - x\,g_n$. The following is verified by the straightforward computation.

Proposition 6.2. *After the simple blow-up of the zero section* $\{x = y = 0\}$ *the pasted in projective line on the fiber* $\varepsilon = 0$ *consists of singularly perturbed points if and only if*:

(1) $r_{n+1}(\cdot\,,\,\cdot\,;\,\cdot) = 0$, $f_n(\cdot\,,\,\cdot\,;\,0) = g_n(\cdot\,,\,\cdot\,;\,0) = 0$, *or*

(2) $r_{n+1}(\cdot\,,\,\cdot\,;\,\cdot) \neq 0\,,\ r_{n+1}(\cdot\,,\,\cdot\,;\,0) = 0\,.$ □

Thus we see that after the blow-up the family exhibits a singular perturbation in two cases: either the order of the singularity (the degree of the principal terms) is discontinuous at the given point $\varepsilon = 0$ as an integer-valued function of the parameter ε, or the diacriticity occurs for $\varepsilon = 0$ and not for generic small ε.

In particular, if a generic point of the blown-up section is nonsingular, while at $\varepsilon = 0$ a singularity occurs, then one necessary falls into the first category, since the order is increased from 0 to 1. Hence singular perturbations appear after the very first blowing of such section.

6.4. Connections with the Hilbert 16th problem. In the papers [6], [12] the program of attacking the existential form of the Hilbert 16th problem was suggested (for the detailed survey of this problem see the introductory paper in this volume). This program consisted of two steps.

- Prove that in a family of vector fields with only elementary singularities a bounded number of limit cycles can be born from a limit periodic set.
- Suggest a procedure of parameter blow-up that would reduce an arbitrary family of vector fields to another family having only elementary singular points.

The first step of the program was recently partially implemented: in the first paper from this volume it is proved that a *generic smooth finite-parameter* family of vector fields has a finite cyclicity in the above sense.

The main goal of this paper was to establish a procedure required for the second step. But from what was written above it should be clear that if one uses the parameter desingularization in the form described in the Main Theorem, then it is necessary to add the third step.

- Investigate limit cycles born after singular perturbations in analytic families of line fields on two-dimensional surfaces.

This step is fairly difficult. As a simple example of the perturbation giving rise to a limit cycle we mention the Van der Pol system describing relaxation oscillations,

$$\begin{cases} \dot{x} = \varepsilon y\,, \\ \dot{y} = x - y(y^2 - 1). \end{cases}$$

In any case, we do not attempt to go deep into the subject in this paper.

Appendix A. Uniform bound for the multiplicity of singularities in analytic families

A.1. Generalized multiplicity. Let $\mathbf{F}\colon (\mathbb{C}^2\,,\,p) \to (\mathbb{C}^2\,,\,0)$ be an analytic germ. We define its *generalized geometric multiplicity* $\overline{\mu}(\mathbf{F})$ as follows. Take a representative F of the germ $\mathbf{F}$ and choose two systems of shrinking neighborhoods $\{U_i\}$, $\{V_j\}$,

$$U_1 \supset U_2 \supset \cdots \supset U_i \supset \cdots\,, \qquad \bigcap_i U_i = \{p\}\,,$$

and the same for the family V_j centered around the point $0 \in \mathbb{C}^2$, and let $N(U_i, V_j)$ be the maximal number of *isolated* preimages of a point $q \in V_j$ by F in U_i:

$$N(U_i, V_j) = \sup_{q \in V_j} \#\{x \in U_i \mid F(x) = q, \quad x \text{ is an isolated preimage of } q\}.$$

Then the number N is nonincreasing with i, j growing, so the limit

$$\overline{\mu}(\mathbf{F}) = \lim_{i, j \to \infty} N(U_i, V_j) = \inf_{i, j \in \mathbb{N}} N(U_i, V_j)$$

is well defined and independent of the choice of the families U_i, V_j.

REMARK. If p is an isolated preimage of the point $0 \in \mathbb{C}^2$, then this definition coincides with the usual multiplicity understood as the index of the vector field F on a small sphere centered at p:

$$\mu(\mathbf{f}, \mathbf{g}) < \infty \implies \overline{\mu}(\mathbf{f}, \mathbf{g}) = \mu(\mathbf{f}, \mathbf{g}).$$

But in the case when $\dim F^{-1}(0) > 0$ the value $\overline{\mu}(F)$ is finite unlike the value $\mu(F)$ which (according to the definitions chosen) is either undefined or $+\infty$.

The notion of the generalized geometric multiplicity is close to the Khovanskiĭ and Gabrielov numbers introduced in [18] for differentiable maps in the real category.

If $F = (f, g)$ is the coordinate representation of F, then we use the notation $\overline{\mu}(f, g)$. Usually the point p (the source of the germ) will be at the origin in $\mathbb{C}^2$.

Recall that a *uniformization* of the germ of an irreducible curve $\gamma \subset (\mathbb{C}^n, 0)$ is a proper analytic map $\lambda: (\mathbb{C}, 0) \to (\mathbb{C}^n, 0)$ whose image is γ and which is one-to-one at a generic point of γ. Any germ of an analytic irreducible curve admits uniformization [5].

LEMMA A.1 (see [5, §10.2, §11.5] or [8, Appendix 1]). *For the germ of a map* $\mathbf{F} = (\mathbf{f}, \mathbf{g})$ *with relatively prime analytic germs* $\mathbf{f}, \mathbf{g}: (\mathbb{C}^2, p) \to (\mathbb{C}, 0)$ *the following assertions hold*:

(1) *There exists a representative* $F = (f, g)$ *of the germ* $\mathbf{F}$ *which is a proper map on its domain and range,* $U \xrightarrow{F} V$.

(2) *The set* $D = D_F$ *of critical values of* F *is an analytic subset in* $(\mathbb{C}^2, 0)$ *of dimension* $\leqslant 1$.

(3) *For any regular value* $q \in V \setminus D$ *the full preimage* $F^{-1}(q) \cap U$ *consists of the same number of points equal to the multiplicity* $\mu(F) = \mu(f, g)$ *of the germ* F.

(4) *Let* $\mathbf{f} = \mathbf{f}_1 \cdots \mathbf{f}_k$ *be the irreducible factorization of the germ* $\mathbf{f}$ *in the ring* $\mathbb{C}_p\{x, y\}$ *of analytic germs at* p, $\lambda_i: (\mathbb{C}, 0) \to (\mathbb{C}^2, p)$ *a uniformization of the irreducible set* $\{f_i = 0\}$ *and* $\mathbf{g}_i = \mathbf{g} \circ \lambda_i: (\mathbb{C}, 0) \to (\mathbb{C}, 0)$. *Denote by* $\operatorname{ord}_0 \mathbf{g}_i$ *the order of zero at the origin of the germ* $\mathbf{g}_i$. *Then the multiplicity* $\mu(F)$ *admits the representation*

$$\mu(\mathbf{F}) = \mu(\mathbf{f}, \mathbf{g}) = \sum_{i=1}^{k} \operatorname{ord}_0 \mathbf{g}_i.$$

COROLLARY A.2. *If both pairs of germs* $(\mathbf{f}, \mathbf{h})$ *and* $(\mathbf{g}, \mathbf{h})$ *are relatively prime, then*

$$\mu(\mathbf{fg}, \mathbf{h}) = \mu(\mathbf{f}, \mathbf{h}) + \mu(\mathbf{g}, \mathbf{h}).$$

From now on for simplicity we disregard the difference in the notation and terminology between germs and their representatives.

LEMMA A.3. *For a pair of relatively prime germs* (f, g) *and any analytic germ* $h \not\equiv 0$ *one has*

$$\mu(f, g) = \overline{\mu}(f, g) \leqslant \overline{\mu}(fh, gh).$$

PROOF. 1. If $h(0, 0) \neq 0$, then $\overline{\mu}(fh, gh) = \overline{\mu}(f, g)$, as it trivially follows from assertion (4) of Lemma A.1.

2. The critical locus of the map $F = (f, g)\colon (\mathbb{C}^2, 0) \to \mathbb{C}^2$ is the set of zeros of the function

$$J_{f,g} = \frac{\partial f}{\partial x} \cdot \frac{\partial g}{\partial y} - \frac{\partial f}{\partial y} \cdot \frac{\partial g}{\partial x}.$$

Denote the set of *critical values* of F by $D_{f,g}$.

Then in this notation the critical locus for the map $h \cdot F = (hf, hg)$ is the union of $\{h = 0\}$ and the zero set for the function

$$J_{fh,gh} = h\, J_{f,g} + f\, J_{g,h} + g\, J_{f,h}$$

and hence has dimension $\leqslant 1$ in $U = (\mathbb{C}^2, 0)$. Indeed, otherwise the zero level curves of functions fh and gh would be tangent at a generic point, hence coinciding in U. But then the same would happen for the zero level curves of f and g, which contradicts the assumptions of the lemma.

Thus the set of critical values of hF also has dimension $\leqslant 1$ in $V = (\mathbb{C}^2, 0)$, therefore it is nowhere dense in V.

Next, one can easily see that for almost all $\beta \in \mathbb{C}$ except at most a finite number of values, the functions $f - \beta g$ and h are relatively prime. Indeed, h is a product of a finite number of prime factors, so if for two different values of β the functions $f - \beta g$ are divisible by the same prime factor, then f and g are not relatively prime.

Those arguments show that in an arbitrarily small neighborhood V of the origin there are regular values of F of the form $(\varepsilon\beta, \varepsilon) \in V \setminus D_{fh,gh}$ with $\varepsilon \neq 0$, $\beta \neq 0$ such that $f - \beta g$ and h are relatively prime.

The set $P = (hF)^{-1}(\beta\varepsilon, \varepsilon) \subset U$ coincides with the preimage of the point $(0, \varepsilon)$ by the map $(f - \beta g, gh)$. The pair of germs $(f - \beta g, gh)$ is relatively prime due to the choice of β. By virtue of Lemma A.1 if $(0, \varepsilon)$ is a regular value for $(f - \beta g, gh)$, then the number of points in P coincides with the multiplicity of the latter map. But by Corollary A.2,

$$\mu(f - \beta g, gh) = \mu(f - \beta g, g) + \mu(f - \beta g, h) > \mu(f, g).$$

To finish the proof, it remains to show that for any point $p \in P$ one has $J_{f-\beta g, gh}(p) \neq 0$, in other words, that the point p is a regular point for the map $(f-\beta g, gh)$. But due to the choice of β and ε,

$$f(p)=-\beta g(p), \quad (gh)(p)=\varepsilon\neq 0, \quad (hF)(p)\notin D_{fh,gh},$$
$$J_{f-\beta g, gh}(p)=\left(J_{f,g}-\beta J_{g,h}+gJ_{h,f}\right)(p)=\frac{J_{fh,gh}(p)}{h(p)}\neq 0.$$

Thus in fact we proved that the inequality stated in the formulation of the lemma, is strict when $h(0,0)=0$. □

COROLLARY A.4. *An essential singularity of a line field spanned by the vector field* $f\,\partial/\partial x+g\,\partial/\partial y$ *has the multiplicity* $\leqslant \overline{\mu}(f, g)$.

A.2. Proof of Lemma 4.5. The basic tool in the proof is the Gabrielov–Teissier finiteness principle for analytic maps. Let $K\subset\mathbb{R}^{n+k}$ be a compact set (usually a cube) and $A\subset K$ an analytic set. Denote by $\mathrm{pr}\colon\mathbb{R}^{n+k}\to\mathbb{R}^k$ the standard projection.

GABRIELOV THEOREM [2]. *The number of connected components of the preimage* $\mathrm{pr}^{-1}(q)\cap A$ *is uniformly bounded over* $q\in\mathrm{pr}(K)$.

PROOF OF LEMMA 4.5. Consider a precompact set U in the total space of the fibration (M, B, π) endowed with an adapted coordinate system (x, y, ε). Let

$$f(x, y, \varepsilon)\,\partial/\partial x+g(x, y, \varepsilon)\,\partial/\partial y+0\cdot\partial/\partial\varepsilon$$

be a family of vector fields spanning the line field α. We will prove that the generalized geometric multiplicity of the pair (f, g) is uniformly bounded over all parameter values ε and all singular points of the field. Then, by Corollary A.4, this would imply the existence of a uniform upper bound for the multiplicity of all essential singularities of the line field.

Consider the set $A=\{z=f(x, y, \varepsilon),\ w=g(x, y, \varepsilon)\}\subset U\times\mathbb{C}^2_{z,w}\subset\mathbb{C}^{n+4}\simeq\mathbb{R}^{2n+8}$. Since f, g are analytic, then A is also analytic and projecting this set onto the space $\mathbb{C}^{n+2}_{z,w,\varepsilon}\simeq\mathbb{R}^{2n+4}$ satisfies the assumptions of the Gabrielov theorem with respect to the projection $\mathrm{pr}\colon\mathbb{R}^{2n+8}\to\mathbb{R}^{2n+4}$ which takes a point $(\mathrm{Re}\,x, \mathrm{Re}\,y, \mathrm{Re}\,\varepsilon, \mathrm{Re}\,z, \mathrm{Re}\,w, \mathrm{Im}\,x, \dots, \mathrm{Im}\,w)$ into the point $(\mathrm{Re}\,\varepsilon, \mathrm{Re}\,z, \mathrm{Re}\,w, \mathrm{Im}\,\varepsilon, \mathrm{Im}\,z, \mathrm{Im}\,w)$.

The number of isolated preimages of a point (z, w) is uniformly bounded over all admissible ε, hence the conclusion on the boundedness of the multiplicity. □

Appendix B. Rectification lemma in the complex analytic case

We prove here the basic Lemma R for any dimension of the base and fiber of a fibration. The exposition is independent of the main body of the paper. Recall that we are studying an analytic subset $\Omega\subseteq M$ of the total space of an analytic fibration (M, B, π) with an arbitrary dimension of the fibers.

In what follows we denote by $\mathrm{reg}\,\Omega$ the subset of regular points of an analytic set Ω. The complement is the set $\mathrm{sng}\,\Omega=\Omega\setminus\mathrm{reg}\,\Omega$.

B.1. Proof of Proposition 2.1. We need to prove that the subset of points at which the set Ω is not π-isolated, is analytic and its image is less than ν-dimensional, $\nu = \dim \Omega$.

We represent Ω as the finite union of analytic manifolds of different dimensions (strata),

$$\Omega_0 = \operatorname{reg} \Omega, \quad \Omega_1 = \operatorname{reg}(\operatorname{sng} \Omega), \quad \Omega_2 = \operatorname{reg}(\operatorname{sng}(\operatorname{sng} \Omega)), \quad \ldots$$

such that the closure of each Ω_i is analytic and $\Omega_i = \overline{\Omega}_i \setminus \overline{\Omega}_{i+1}$. Let ω_i be the set of π-nonisolated points of Ω_i. Then the closure $\overline{\omega}_i$ is an analytic subset of Ω. Let us show that the union ω of all $\overline{\omega}_i$ coincides with the set $\Omega \setminus \Omega_{\text{is}}$ of π-nonisolated points of Ω.

Let $p_0 \in \Omega$ be a point at which Ω is not π-isolated, $\pi(p_0) = \varepsilon_0$ being its projection on the base B. Then in any neighborhood of p_0 the corresponding fiber M_{ε_0} will intersect Ω by a nonisolated set whose generic point belongs to some Ω_i. Evidently, this implies that such generic points belong to ω_i and hence $p_0 \in \overline{\omega}_i$.

It remains to show that at all points of ω the set Ω is not π-isolated. This follows from a general property of analytic sets [5]: for any point $p \in \Omega_{\text{is}}$ all sufficiently close points are also π-isolated for Ω, hence a generic point $p \in \operatorname{reg} \Omega$ falls into this class.

Now we need to prove the inequality for the dimension. Without loss of generality one may assume Ω to be irreducible. If $\dim \omega < \dim \Omega$, then the assertion is evident. Otherwise, if $\dim \omega = \dim \Omega$, then $\omega = \overline{\omega}_0 = \Omega$, since Ω is irreducible, hence $\pi|_{\Omega_1}$ has not the full rank and

$$\dim \pi(\Omega) = \dim \overline{\pi(\omega_0)} < \dim \Omega.$$

This proves the assertion of the lemma. $\square$

REMARK AND NOTATION. For any analytic Ω denote by $\operatorname{br}_\pi \Omega$ the *ramification set*, that is the union $\operatorname{sng} \Omega \cup (\Omega \setminus \Omega_{\text{is}})$. The assertion proved above means that

$$\dim \pi(\operatorname{br}_\pi \Omega) < \dim \Omega.$$

REMARK. The above proof shows also that if the set $\Omega \setminus \Omega_{\text{is}}$ is nowhere dense in Ω, then $\dim \pi(\Omega) = \dim \Omega$.

B.2. Proper preimages of analytic sets and covering multiplicity. Let $\Omega \subset M$ be an analytic subset of pure dimension n in the total space of an analytic connected fibration (M, B, π) over an n-dimensional base B. Then generically the intersection of Ω with a fiber of the fibration is zero-dimensional (eventually empty), and for apparent reasons the number of points in this intersection is the same for almost all fibers. We call this number the *covering multiplicity* of the projection $\pi|_\Omega$. Below we always assume that $\dim \Omega = \dim B$ unless explicitly stated otherwise.

It may happen that the set Ω contains one or more analytic sections $S_i \colon B \to M$, $S_i(B) \subseteq \Omega$. The *reduced covering multiplicity* is the difference between the covering multiplicity and the maximal number of different analytic sections contained in Ω. The following proposition is standard.

PROPOSITION B.2. *If $U \subset B$ is a simply connected subset of the base, which does not intersect the image of the ramification set $\pi(\mathrm{br}_\pi \Omega)$, then the reduced covering multiplicity of the set $\Omega \cap \pi^{-1}(U)$ over U is zero, which means that this set is the union of analytic sections.* $\square$

Now let $(H, \rho)\colon \widetilde{M} \times \widetilde{B} \to M \times B$ be an unfolding of the base of the fibration. Then the closed set

$$\widetilde{\Omega} = \overline{H^{-1}(\Omega_{\mathrm{is}})} \subset \widetilde{M}$$

is well defined. For a moment we will call it a *proper unfolding* of the set Ω.

PROPOSITION B.3. *Let $\widetilde{\Omega}$ be a proper unfolding of an analytic set Ω of uniform dimension. Then this unfolding is an analytic set, and*

(1) *if* $\dim \Omega < \dim B$ *then* $\dim \widetilde{\Omega} < \dim B$;
(2) *if* $\dim \Omega = \dim B$, *then $\widetilde{\Omega}$ is of the same uniform dimension $n = \dim B$, and both the total and the reduced covering multiplicities for $\widetilde{\pi}|_{\widetilde{\Omega}}$ do not exceed the corresponding values for $\pi|_\Omega$.*

PROOF. The proper unfolding is the closure of the difference of two analytic sets, hence analytic:

$$\widetilde{\Omega} = \overline{H^{-1}(\Omega_{\mathrm{is}})} = \overline{H^{-1}(\Omega) \setminus H^{-1}(\Omega \setminus \Omega_{\mathrm{is}})}.$$

A generic point of $\widetilde{\Omega}$ is π-isolated, and by the remark at the end of B.1, $\dim \widetilde{\Omega} = \dim \widetilde{\pi}(\widetilde{\Omega})$. If $\dim \Omega < \dim B$, then the set $\pi(\Omega)$ is nowhere dense in B so the same is true for the set $\widetilde{\pi}(\widetilde{\Omega})$ in $\widetilde{B}$, and $\dim \widetilde{\Omega} < \dim B$.

Next, by definition of the unfolding of the base, the map $\rho\colon \widetilde{B} \to B$ is biholomorphic near a generic point $\widetilde{\varepsilon} \in \widetilde{B}$ and in a neighborhood of the corresponding fiber $\widetilde{M}_{\widetilde{\varepsilon}}$. The remaining part of the Proposition trivially follows from that observation. $\square$

The notion of a proper unfolding allows us to reformulate Lemma R as the assertion that *after an appropriate unfolding of the base, the (induced) proper unfolding of an analytic set Ω in each connected component of the new fibration belongs to the union of several analytic sections.*

B.3. Examples. Before proceeding with the formal proof of Lemma R, we give several examples of effective construction of unfoldings solving this problem.

EXAMPLE 1. Let

$$M = \mathbb{C}^2 \times \mathbb{C}^1, \quad B = \mathbb{C}^1, \quad \pi(x, y, \varepsilon) = \varepsilon; \qquad \Omega = \{(x, y, \varepsilon) \mid x^2 = \varepsilon, y = 0\}.$$

Consider the unfolding of the base

$$\rho\colon \mathbb{C} \to \mathbb{C}, \qquad \widetilde{\varepsilon} \overset{\rho}{\mapsto} \widetilde{\varepsilon}^2.$$

Then the pullback fibration is

$$\widetilde{M} = \mathbb{C}^2 \times \mathbb{C}^1, \quad \widetilde{B} = \mathbb{C}^1, \quad \widetilde{\pi}(x, y, \widetilde{\varepsilon}) = \widetilde{\varepsilon},$$

and the full preimage of the set Ω splits into two analytic sections:

$$H^{-1}(\Omega) = \{(x, y, \varepsilon) \mid x = \varepsilon, y = 0\} \cup \{(x, y, \varepsilon) \mid x = -\varepsilon, y = 0\}.$$

EXAMPLE 2 (proof of the general statement in the particular case of algebraic curve B). Let

$$M = \mathbb{C}P^2_p \times \mathbb{C}P^1_\varepsilon, \quad B = \mathbb{C}P^1_\varepsilon, \quad \pi(p, \varepsilon) = \varepsilon,$$

and consider a one-dimensional irreducible set (an algebraic curve by virtue of general results) $\Omega \subset \mathbb{C}P^2$, which is not a fiber of the fibration. Then Ω admits an algebraic uniformization $\varphi\colon \mathbb{C}P^1 \to \mathbb{C}P^2$, $\varphi(\mathbb{C}P^1) = \Omega$, of rank 1 at a generic point of $\mathbb{C}P^1$.

The map $\rho = \pi \circ \varphi$ yields an unfolding of the base $B = \mathbb{C}P^1$. By Proposition 1.2, there exists at least one section of the pullback fibration, entirely belonging to $H^{-1}(\Omega)$.

If Ω is reducible and the covering multiplicity k is positive, then one may choose an irreducible component which is not a section and apply the above construction. Evidently, this would increase the number of different analytic sections entirely belonging to the proper unfolding, thus decreasing the reduced multiplicity (since the total multiplicity remains the same). Therefore, after no more than k steps the process will terminate, and we obtain the required unfolding.

EXAMPLE 3. Looking at the previous example, one might think that after one step of the construction an irreducible set is transformed into the union of smooth sections. However, in general this is not the case. For example, if the original set Ω has a pair of points p_2, p_3 with $\pi(p_2) = \pi(p_3) = \varepsilon$, and the germs $\pi|_\Omega\colon (\Omega, p_k) \to (B, \varepsilon)$ for $k = 2, 3$ have the multiplicities 2 and 3 respectively, then after the unfolding described in Example 2, one gets two new points $\widetilde{p}_1, \widetilde{p}_2$ in the total space of the new fibration, such that $\widetilde{\pi}(\widetilde{p}_1) \neq \widetilde{\pi}(\widetilde{p}_2)$ and over which the preimage $H^{-1}(\Omega)$ would have branching of orders 2 and 3 respectively. However, this preimage is reducible and contains at least one section.

REMARK. The example in 3.2 shows that blowing-ups of the base are very important for resolution of singularities in families. But in general unfoldings of the base increase the dimension of proper preimages (cf. with the statement (1), Proposition B.3) of analytic sets over blown-up points of the base.

B.4. Hironaka theorem. In general, one cannot hope for the uniformization so simple as in the case of algebraic curves. Still, it is sufficient to use the following very powerful tool from the local theory of analytic sets.

HIRONAKA THEOREM (on desingularization of analytic sets) [9], [14]. *Let $\Omega \subset M$ be an analytic subset of uniform dimension. Then for any point $p \in \Omega$ one may find a neighborhood $U \ni p$, an analytic nonsingular manifold $\widetilde{\Omega}$ and a proper map $\varphi\colon \widetilde{\Omega} \to M$ such that $\varphi(\widetilde{\Omega}) \subset \Omega \cap U$ and φ is biholomorphic at a generic point of $\widetilde{\Omega}$.*

B.5. The case of pure (uniform) dimension n. In this section we show that Lemma R holds for an analytic set Ω (eventually, reducible) of the same uniform dimension as that of the base B.

LEMMA B.4.

$$\dim \Omega = \dim B \implies \text{Lemma R holds for } \Omega.$$

PROOF. 1. Without loss of generality we may assume that the fibration is connected and the set Ω does not have irreducible components without π-isolated points. Then $\dim \operatorname{br}_\pi \Omega < n = \dim B$. Denote by k the reduced covering multiplicity of the projection $\pi|_\Omega$. If $k = 0$, then the assertion of the lemma trivially holds.

2. We construct an unfolding of the base in such a way that the reduced covering multiplicity will strictly decrease in any connected component of the pullback fibration. Denote by Ω' the set of points $p \in M$ such that in the restriction of the fibration to any small neighborhood of p the set Ω cannot be represented as a union of sections. Obviously, $\Omega' \subseteq \operatorname{br}_\pi \Omega$ so the exceptional set $M \setminus \Omega'$ is open and the set Ω' is closed. According to Proposition B.2 the reduced covering multiplicity of $\pi|_\Omega$ is zero over any connected subset $U \subseteq B$ disjoint with Ω'. Thus we need to construct the unfolding only for a small neighborhood of the set $\pi(\Omega')$.

Choose a locally finite covering of $\pi(\Omega')$ by open neighborhoods for which the assertion of Hironaka theorem holds: thus there exists a family of uniformizing maps $\varphi_i : \widetilde{\Omega}_i \to M$, $\varphi_i(\widetilde{\Omega}_i) \subseteq \Omega$. Shrinking, if necessary, the manifolds $\widetilde{\Omega}_i$, we can assume that the images $\varphi(\widetilde{\Omega}_i)$ coincide with irreducible components of the intersection of Ω with the corresponding neighborhood of the covering. We may also suppose that neither of these sets $\varphi(\widetilde{\Omega}_i)$ is a section of the restricted fibration.

3. Since $\dim \pi(\Omega') \leqslant \dim \operatorname{br}_\pi \Omega < n$, for any i the map $\pi \circ \varphi_i : \widetilde{\Omega}_i \to B$ is locally biholomorphic at a generic point. Therefore the sets $U_i = (\pi \circ \varphi_i)(\widetilde{\Omega}_i)$ are open and together constitute a covering of the set $\pi(\Omega')$. Choose from that covering a locally finite covering (to simplify notation, assume that it was such from the very beginning) and construct the disjoint sum $\widetilde{B}_1 = \coprod_i \widetilde{\Omega}_i$ together with its projection onto the base $\rho_1 = \coprod \pi \circ \varphi_i$. This is evidently an unfolding of the neighborhood of $\pi(\Omega')$.

For any i the set $\varphi_i(\widetilde{\Omega}_i)$ is not a section, but by Proposition 1.2, the pullback of this set under the unfolding φ_i would contain at least one section. So in any connected component of $\rho^* M$ the reduced covering multiplicity was in fact decreased.

4. Inductive iteration of the steps 2–3 proves the Lemma in the case of uniform dimension n. $\square$

B.6. Proof of Lemma R in the general case. We treat the general case using induction in the dimension of the base $n = \dim B$. If $\dim B = 0$, then the base is a discrete set and there is nothing to prove. The induction step is justified by the following assertion.

LEMMA B.5. *If Lemma* R *holds for any subset of any fibration with* $(n-1)$*-dimensional base, then it holds for subsets of all fibrations with* n*-dimensional bases.*

PROOF. Consider the collection of analytic subsets Ω_i whose π-isolated points are to be placed on sections in the total space of a fibration (M, B, π) with n-dimensional base B. Assume that the conditions of Lemma R are satisfied: $\dim \Omega_i \leqslant \dim B$ and the restriction $\pi|_{\Omega_i}$ is proper for any i.

Proposition B.3 and Lemma B.4 reduce the considerations to the situation when $\dim \Omega_i < n$; indeed, all n-dimensional subsets can be unfolded as required by virtue of Lemma B.4, while the other sets should remain no more than $(n-1)$-dimensional by Proposition B.3.

Since the restriction of π to Ω_i is proper, the image cannot be more than $(n-1)$-dimensional. Denote by N the union of all these images. Since there are no points of Ω_i over $B \setminus N$, we may consider only a small neighborhood of the set N and construct an unfolding of the base only for that neighborhood.

By the theorem on existence of proper projections from [5, Lemma 1 in §3.4], in a small neighborhood of any point $\varepsilon \in N$ there exists a projection onto an $(n-1)$-dimensional hyperplane whose restriction to N is proper. Thus we can choose a locally finite covering of N by open neighborhoods with this property and work with only one such neighborhood.

The base locally has the product structure $B = B' \times B''$ with $\dim B' = n-1$, $\dim B'' = 1$ and the projection $\xi\colon B' \times B'' \to B'$ restricted to N is proper. Consider an auxiliary fibration with the base B', the same (locally) total space M and the projection $\pi' = \xi \circ \pi$.

One can easily see that the sets Ω_i (or more precisely their parts over the neighborhood B') satisfy the conditions of Lemma R for the newly constructed fibration. By the induction assumption, there exists an unfolding $\rho'\colon \widetilde{B}' \to B'$ of the base B' whose pullback places preimages of the sets Ω_i on the union of sections $\widetilde{S}'_j$. Define the unfolding of the base B as

$$\rho = \rho' \times \mathrm{id}\colon (\widetilde{\varepsilon}', \varepsilon'') \mapsto (\rho'(\widetilde{\varepsilon}'), \varepsilon'').$$

In fact, the total spaces of both the main and the auxiliary fibrations after the unfoldings coincide: one can easily see that implementing the pullback construction from 1.1.3, in both cases we obtain

$$\widetilde{M} = \{(p, \widetilde{\varepsilon}', \varepsilon'') \mid \xi \circ \pi(p) = \rho'(\widetilde{\varepsilon}')\} \subset E = M \times B' \times B''.$$

It is sufficient to prove that any section of the unfolded auxiliary fibration belongs to a section of the unfolded principal fibration.

Denote by $P\colon E \to M$ the Cartesian projection. Then for any section $\widetilde{S}'\colon \widetilde{B}' \to \widetilde{M} \subset E$ of the unfolded auxiliary fibration, one may define the section

$$\widetilde{S}\colon \widetilde{B} \to \widetilde{M}, \qquad (\widetilde{\varepsilon}', \varepsilon'') \mapsto (P \circ \widetilde{S}'(\widetilde{\varepsilon}'), \widetilde{\varepsilon}', \varepsilon'').$$

Evidently, $\widetilde{S}'(\widetilde{B}') \subseteq \widetilde{S}(\widetilde{B}' \times B'')$. This finishes the proof of Lemma B.5. At the same time we proved the principal Lemma R in all generality. $\square$

Appendix C. Constructions in the real case

In this appendix we discuss the real counterparts of the principal tools of the desingularization. Since the basic constructions remain essentially the same, we mostly stress the differences between the real and the complex case and make necessary modifications.

C.1. Properties of real subanalytic sets. Here we use the category of closed real subanalytic sets instead of that of complex analytic sets. In a sense, the former constitute a "real analog" of the latter, allowing for the reproduction of the main constructions. Considering subanalytic sets we follow the papers [2], [10], [14] discussing equivalent properties. The paper [2] of Gabrielov appeared prior to the work [14] of Bierstone and Milman, the latter being more clear and understandable. However, we prefer to use formulations of Hironaka [10] because they are more close to the context of this paper.

DEFINITION [10]. A subset θ of a real analytic manifold B is called (real) *subanalytic*, if near each $q \in \theta$ it can be represented as the difference of two embedded analytic sets: there exists a neighborhood $U \subset B$ containing q, and two real analytic manifolds W_1, W_2 with two proper real analytic mappings $\lambda_i\colon W_i \to U$ such that

$$\theta \cap U = \lambda_1(W_1) \setminus \lambda_2(W_2).$$

REMARKS. 1. In this definition the manifolds W_i should not necessarily be connected; they must be only countably compact as all other topological spaces occurring in this paper. Equivalent definitions are given in the papers [2], [10], [14].

2. Any analytic set is closed subanalytic. From the definition it is clear that a proper projection of closed subanalytic set is also closed subanalytic—the main property that we use in this paper.

We also use the following fundamental properties of subanalytic sets. The notation $\dim_H \theta$ is used for the Hausdorff dimension of the set θ.

C.1.1. *Stratification* [2]. Any subanalytic set can be represented as a countable union of local analytic sets (i.e. the sets that are analytic near each their point). Thus for any real subanalytic set Ω the set $\operatorname{sng}\Omega$ of points near which Ω is not a regular submanifold, is a set of Hausdorff dimension less than that of Ω: $\dim_H \operatorname{sng}\Omega < \dim_H \Omega$.

C.1.2. *Proper projections of subanalytic sets* [2]. Any closed subanalytic set of Hausdorff dimension $< n$ locally near each point admits a proper projection onto some $(n-1)$-dimensional hyperplane.

C.1.3. *Closure of the difference* [2], [10], [14]. The category of real subanalytic sets is closed under taking unions, intersections, preimages under analytic mappings, differences and closures. Thus for any pair Ω_1, Ω_2 of real subanalytic sets the set $\overline{\Omega_1 \setminus \Omega_2}$ is closed subanalytic.

C.1.4. *Desingularization of subanalytic sets.* In [10] Hironaka formulated and proved the the following theorem.

THEOREM (on representation of subanalytic sets, see [2], [10], [14]). *For a subanalytic set $\theta \subseteq B$ of integral Hausdorff dimension $k \in \mathbb{Z}$ there exists a real analytic manifold D, $\dim D = k$, and a proper analytic map $\lambda\colon D \to B$ such that $\overline{\theta} = \lambda(D)$.*

REMARK. It is easy to see that the map λ is a local analytic diffeomorphism at a generic point of D: the exceptional set may be chosen as a real analytic discriminant set of the analytic mapping. So, this theorem may be considered as a real subanalytic analog of Hironaka Theorem from B.4 with D replacing the set $\widetilde{\Omega}$.

C.2. Maximal and reduced multiplicity of a covering. Consider a real analytic family of surfaces (M, B, π) and a closed subanalytic subset $\Omega \subset M$, $\dim \Omega = \dim B = n$.

Define the set $\mathrm{br}_\pi \Omega$ consisting of points from Ω, such that Ω is not a local section near these points (i.e., cannot be represented as the graph of an analytic map $B \to M$). One can easily see that the projection $\pi(\mathrm{br}_\pi \Omega)$ is a closed subset of the base B of Hausdorff dimension $< n$. Indeed, by the theorem on rank, the dimension of the set $\pi((\Omega \setminus \mathrm{sng}\,\Omega) \cap \mathrm{br}_\pi \Omega)$ is less than n, and by the property C.1.1 of stratifications, $\dim \mathrm{sng}\,\Omega < n$. Thus the following definitions are self-consistent; they generalize the notions of (reduced) covering multiplicity introduced in B.2, for the real case.

DEFINITION. The *maximal covering multiplicity* is the maximal number of the points in the intersection $M_\varepsilon \cap \Omega$, over $\varepsilon \in B \setminus \pi(\mathrm{br}_\pi \Omega)$. The *reduced covering multiplicity* is the difference between the maximal multiplicity of the covering and the number of different sections contained in Ω.

For the multiplicities thus introduced, the real analogs of Propositions B.2 and B.3 hold. The proof for the real case differs from that for the complex case at the following point: instead of (complex) dimensions in the complex analytic category one should use the Hausdorff dimension in the real topological category.

C.3. Lemma on sections in the real subanalytic category. Before analyzing the general case, consider an example which is a real counterpart of Example 1 from B.3.

EXAMPLE. Let

$$M = \mathbb{R}^2 \times \mathbb{R}^1, \quad B = \mathbb{R}^1, \quad \pi(x, y, \varepsilon) = \varepsilon, \qquad \Omega = \{(x, y, \varepsilon) \mid x^2 = \varepsilon, y = 0\}.$$

To construct the unfolding of the base which would place the set Ω on the union of sections, consider not one but rather *two unfoldings*,

$$\rho_1, \rho_2 \colon \mathbb{R} \to \mathbb{R}, \qquad \rho \colon \tilde{\varepsilon}_1 \mapsto \tilde{\varepsilon}_1^2, \quad \rho_2 \colon \tilde{\varepsilon}_2 \mapsto -\tilde{\varepsilon}_2^2.$$

Each unfolding is the family of surfaces $\mathbb{R}^2 \times \mathbb{R}^1_i \xrightarrow{\pi_i} \mathbb{R}^1_i$, $i = 1, 2$, and while in the first family the pullback of Ω is the union of two linear sections $\{y = 0, x = \pm\tilde{\varepsilon}_1\}$, the second unfolding of Ω yields just one point at the origin; there are many different sections containing this single point, and we can choose any of them.

The real analog of Lemma B.4 looks as follows. Let (M, B, π) be a real analytic family of surfaces.

LEMMA C.1. *Let $\Omega \subset M$ be a closed real subanalytic set of dimension $\Omega = \dim B$ and the restriction $\pi|_\Omega$ be proper. Then there exists an unfolding ρ such that the proper preimage of Ω in any connected component of the pullback family $\rho^* M$ belongs to the union of a finite number of real analytic sections and a subanalytic subset of Hausdorff dimension less than* $\dim B$.

THE IDEA OF THE PROOF OF LEMMA C.1. Define the subset Ω' of the set Ω in the same way as in the complex case: Ω' is the set of points near which Ω is not a union of sections. This set is closed and its projection $\pi(\Omega')$ is a closed subset of B of Hausdorff dimension less than n.

Similarly to the complex case, it is sufficient to prove the assertion of the lemma for a small neighborhood of an arbitrary point from the projection $\pi(\Omega')$ on the base. By the Gabrielov theorem (see Appendix A) for the restriction of the original family to such a neighborhood, both the maximal and the reduced multiplicities of the covering $\Omega \to B$ are finite, and we want to construct an unfolding of the base which would diminish the maximal multiplicity of the covering.

Using the real counterpart of the Hironaka theorem (property C.1.4), we can reproduce the construction of the set $\widetilde{B}_1$ as the disjoint union of the corresponding components $\widetilde{\Omega}_i$ from the step 3 in the proof of Lemma B.4, together with the projection ρ_1.

In the notations of Lemma B.4 and its proof, the set $B_1 = \rho_1(\widetilde{M})$ is *not an open neighborhood* of the discriminant set $\pi(\mathrm{br}_\pi \Omega)$. However, the Hironaka theorem from C.1.4 implies that $B_2 = \overline{B \setminus B_1}$ is subanalytic, thus there exists a local parametrization $\lambda: D \to B_2$ by a real manifold D such that the pair (ρ_1, λ) is an unfolding of the full neighborhood of the discriminant. In every connected component of the pullback family induced by λ, the reduced covering multiplicity will be decreased. Indeed, in this component the preimage of at least one non-section piece of Ω must belong to the preimage of $\partial B_1 \subset \mathrm{br}_\pi \Omega$ (the topological boundary). Therefore it will be of dimension smaller than $\dim \Omega$ (in the same way as this happened with the map ρ_2 from the example in the beginning of this subsection). □

Remarks on the proof of Lemma R in the real case. In fact, we prove Lemma R not only for real analytic sets but for the extended category of closed subanalytic sets. In this context Lemma C.1 is the complete real analog of the proof of Lemma B.4.

The final part of the proof of Lemma R for the real case repeats that for the complex case with just one modification: instead of referring to the existence of a locally proper projection for a complex analytic set, we use the property C.1.2. □

References

1. V. I. Arnold and Yu. S. Ilyashenko, *Ordinary differential equations* Dynamical systems, I, Itogi Nauki. Sovremennye Problemy Matematiki: Fundamental′nye Napravleniya., vol. 1 7–149, VINITI, Moscow, 1985; English transl., Encyclopaedia of Math. Sci., vol. 1, Springer-Verlag, Heidelberg, 1988.
2. A. M. Gabrielov, *Projections of semianalytic sets*, Functional Anal. Appl. **2** (1968), 282–291.
3. P. Griffits and J. Harris, *Principles of algebraic geometry*, Wiley, New York, 1978.
4. O. Forster, *Riemannsche Fläschen*, Springer-Verlag, Berlin, Heidelberg, and New York, 1977.
5. E. M. Chirka, *Complex analytic varieties*, "Nauka", Moscow, 1985. (Russian)
6. V. I. Arnold, M. I. Vishik, Yu. S. Ilyashenko, A. S. Kalashnikov, V. A. Kondratyev, S. N. Kruzhkov, E. M. Landis, V. M. Millionshchikov, O. A. Oleinik, A. F. Filippov, and M. A. Shubin, *Some unsolved problems from the theory of differential equations ant mathematical physics*, Uspekhi Math. Nauk **44** (1989), no. 4, 191–202; English transl. in Russian Math. Surveys **44** (1989).
7. F. Dumortier, *Singularities of vector fields on the plane*, J. Differential Equations **23** (1977), no. 1, 55–106.
8. J.-F. Mattei and R. Moussu, *Holonomie et integrales premiéres*, Ann. Sci. École Norm. Sup. (4) **13** (1980), 469–523.

9. H. Hironaka, *Resolution of singularities of an algebraic variety over a field of characteristic zero*, Ann. of Math. **79** (1964), no. 1–2, 109–326.
10. ———, *Subanalytic sets*, Number Theory, Algebraic Geometry and Commutative Algebra, in honor of Akizuki, Kinokunia Bookstore Co., Tokio, 1973.
11. ———, *Bimeomorfic smoothing of a complex-analytic space* (*summary*), Math. Inst. Warwick Univ.
12. R. Roussarie, *A note on finite cyclicity and Hilbert's 16th problem*, Dynamical systems, Proc. of the Chilean symposium, (Valparaiso, 1986) (R. Bamon, R. Labarca, and J. Palis, eds.), Lecture Notes Math., vol. 1331, Springer-Verlag, Berlin, 1988, pp. 161-188.
13. A. Seidenberg, *Reduction of singularities of differential equation* $Ady = Bdx$, Amer. J. Math. **90** (1968), no. 1, 248–269.
14. E. Bierstone and P. Milman, *Semianalytic and subanalytic sets*, Inst. Hautes Études Sci. Publ. Math. **67** (1988), 1–42.
15. Z. Denkowska and R. Roussarie, *A method of desingularization for analytic two-dimensional vector field families*, Bol. Soc. Brasil. Mat. **22** (1991), no. 1, 93–126.
16. A. Van den Essen, *Reduction of singularities of the equation* $A\,dy = B\,dx$, Equations differentielles et systemes de Pfaff dans le champ complexe (R. Gerard and J.-P. Ramis, eds.), Lecture Notes Math., vol. 712, Springer-Verlag, Berlin and New York, 1979, pp. 44-60.
17. O. Kleban, *The order of topologically sufficient jet of a smooth vector field on the real plane at a singular point of finite multiplicity*, in this volume.
18. Yu. S. Ilyashenko and S. Yu. Yakovenko, *Finite cyclicity of elementary polycycles in generic families*, in this volume.
19. S. Trifonov, *Resolution of singularities in one-parametric analytic families of differential equations*, Amer. Math. Soc. Transl. (2), vol. 151, Amer. Math. Soc., Providence, RI, 1992, pp. 135–145.
20. ———, *Analytic families of differential equations on two-dimensional manifolds, resolution of singularities and singular perturbations*, Dokl. Akad. Nauk **331** (1993), no. 5, 559–562; English transl. in Russian Math. Doklady (1994) (to appear).

Translated by S. YAKOVENKO

Moscow State University
E-mail address: strif@arcady.msk.su, trifon@mmcentre.msk.su

Amer. Math. Soc. Transl.
(2) Vol. **165**, 1995

Order of the Topologically Sufficient Jet of a Smooth Vector Field on the Real Plane at a Singular Point of Finite Multiplicity

O. KLEBAN

§1. Formulation of the main result

In the paper we consider germs of smooth vector fields having an isolated singular point at the origin on the real plane. The term "smooth" here and below means C^∞-differentiability. The Dumortier theorem [1] claims that if a germ $\mathbf{v}$ has a characteristic orbit (a phase trajectory tangent to a certain direction when tending to the singular point) and satisfies the Łojasiewicz condition [2], then such a germ is topologically finitely determined: there exists a natural N such that all the representatives of the N-jet $\mathbf{v}^{(N)}$ of the germ $\mathbf{v}$ are topologically equivalent. Such a jet is called *topologically sufficient*, and the minimal integer N is the *order of the topologically sufficient jet*.

DEFINITION [3]. Multiplicity of the singular point $O = (0, 0)$ for the germ of a smooth vector field $v = (v_1, v_2)$ is the dimension of the quotient ring $Q_v = \mathbb{R}[[x]]/(\hat{v})$, where $\mathbb{R}[[x]]$ is the ring of Taylor series in two variables, and $(\hat{v})$ is the ideal in this ring, generated by the formal power series $\hat{v}_1, \hat{v}_2$.

A singular point of finite multiplicity is isolated, and the corresponding germ satisfies the Łojasiewicz condition [4]. Thus the germ of a smooth vector field at a singular point of finite multiplicity is finitely topologically determined provided that there exists a characteristic orbit. *Throughout this paper we deal only with singular points of finite multiplicity*, sometimes not stating this fact explicitly.

It turns out that it is possible to estimate the order N of the topologically sufficient jet in terms of the multiplicity of the singular point.

THEOREM. *The order of the topologically sufficient jet for a vector field possessing a characteristic orbit at the singular point of multiplicity* μ_0, *does not exceed* $2\mu_0 + 2$.

The proof is based on the desingularization technique [3].

REMARK 1. The above theorem and its proof may be considered as the new demonstration of the Dumortier theorem [1] on existence of a nice desingularization

1991 *Mathematics Subject Classification.* Primary 34C05.

for a singular point of a smooth vector field. In our proof we use the ideas and the technique suggested by Van den Essen [5], [7] for the complex analytic case.

REMARK 2. In some degenerate cases the theorem allows us to reduce investigation of stability for smooth vector fields to that for polynomial fields.

§2. Elementary singularities and points of contact

DEFINITION. A singular point of a planar vector field is said to be *elementary* if at least one eigenvalue of the linear part of the field at this point is nonzero.

Throughout this paper we will use the following names for different types of singular points. These types are defined in terms of the eigenvalues λ_1, λ_2 which are either real or complex conjugate:

Elliptic point:	$\lambda_{1,2} = \alpha \pm i\omega, \quad \omega \neq 0$,
Resonant saddle:	$\lambda_1/\lambda_2 = -p/q, \quad p, q \in \mathbb{N}$,
Nonresonant saddle:	λ_1/λ_2 negative irrational,
Resonant node:	λ_1/λ_2 or $\lambda_2/\lambda_1 \in \mathbb{N}$,
Nonresonant node:	λ_2/λ_1 positive, not an integer or an inverse integer,
Diacritical node:	$\lambda_1/\lambda_2 = 1$, linearization is diagonal,
Jordanian node:	$\lambda_1 = \lambda_2 = \lambda \neq 0$,
	the linear part is $\lambda(x\partial/\partial x + y\partial/\partial y) + y\partial/\partial x$,
Cuspidal point:	$\lambda_1 = \lambda_2 = 0$, the linear part is $y\partial/\partial x$.

Out of this list, only the last type corresponds to a nonelementary singularity.

DEFINITION. A *point of contact* between a vector field and a smooth curve L is the point on the curve at which the direction of the filed is tangent to the curve. By definition, the zero vector is tangent to any curve passing through the point.

From now on we will be interested only in isolated points of contact. Let C be a point of contact between the field $v(x, y)$ and a curve L, and $\gamma\colon (-\delta, \delta) \to \mathbb{R}^2$ a smooth parameterization of the curve L, regular at the origin: $C = \gamma(0)$, $\dot\gamma(0) \neq 0$. Denote by $n(t)$ the unit vector normal to L at the point $\gamma(t)$. Consider the projection $\Pi_n v = (v(t), n(t)) = K(t)$. The function $K(t)$ is smooth in a neighborhood of zero. If its Taylor expansion has the form $K(t) = c_k t^k + \cdots$, where $c_k \neq 0$, then k is the order of contact at the point C. We denote the order of contact by $k_C(v)$. If the function $K(\cdot)$ is flat at the origin (having identically zero Taylor series), then C is the point of contact of infinite order. The definition of the order of contact is self-consistent, i.e., independent of the choice of parameterization of the curve.

Let us establish a relationship between the order of contact of the field v with the curve L and the order of tangency of phase trajectories to this curve. Without loss of generality we may assume that the field v corresponds to the system of differential equations

$$\dot x = f(x, y), \qquad \dot y = g(x, y), \tag{1}$$

while the curve L is the x-axis and the origin is the point of contact. If

$$g(x, 0) = c_k x^k + o(x^k), \qquad c_k \neq 0, \ k \geqslant 1, \tag{2}$$

then, by definition, k is the order of contact at the point $(0, 0)$. If $f(0, 0) = \lambda \neq 0$, then the point of contact is *nonsingular*, otherwise it is *singular*.

LEMMA 1. *If the origin O is a nonsingular point of contact of the field with the x-axis, then the order of contact is equal to the order of tangency of a phase curve with the axis at the origin.*

PROOF. Let $y = \varphi(x)$ be the phase curve of (1) passing through the point O (that is, $\varphi(0) = 0$). Differentiating $y - \varphi(x)$ along the field, we obtain

$$g(x, \varphi(x)) - \varphi'(x) f(x, \varphi(x)) \equiv 0.$$

By condition (2), in a nonsingular point of contact of order k the right-hand sides of equations in (1) have the form

$$f(x, y) = \lambda + f_1(x, y), \qquad g(x, y) = c_k x^k + o(x^k) + y g_1(x, y),$$

where $\lambda \neq 0$, $c_k \neq 0$, $f_1(0, 0) = 0$. The first nonzero coefficient in the Taylor expansion of $\varphi(x)$ at the origin occurs at x^{k+1} and it is equal to $c_k/(k+1)\lambda$. □

LEMMA 2. *Let O be an elementary singular point with real eigenvalues and one of the eigenvectors collinear to the x-axis. In this case there exists an invariant one-dimensional manifold (a curve) tangent to the axis with a certain order k.*

If the singular point is not a resonant node (the ratio of eigenvalues λ_1/λ_2 is not in the set $\{n, 1/n \mid n \in \mathbb{N}\}$), then the order of contact between the field $v(x, y)$ and the x-axis at the origin is $k + 1$.

PROOF. Without loss of generality we may assume that the right-hand sides of equations in the system (1) have the form

$$f(x, y) = \lambda_1 x + f_2(x, y), \qquad g(x, y) = \lambda_2 y + c_k x^k + y g_2(x, y) + o(x^k),$$

where $\lambda_1^2 + \lambda_2^2 \neq 0$, $g_2(0, 0) = 0$, $f_2(x, y) = o\big(\sqrt{x^2 + y^2}\big)$. Let $y = \varphi(x)$ be the invariant manifold of the system (1), tangent to the x-axis at the origin. Differentiating $y - \varphi(x)$ along v, one obtains

$$\lambda_2 \varphi(x) + \varphi(x) g_2(x, \varphi(x)) + c_k x^k + o(x^k) - \lambda_1 x \varphi'(x) - f_2(x, \varphi(x)) \varphi'(x) \equiv 0.$$

Therefore, the Taylor coefficients for φ at the origin may be found:

$$\varphi_1 = \cdots = \varphi_{k-1} = 0, \qquad \varphi_k = c_k/(\lambda_1 k - \lambda_2) \neq 0. \quad \square$$

§3. Nice desingularization

Recall the scheme of resolution of singularities (desingularization), see [1]–[3]. We begin with the *polar blow-up*. Let $v = v(x, y)$ be a smooth vector field in a neighborhood U of the singular point $O = (0, 0)$ on the plane $\mathbb{R}^2$. As usual, we assume that the singularity of the vector field at the origin is of finite multiplicity. In particular, the field satisfies the Łojasiewicz condition at the origin. Sometimes we say that the vector field is *nonflat* at the origin.

Consider the map $U \setminus O \to \mathbb{R}^2$, which in the polar coordinates (r, φ) has the form $(r, \varphi) \mapsto (r + 1, \varphi)$. The punctured neighborhood $\dot{U} = U \setminus O$ is mapped into an annulus with the inner boundary $r = 1$. The pullback of the original vector field to this annulus extends smoothly to a full neighborhood of the circle $r = 1$. After dividing by an appropriate power of $r-1$, the extended field has only isolated singularities on $r = 1$, while still remaining smooth. All these singularities also satisfy the Łojasiewicz condition, and if the curve $r = 1$ is not invariant by the pullback vector field after the above division, then there is only a finite number of points of contact.

This procedure is called the polar blow-up of the singular point at the origin. The circle $r = 1$ will be referred to as the *pasted in circle*. Each of the singular points on the pasted in circle is in some sense less degenerate than the original singularity O. For further simplification of singularities the process may be iterated: each new singularity can in turn be blown up. After a finite number of polar blow-ups the punctured neighborhood $\dot{U}$ will be transformed into a topological annulus A, and the pullback of the original field $v|_U$ is a smooth vector field $\tilde{v}$ defined in A. The outer boundary of A is smooth (provided that the boundary of U was smooth, which one may assume without loss of generality). The inner boundary of A is a piecewise smooth curve which consists of the union of arcs of pasted in circles and their preimages by subsequent blow-ups. We denote this curve by γ and call it the *pasted in curve*. By construction, γ is homeomorphic to the circle $\mathbb{S}^1$ and is endowed with the natural orientation. Interior points of smooth components of γ are called *interior points*, and the rest are *corner points*.

As before, the pullback $\tilde{v}$ extends smoothly across γ, and after an appropriate division has only a finite number of singularities and points of contact on γ.

There exists an algebraic analog of the polar blow-up, known as *σ-process*. Consider the natural map $\mathbb{R}^2 \setminus O \to \mathbb{R}P^1$, $(x, y) \mapsto (x : y)$. The graph of this map in $\mathbb{R}^2 \times \mathbb{R}P^1 \simeq \mathbb{R}^2 \times \mathbb{S}^1$ denote by M. The closure $\overline{M}$ is diffeomorphic to the Möbius band. The projection $\pi\colon \mathbb{R}^2 \times \mathbb{R}P^1 \to \mathbb{R}^2$ along $\mathbb{R}P^1$ takes $\overline{M}$ into $\mathbb{R}^2$, and the preimage of the origin is the whole projective line $\mathbb{R}P^1$, which we call the *pasted in projective line*. The restriction of π on $\overline{M} \setminus \mathbb{R}P^1$ is a diffeomorphism of that open domain onto $\mathbb{R}^2 \setminus O$.

After the σ-process, the line field spanned by the vector field v in $\dot{U}$ is pulled back to M as a smooth line field which extends smoothly to a line field on the whole of $\overline{M}$ except for a finite number of singular points on the pasted in projective line. In a small neighborhood of each such point the line field is spanned by a smooth vector field which is nonflat at that point. This allows for iterations of the process.

In what follows we will refer to both the polar blow-up and the σ-process as the blow-up of a singularity. The *step of desingularization* is the blow-up of a nonelementary singularity or a point of contact of a vector field with the pasted in curve. Note that each step implicitly involves maximal extension of the line field (for σ-process) or division by a factor of $(r-1)^\nu$ (for the polar blow-up), so that the result will again be a (line, vector) field.

The following terminology is introduced inductively. A *stage* of desingularization is the superposition of steps (blow-ups) at all nonelementary singular points and points of contact, obtained on the previous stage.

The first stage of desingularization is blow-up of the singularity at the origin $O = (0, 0)$.

The order of consecutive blow-ups at any given stage is not important: the vector field obtained as the result of superposition of the same blow-ups but in a different order, is orbitally equivalent to the result of the initial desingularization near the pasted in curve.

DUMORTIER THEOREM [4]. *The germ* $\mathbf{v}$ *of a smooth planar vector field satisfying the Łojasiewicz condition at a nonelementary singular point, after a finitely many stages can be taken into the germ* $\tilde{\mathbf{v}}$ *of a smooth vector field along the pasted in curve* γ *and having the following properties*:

1. *All the singular points of the germ* $\tilde{\mathbf{v}}$ *belong to the curve* γ. *They are elementary and of finite multiplicities. Their number is finite and they cannot be of the following types*: *an elliptic point, a diacritical node, a Jordanian node, a resonant node.*
2. *The germ* $\tilde{\mathbf{v}}$ *does not have points of contact with* γ. *Moreover, at any point of the line the pair* $(\tilde{\mathbf{v}}, \gamma)$ *is orbitally topologically equivalent to one of the following normal forms*:

A. *At a noncorner point,* γ *is a part of the* x*-axis, and the germ* $\tilde{\mathbf{v}}$ *is of one of the four types,*

$$1.\ \frac{\partial}{\partial x}, \quad 2.\ \pm\frac{\partial}{\partial y}, \quad 3.\ \pm x\frac{\partial}{\partial x} \pm y\frac{\partial}{\partial y}, \quad 4.\ x^2\frac{\partial}{\partial x} \pm y\frac{\partial}{\partial y}.$$

B. *At a corner point,* γ *is the union of positive semi-intervals on* x*- and* y*-axes and the origin, while the germ* $\tilde{\mathbf{v}}$ *is of one of the three types,*

$$5.\ \pm x\frac{\partial}{\partial x} + y\frac{\partial}{\partial y}, \quad 6.\ -x\frac{\partial}{\partial x} - y\frac{\partial}{\partial y}, \quad 7.\ \pm\frac{\partial}{\partial y}$$

(*in the third case any combination of signs is admissible*).

The desingularization described in the theorem is called the *nice desingularization*.

Let the nice desingularization be achieved in m stages and $\sigma\colon \mathbb{R}^2 \setminus O \to \mathbb{R}^2$ be the corresponding map: $\sigma = \sigma_m \circ \cdots \circ \sigma_1$, where σ_i are consecutive stages of the desingularization. Then for any representative $v(x)$, $x \in \mathbb{R}^2$, of the germ $\mathbf{v}$ defined in some neighborhood U of the origin O in $\mathbb{R}^2$ there exists a representative $\tilde{v}(y)$, $y = \sigma(x)$, defined in a certain neighborhood of the pasted in curve γ, such that the condition

$$\tilde{v}(y) = (F\sigma_* v) \circ \sigma^{-1}(y), \qquad y \in \tilde{U} = \sigma(U \setminus O) \tag{3}$$

holds. Here σ_* stands for the derivative of the map σ and $F(x)$ is a certain positive function equal to the product of m functions by which the field is multiplied on each stage of the desingularization.

The precise definition of the function $F(x)$ and related estimates will be given in Section 10.

We will need smooth classification of singular points appearing in the nice desingularization of a singular point. The corresponding theorem is formulated in the next section.

§4. The classification theorem

The definition of smooth orbital equivalence of germs of vector fields is obtained from the definition of topological orbital equivalence by replacing the term "homeomorphism" by the term "smooth diffeomorphism".

THEOREM [3], [4]. *At each singular point of the pasted in curve* γ *the germ* $\tilde{\mathbf{v}}$ *obtained after a desingularization of a singular point of a vector field is smoothly orbitally equivalent to one of the following germs*:

(1) *nonresonant singular point*

$$\lambda_1 x\frac{\partial}{\partial x} + \lambda_2 y\frac{\partial}{\partial y};$$

(2) *resonant saddle*

$$x\frac{\partial}{\partial x} - y\left(p/q + \varepsilon(u^n + au^{2n})\right)\frac{\partial}{\partial y},$$

where $\varepsilon = \{0, \pm 1\}$, p/q *is an irreducible fraction, and* $u = x^p y^q$ *the resonant monomial*;

(3) *degenerate elementary singularity* (*semihyperbolic point*)

$$(x^\mu + ax^{2\mu-1})\frac{\partial}{\partial x} - y\frac{\partial}{\partial y} \quad \text{or} \quad (y^\mu + ay^{2\mu-1})\frac{\partial}{\partial y} - x\frac{\partial}{\partial x},$$

where μ *is the multiplicity of the degenerate elementary singularity*;

(4) *resonant node*

$$(nx + \varepsilon y^n)\frac{\partial}{\partial x} + y\frac{\partial}{\partial y}.$$

The conjugacy can be chosen in such a manner that the germ of the pasted in curve corresponds either to the axis $\{y = 0\}$ *or to the part of the coordinate cross* $\{xy = 0,\ x, y \geqslant 0\}$ *depending on whether the point is noncorner or corner.*

REMARK. In the current paper we use the term "nice desingularization" in a weaker sense that that suggested by Dumortier: a similar list from [3] should not contain resonant nodes that can be further resolved to one of the above types (1)–(3). Since we are minimizing the number of stages in the desingularization process, we will not blow up resonant nodes unless they are points of contact.

§5. Scheme of desingularization

Let $\mathbf{v}$ and $\tilde{\mathbf{v}}$ be the same as in the above Dumortier theorem. With each singularity p of the germ $\tilde{\mathbf{v}}$ on the pasted in curve γ one can associate the germ of a standard vector field, topologically orbitally equivalent to the germ of $\tilde{\mathbf{v}}$ at this point. The curve γ is at the same time taken into either the germ $\{y = 0\}$ or the germ $\{xy = 0,\ x, y \geqslant 0\}$ at the origin (for simplicity we call these germs the *standard germs* of curves).

The *scheme of desingularization* of the germ $\mathbf{v}$ is the ordered collection of pairs

(germ of a standard vector field, standard germ of oriented curve),

each pair corresponding to a singular point of the germ $\tilde{\mathbf{v}}$ along the entire pasted in curve γ.

THEOREM [1]. *Two germs of planar vector fields at singular points, both satisfying the Łojasiewicz condition and possessing characteristic orbits, are orbitally topologically equivalent if and only if the corresponding schemes of desingularization coincide, possibly after a cyclic renumbering of the elements and/or the reversal of the order.*

In the next three sections we study the nice desingularization of a singular point for an analytic vector field, estimating the number of stages necessary to resolve it into singularities from the above list.

§6. Behavior of multiplicity under desingularization

Let $\mathbf{v}$ be the germ of an analytic vector field on $\mathbb{R}^2$ at the origin O, which is a singular point. The field can be extended to a neighborhood of the origin in $\mathbb{C}^2$. The σ-process in $\mathbb{C}^2$ is the same as its real counterpart, with the only difference that instead of $\mathbb{R}P^1$ one has to paste in the complex projective line $\mathbb{C}P^1$.

Let $\tilde{\mathbf{v}}$ be the germ obtained after the blow-up of the origin. The step is called *nondiacritical* if the pasted in projective line $\mathbb{C}P^1$ is invariant for $\tilde{\mathbf{v}}$. Otherwise the step is called *diacritical*. In the diacritical case on the pasted in projective line there may appear isolated points of contact of a finite order. In what follows ν_0 denotes the order of the lowest terms in the Taylor expansion of $\mathbf{v}$ at O, and μ_0 denotes the multiplicity of the point O. Also, let $\mu_p(\tilde{\mathbf{v}})$ be the multiplicity of the the field $\tilde{\mathbf{v}}$ at a point p and $k_q(\tilde{\mathbf{v}})$ the tangency order between the field and the projective line at a point q, $p, q \in \mathbb{C}P^1$.

LEMMA 3. *In the nondiacritical case the the multiplicities of all singular points of the resolution on $\mathbb{C}P^1$ satisfy the equality*

$$\sum_{p\in\mathbb{C}P^1} \mu_p(\tilde{\mathbf{v}}) = \mu_0(\mathbf{v}) - \nu_0^2 + \nu_0 + 1. \tag{4}$$

In the diacritical case the following holds:

$$\sum_{p\in\mathbb{C}P^1} \mu_p(\tilde{\mathbf{v}}) + \sum_{q\in\mathbb{C}P^1} k_q(\tilde{\mathbf{v}}) = \mu_0(\mathbf{v}) - \nu_0^2, \tag{5}$$

where the first (the second) summation is taken over all singular points (resp., all points of contact).

REMARK 1. In the paper [5] an analogous assertion is proved for holomorphic differential forms, but without taking into account points of contact. Moreover, the proof from [5] requires some additional remarks given below.

REMARK 2. Below we study the desingularization at an elementary singular or a nonsingular point of contact: nonelementary singular points are to be blown up regardless of whether they are points of contact or not.

COROLLARY. *Let $\mathbf{v}$ be the germ at the origin of a real analytic field. Then, after the σ-process one gets the inequalities*

$$\sum_{p\in\mathbb{R}P^1} \mu_p(\tilde{\mathbf{v}}) \leqslant \mu_0(\mathbf{v}) - \nu_0^2 + \nu_0 + 1, \tag{4'}$$

$$\sum_{p\in\mathbb{R}P^1} \mu_p(\tilde{\mathbf{v}}) + \sum_{q\in\mathbb{R}P^1} k_q(\tilde{\mathbf{v}}) \leqslant \mu_0(\mathbf{v}) - \nu_0^2, \tag{5'}$$

respectively for the nondiacritical and the diacritical case.

PROOF. The corollary immediately follows from Lemma 3.

PROOF OF LEMMA 3. For the proof we use the definition of the multiplicity of an isolated root for an analytic map, equivalent to that from §1. Let $F\colon (\mathbb{C}^2, 0) \to (\mathbb{C}^2, 0)$ be the germ of a map holomorphic at the origin,

$$F(z, w) = (f(z, w), g(z, w)), \qquad F(0, 0) = (0, 0).$$

If $\frac{\partial f}{\partial w}(0, 0) \neq 0$, then by the implicit function theorem the germ of the set $\{f = 0\}$ at the origin is the graph of a function $w = \varphi(z)$. Then the multiplicity of the map $F = (f, g)$ at O is equal to the order of the zero of the function $\psi(z) = g(z, \varphi(z))$ at the origin or, in other words, to the order of the lowest term in the Taylor series for $\psi(z)$ at the origin:

$$\mu_0(F) = \mu_0(f, g) = \operatorname{ord}_0 g(z, \varphi(z)).$$

This definition can be generalized also for the case when $df(0, 0) = 0$ and the implicit function theorem does not apply. It is known that an irreducible germ of the curve $\{f(z, w) = 0\}$ admits an analytic uniformization

$$\gamma\colon (\mathbb{C}, 0) \to (\mathbb{C}^2, 0), \ \gamma\colon t \mapsto (z(t), w(t)), \qquad f(z(t), w(t)) \equiv 0.$$

This uniformization is minimal in the following sense: the orders of principal (lowest) terms of $z(\cdot)$ and $w(\cdot)$ are relatively prime.

The multiplicity of the map $F = (f, g)$ at the origin is the order of zero of the function $g \circ \gamma$:

$$\mu_0(F) = \operatorname{ord}_0 g(z(t), w(t)).$$

If $\{f = 0\}$ is a reducible germ and $f = f_1^{k_1} \cdots f_l^{k_l}$ the corresponding primary decomposition, then

$$\mu_0(F) = \sum_{j=1}^{l} k_j \mu_0(f_j, g).$$

Below follows a list of elementary properties of the multiplicity.

(1) $\mu_0(f, g) = 0$, if $(f, g)(0, 0) \neq (0, 0)$;
(2) $\mu_0(f_1 f_2, g) = \mu_0(f_1, g) + \mu_0(f_2, g)$;
(3) if A is a nondegenerate square 2×2-matrix, and $(\tilde{f}, \tilde{g}) = (f, g) \cdot A$, then $\mu_0(\tilde{f}, \tilde{g}) = \mu_0(f, g)$. In particular, $\mu_0(f, g) = \mu_0(g, f)$.

Now we turn to the investigation of the behavior of multiplicity under σ-processes. In the two affine charts on the projective line the formulas for the coordinate transformations associated with the standard σ-process look as follows:

$$(z, w) \mapsto (z, u = w/z) \quad \text{for } z \neq 0,$$
$$(z, w) \mapsto (v = z/w, w) \quad \text{for } w \neq 0.$$

Let ν_1 and ν_2 be the orders of lowest terms of the components f, g at the origin. Denote

$$\tilde{f}(z, u) = f(z, zu)/z^{\nu_1}, \quad \tilde{g}(z, u) = g(z, zu)/z^{\nu_2},$$
$$\check{f}(v, w) = f(vw, w)/w^{\nu_1}, \quad \check{g}(v, w) = g(vw, w)/w^{\nu_2}.$$

Then the following identity holds [5]:

$$\mu_0(f, g) = \nu_1\nu_2 + \sum_{\substack{c\in\mathbb{C}P^1\\ z(c)\neq 0}} \mu_c(\tilde{f}, \tilde{g}) + \mu_0(\check{f}, \check{g}).$$

The second term in the expression is the sum of multiplicities of the map $(\tilde{f}, \tilde{g})$ at critical points on the line $\mathbb{C}P^1$ in the chart (z, u).

Now we proceed with the proof of Lemma 3. Let $v(z, w)$ be a representative of the germ $\mathbf{v}$ defined in a neighborhood $(\mathbb{C}^2, 0)$. The corresponding system of differential equations has the form

(6) $$\dot{z} = f(z, w), \qquad \dot{w} = g(z, w)$$

(we assume that $f(0, 0) = g(0, 0) = 0$). One can always put the orders of lowest terms of f and g be equal to $\nu = \nu_1 = \nu_2$ (this can be achieved by a linear transformation, if necessary). After one σ-process in the chart (z, u) the system (6) takes the form

$$\dot{z} = f(z, zu), \qquad \dot{u} = z^{-2}[zg(z, zu) - zuf(z, zu)].$$

Using the above notation, we rewrite it in the form

$$\dot{z} = z^{\nu}\tilde{f}(z, u), \qquad \dot{u} = z^{\nu-1}[\tilde{g}(z, u) - u\tilde{f}(z, u)].$$

After division by $z^{\nu-1}$ (this does not change the phase portrait outside the line $z = 0$) we obtain

(7) $$\dot{z} = z\tilde{f}(z, u), \qquad \dot{u} = \tilde{g}(z, u) - u\tilde{f}(z, u).$$

Denote

$$P(z, w) = zg(z, w) - wf(z, w),$$
$$\tilde{P}(z, u) = \tilde{g}(z, u) - u\tilde{f}(z, u),$$
$$P_{\nu+1}(z, w) = zg_{\nu}(z, w) - wf_{\nu}(z, w),$$

where f_ν and g_ν are homogeneous lower terms of f and g respectively. There are two possibilities.

1. If $P_{\nu+1} \not\equiv 0$ (the nondiacritical case), we may assume that singular points of the resolved map on $\mathbb{C}P^1$ are all covered by the chart $u = w/z$, or, what is the same, that the polynomial $P_{\nu+1}$ is not divisible by z and therefore contains a

term proportional to $w^{\nu+1}$ (this may be achieved by a linear transformation of the coordinates).

Let us compute the sum of multiplicities of all singular points of the system (7) on $\mathbb{C}P^1$:

$$\sum_p \mu_p(z\tilde{f}, \tilde{P}) = \sum_p \mu_p(z, \tilde{P}) + \sum_p \mu_p(\tilde{f}, \tilde{P}) = \nu + 1 + \sum_p \mu_p(\tilde{f}, \tilde{g}).$$

Now the identity

$$\mu_0(f, g) = \nu^2 + \sum_{c \in \mathbb{C}P^1} \mu_p(\tilde{f}, \tilde{g}),$$

implies (4).

2. If $P_{\nu+1}(z, w) \equiv 0$ (the diacritical case), then the right-hand sides of equations in the system (7) are divisible by one extra factor z. After such a division, one obtains

$$\dot{z} = \tilde{f}(z, u), \qquad \dot{u} = \tilde{\mathcal{P}}(z, u), \tag{8}$$

where $\mathcal{P}(z, u) = P(z, zu)/z^{\nu+2}$. Let us point out that in the diacritical case

$$f_\nu(z, w) = zh_{\nu-1}(z, w), \qquad g_\nu(z, w) = wh_{\nu-1}(z, w),$$

where $h_{\nu-1}$ is a homogeneous polynomial of degree $\nu - 1$.

Since the singular point of $v(z, w)$ is isolated, there exists a natural l such that

$$P_{\nu+j+1} \equiv 0 \text{ for all } j < l,, \qquad P_{\nu+l+1} \not\equiv 0.$$

Without loss of generality one can assume that the polynomials $h_{\nu-1}(z, w)$ and $P_{\nu+l+1}(z, w)$ contain the terms $w^{\nu-1}$ and $w^{\nu+l+1}$ respectively (otherwise an appropriate linear transformation can be applied). This means that all singular points and points of contact are within the domain of the chart $u = w/z$ on $\mathbb{C}P^1$. From the definition it follows that $\mu_0(f, z) = \operatorname{ord}_0 f(0, w) = \nu + l$. Hence we conclude that

$$\mu_0(f, P) = \nu + l + \mu_0(f, g). \tag{9}$$

On the other hand, by virtue of the property of the multiplicity proved above,

$$\mu_0(f, P) = \nu(\nu + l + 1) + \sum_{p \in \mathbb{C}P^1} \mu_p(\tilde{f}, \check{\mathcal{P}}), \tag{10}$$

where $\check{\mathcal{P}}(z, u) = P(z, zu)/z^{\nu+l+1}$. Next, since

$$\tilde{\mathcal{P}}(z, u) = z^{l-1}\check{\mathcal{P}}(z, u),$$

one has

$$\begin{aligned} \sum_{p \in \mathbb{C}P^1} \mu_p(\tilde{f}, \tilde{\mathcal{P}}) &= \sum_p \mu_p(\tilde{f}, z^{l-1}) + \sum_p \mu_p(\tilde{f}, \check{\mathcal{P}}) \\ &= (l-1)(\nu-1) + \sum_p \mu_p(\tilde{f}, \check{\mathcal{P}}). \end{aligned} \tag{11}$$

Combining (9), (10) and (11), one concludes that

$$\sum_{c\in\mathbb{C}P^1} \mu_p(\tilde{f}, \tilde{\mathcal{P}}) = \mu_0(f, g) - \nu^2 - \nu + 1. \tag{12}$$

The sum of orders of contacts for the field $\tilde{v}(z, u)$ corresponding to the system (8) on the pasted in projective line, coincides with the number of roots (counted with the multiplicities) of the function $\tilde{f}(0, u) = h_{\nu-1}(1, u)$, the latter being $\nu - 1$. Therefore (12) implies (5). □

COROLLARY 1. *If the linear part of a vector field at a singular point of a finite multiplicity is identically zero (i.e., $\nu > 1$), then the sum of multiplicities of all points obtained after one step of desingularization, is smaller than the multiplicity of the initial singularity.*

COROLLARY 2. *After a finitely many stages of desingularization an analytic singular point of a finite multiplicity is resolved into finitely many singular points with nonzero linear parts.*

In the next section we study desingularization of singularities with nonzero linear parts. Section 8 contains the results concerning points of contact.

§7. Cuspidal points

It is known [6] that an analytic vector field on the real plane at a nonelementary singular point but with nonzero linear part is smoothly orbitally equivalent to the vector field associated with the equation

$$\begin{aligned} \dot{x} &= y, \\ \dot{y} &= ax^k + bx^n y + g(x, y), \end{aligned} \tag{13}$$

where $k \geqslant 2$, $n \geqslant 1$, $a \neq 0$, $b \neq 0$ and

$$g(x, y) = x^{k+1}h_1(x, y) + x^{n+1}y\, h_2(x, y) + y^2 h_3(x, y).$$

REMARK. The number k in (13) is the multiplicity of the singular point O at the origin $(0, 0)$.

LEMMA 4. *After the nice desingularization of the singularity O for the system (13), the following possibilities may occur:*

1. *If $k \leqslant 2n$ is even, then in $m = k/2 + 2$ steps the singularity is resolved into m nondegenerate saddles.*
2. *If $k \leqslant 2n + 1$ is odd, then in $m = (k+1)/2$ steps the singularity is resolved either into*
   - 2a. *$(k+3)/2$ nondegenerate saddles, or*
   - 2b. *$(k+1)/2$ nondegenerate saddles and one nondegenerate node, or*
   - 2c. *$(k-1)/2$ nondegenerate saddles and one saddle-node of multiplicity 2.*
3. *If $k > 2n+1$, then in $m = n+1$ steps the singularity is resolved into m nondegenerate saddles and one degenerate elementary singularity of multiplicity $\mu = k - 2n$.*

Moreover, all the steps of the nice desingularization are not diacritical, i.e., no points of contact appear, and the order of lowest terms of the vector field at all intermediate singularities (along the process) is equal to two except for the first step for all three cases 1–3 *and* $(m-1)$ *st step for the case* 3, *when this order is equal to one.*

PROOF. The demonstration goes on in three stages.

Stage 1. Make one σ-process at the origin $O=(0,0)$. Since the order ν of lowest terms is 1 and the polynomial $P_{\nu+1}(x,y)=P_2(x,y)=-y^2$ is not divisible by x, all the singular points of the first resolution are covered by the chart $u=y/x$ and the blow-up is not diacritical. In the coordinates (x,u) the system (13) takes the form

$$\begin{aligned} \dot{x} &= xu, \\ \dot{u} &= ax^{k-1}+bx^n u-u^2+\cdots. \end{aligned} \tag{14}$$

The point $x=u=0$ is the only singularity on the pasted in u-axis, and its multiplicity is $k+1$ by Lemma 3.

Stage 2. For simplicity we write (14) in the coordinates (x,y),

$$\begin{aligned} \dot{x} &= xy, \\ \dot{y} &= ax^{k-1}+bx^n y-y^2+\cdots. \end{aligned} \tag{15}$$

PROPOSITION 1. *If* $k\gg 3$, $n\gg 1$, *then after* l *blow-ups the singularity* O *is resolved into* l *nondegenerate saddles and one degenerate singularity of multiplicity* $k-2l+1$ *of the form*

$$\begin{aligned} \dot{x} &= xy, \\ \dot{y} &= ax^{k-2l-1}+bx^{n-l}y-(l+1)y^2+\cdots. \end{aligned} \tag{16}$$

PROOF OF THE PROPOSITION. The proof goes on by induction in l.

Base. After one blow-up the system (15) in coordinates y, $v=x/y$ takes the form

$$\begin{aligned} \dot{v} &= 2v-ay^{k-3}v^k-by^{n-1}v^{n+1}+\cdots, \\ \dot{y} &= -y+ay^{k-2}v^{k-1}+by^n v^n+\cdots. \end{aligned}$$

The only singular point of the resolution is the nondegenerate saddle at $v=0$, $y=0$. In the same way in the coordinates x, $u=y/x$ after division by x one obtains

$$\begin{aligned} \dot{x} &= xu, \\ \dot{u} &= ax^{k-3}+bx^{n-1}u-2u^2+\cdots, \end{aligned}$$

and the only additional singularity is at $u=x=0$, which is of multiplicity $k-1$ by Lemma 3.

Induction step. An induction step is implemented in the same way. The proposition is proved. □

Stage 3.

PROPOSITION 2. *Let the integer* l *in the system* (16) *is such that one of the following conditions holds*:

a) $k-2l=2\,,\ n-l\geqslant 1\,,$
b) $k-2l=3\,,\ n-l\geqslant 1\,,$
c) $k-2l>3\,,\ n-l=1\,.$

Then in each of these cases:

a) *after two blow-ups the singularity is resolved into three nondegenerate saddles*;
b) *after one blow-up the singularity is resolved either into three nondegenerate saddles, or into two nondegenerate saddles and one nondegenerate node, or one nondegenerate saddle and one saddle-node of multiplicity* 2;
c) *after one blow-up the singularity is resolved into two nondegenerate saddles and one degenerate elementary point of multiplicity* $k-2l-2$.

PROOF OF THE PROPOSITION. We consider only case c); the remaining cases are treated similarly. The system (16) is of the form

$$\dot{x}=xy\,,\qquad\qquad (17)$$
$$\dot{y}=ax^{k-2n+1}+bxy-ny^2+\cdots\,,$$

and the order ν of lowest terms is 2. The homogeneous polynomial

$$P_{\nu+1}(x\,,\,y)=P_3(x\,,\,y)=bx^2y-(n+1)xy^2$$

is divisible by both x and y, therefore both charts should be introduced. In the first one, $x\,,\ u=y/x$, the system after division by x takes the form

$$\dot{x}=xu\,,$$
$$\dot{u}=ax^{k-2n-1}+bu-(n+1)u^2+\cdots\,,$$

and there are two singularities on the u-axis, $(0\,,\,0)$ and $(0\,,\,b/2)$. The Jacobian matrices at these points are

$$J(0\,,\,0)=\begin{pmatrix}0 & 0\\ * & b\end{pmatrix}\,,\qquad J(0\,,\,b/2)=\begin{pmatrix}b/2 & 0\\ * & -b\end{pmatrix}\,;$$

therefore, the first point is a degenerate elementary singularity, while the second one is a nondegenerate saddle.

In the coordinates $y\,,\ v=x/y$ the system (17) takes the form

$$\dot{v}=2v-2v^{k-2n+2}y^{k-2n-1}-bv^2+\cdots\,,$$
$$\dot{y}=av^{k-2n+1}y^{k-2n}+bvy-y+\cdots\,.$$

The additional singularity at $v=y=0$ (invisible in the first chart) is a nondegenerate saddle, since

$$J(0\,,\,0)=\begin{pmatrix}2 & 0\\ 0 & -1\end{pmatrix}.$$

Note that the multiplicity of the degenerate elementary singularity here is $k-2n$ by virtue of Lemma 3. The proof of Proposition 2 in the case c) is completed. □

Lemma 4 follows from the above results. □

§8. Resolution of points of contact

LEMMA 5. *A nonsingular point of order k contact of a smooth vector field with a smooth curve, after one blow-up turns into a nondegenerate saddle with the same order of contact.*

PROOF. Let $(0, 0)$ be a nonsingular point of contact of order k between a field $v(x, y)$ and the x-axis. The vector field is associated with the system

$$\begin{aligned} \dot{x} &= \lambda + f(x, y), \\ \dot{y} &= bx^k + y\,g(x, y) + o(x^k), \end{aligned} \tag{18}$$

where $\lambda \neq 0$, $b \neq 0$, $k \geqslant 1$, $f(0, 0) = 0$, and f, g are smooth at the origin. After blow-up $(x, y) \mapsto (x, u = y/x)$ the system (18) multiplied by x yields the system

$$\begin{aligned} \dot{x} &= \lambda x + x f(x, ux), \\ \dot{u} &= -\lambda u + bx^k + u[x\,g(x, ux) - f(x, ux)] + o(x^k). \end{aligned}$$

The point $x = u = 0$ is a nondegenerate saddle, which has the contact of order k with the x-axis by (2). There are no other points of contact or singularities in the chart (x, u). In the chart y, $v = x/y$ the vector field after multiplication by y yields

$$\begin{aligned} \dot{v} &= \lambda + f(vy, y) - v[bv^k y^k + y\,g(vy, y) + o(v^k y^k)], \\ \dot{y} &= y[bv^k y^k + y\,g(vy, y) + o(v^k, y^k)], \end{aligned}$$

which has neither singularities nor points of contact on the v-axis. $\square$

LEMMA 6. *A kth order contact with a smooth curve at an elementary singular point disappears no later than after k resolutions. More precisely,*

a) *a nondegenerate saddle with contact of order k after k blow-ups is resolved into $k + 1$ nondegenerate saddles without contact;*
b) *a nondegenerate nondiacritical diagonalizable node with contact of order k is resolved after k blow-ups into k nondegenerate saddles and one nondegenerate node, all of them without contact;*
c) *elliptic singularities (always first order contact) after one blow-up become nonsingular and without points of contact; the same holds for a diacritical node with contact of any order;*
d) *a Jordanian node with contact of order k after k blow-ups is resolved into $k - 1$ nondegenerate saddles and one saddle-node of multiplicity 2 (all of them without contacts);*
e) *a degenerate elementary singularity of multiplicity μ and contact of order k after k blow-ups is resolved into k nondegenerate saddles and one degenerate singularity of multiplicity μ (also without contacts).*

PROOF. Prove the assertion a). Without loss of generality one may assume that the field has kth order contact with the x-axis, and the origin is a nondegenerate saddle. The associated system looks like

$$\begin{aligned} \dot{x} &= ax + by + f_2(x, y), \\ \dot{y} &= dy + \alpha x^k + g_2(x, y), \end{aligned} \tag{19}$$

with $ad < 0$, $\alpha \neq 0$, $k \geqslant 1$ and f_2, g_2 starting with terms of order $\geqslant 2$. Moreover, $g_2(x, y)$ does not contain monomials of the form $\alpha_n x^n$ with $n \leqslant k$.

After the blow-up $(x, y) \mapsto (x, u = y/x)$ (the only chart to be considered) the system (19) takes the form

$$\dot{x} = ax + bxu + \cdots,$$
$$\dot{u} = (d-a)u + \alpha x^{k-1} - bu^2 + \cdots,$$

and there are two singular points on the pasted in line: a nondegenerate saddle at $x = u = 0$ having contact of order $k-1$ and a nondegenerate saddle without contact at $x = 0$, $u = (d-a)/b$. Thus after each blow-up the order of contact is diminished by 1 and a new nondegenerate saddle appears. This remark completes the proof of assertion a). The rest goes on in a similar way. □

§9. Distortions of blow-up maps

Let $\mathbf{v}$ be the germ of a real analytic vector field having a finite multiplicity singular point O on $\mathbb{R}^2$, σ a desingularization (of the singular point) consisting of m polar blow-ups, U a punctured neighborhood of the origin on $\mathbb{R}^2$, $\tilde{U} = \sigma(U)$, γ a curve pasted in when performing the desingularization σ. Next, let $r(x)$ be the distance from x to the origin in $\mathbb{R}^2$ and $d(y)$ the distance to γ in $\tilde{U}$.

LEMMA 7 [3]. *In a sufficiently small neighborhood U the following inequalities hold*:

$$\|\sigma_*(x)\| \leqslant c[r(x)]^{-m},$$
$$c_1 r(x) \leqslant d \circ \sigma(x) \leqslant c_2[r(x)]^{1/2^m},$$

with some positive c, c_1, c_2. □

§10. Proof of the principal theorem

Let $\mathbf{v}$ be the germ of a smooth vector field on $\mathbb{R}^2$ having a singular point of a finite multiplicity μ_0 at the origin and a characteristic trajectory.

As before, denote $N = 2\mu_0 + 2$ and let $\mathbf{v}^{(N)}$ be the N-jet of the germ $\mathbf{v}$ at the origin. One has to prove that all germs extending the same $\mathbf{v}^{(N)}$ are orbitally topologically equivalent.

Take any representative v of $\mathbf{v}$ and let P_N be its Taylor polynomial of degree N centered at the origin:

$$v(x) = P_N(x) + w(x), \qquad w(x) = o\big([r(x)]^N\big) \text{ as } x \to 0. \tag{20}$$

Let σ be a nice desingularization of the polynomial field P_N; denote by $\tilde{P}_N$ the field obtained from P_N after applying σ and defined in a neighborhood of the curve γ. The fields P_N and $\tilde{P}_N$ are related as follows,

$$\tilde{P}_N(y) = (F\sigma_* P_N) \circ \sigma^{-1}(y), \qquad y \in \tilde{U} = \sigma(U - \{0\}). \tag{21}$$

Here F is a positive function which we specify in a moment. Let the nice desingularization σ be achieved by m stages of polar blow-ups, $\sigma = \sigma_m \circ \cdots \circ \sigma_1$, where $\{\sigma_j, j = 1, \dots, m\}$ is the sequence of resolutions. Then F is the product of the positive functions F_j defined as follows. Let v_{j-1} be the field obtained before the jth stage and defined in a neighborhood of the curve γ_{j-1}. Denote by M_{j-1} the collection of points on γ_{j-1} to be blown-up further: it may contain singularities of v_{j-1} on γ_{j-1} and points of contact (recall that only nonsingular and elementary points of contact are considered). Let us order points from M_{j-1} in an arbitrary manner, so that $M_{j-1} = \{p_i \mid i = 1, \dots, k_{j-1}\}$. Then the jth stage of desingularization is the superposition of blow-ups: the first one is centered at p_1, the second one at the image of p_2, etc. In this process the field on each step (of the current stage) is multiplied by a certain power of the distance to the circle pasted in. Denote the product of all these powers by $\tilde{F}_j$. The field v_j obtained on the jth stage, is related to v_{j-1} by the formula

$$v_j(y) = \tilde{F}_j(y)(\sigma_{j*}v_{j-1}) \circ \sigma_j^{-1}(y).$$

Define now the functions

$$F_j(x) = \tilde{F}_j(\tilde{\sigma}_j(x)), \qquad \tilde{\sigma}_j = \sigma_j \circ \cdots \circ \sigma_1,$$

so that

$$v_j(y) = (F_j \sigma_{j*} v_{j-1}) \circ \sigma_j^{-1}(y).$$

REMARK. The order of each factor of the function $\tilde{F}_j$ depends on the order ν_p of lower terms of the field v_{j-1} at the point $p \in M_{j-1}$ as well as on whether the polar blow-up of p is diacritical or not. More precisely, if this order is denoted by $\varkappa_p$, then $\varkappa_p = \nu_p - 1$ if p is nonelementary and its blow-up is not diacritical (see §6) or if p is a point of contact between the field v_{j-1} and γ_{j-1} (§8). Otherwise $\varkappa_p = \nu_p$ (if the point is nonelementary and its blow-up is diacritical).

PROPOSITION. *In a sufficiently small (punctured) neighborhood of the origin on* $\mathbb{R}^2$, *the functions* F_j, $j = 1, \dots, m$, *satisfy the inequalities*

$$F_j(x) \leqslant c_j/[r(x)]^{\alpha_{j-1}}, \tag{22}$$

where

$$\alpha_i = \max\{0, \varkappa_p \mid p \in M_i\}, \qquad i = 0, \dots, m-1. \tag{23}$$

PROOF. Let d_j be the distance to γ_j. From the above description of $\tilde{F}_j$ it is clear that in a sufficiently small neighborhood of γ_j the following estimate holds

$$\tilde{F}_j(y) \leqslant c_j/[d_j(y)]^{\alpha_{j-1}}, \qquad c_j > 0.$$

The second inequality of Lemma 7 implies (22) for F_j. □

LEMMA 8. *In a sufficiently small neighborhood of the origin the function F satisfies the inequality*

$$F(x) \leqslant c/[r(x)]^{\mu_0}. \tag{24}$$

PROOF. The assertion follows from the inequality

$$\sum_{j=0}^{m-1} \sum_{p\in M_j \mid \varkappa_p>0} \varkappa_p \leqslant \mu_0. \tag{25}$$

Indeed, from (23) one concludes that

$$\alpha_j \leqslant \sum_{p\in M_j \mid \varkappa_p>0} \varkappa_p \tag{26}$$

hence

$$\sum_{j=0}^{m-1} \alpha_j \leqslant \mu_0. \tag{27}$$

The function F is the product, $F = F_1 \cdots F_m$, therefore (22) and (27) together imply (24). It remains to prove (25).

The proof of (25) will be given first for the case when linearization of the field is a nilpotent Jordan block. From Lemma 4 it follows that M_j consists of a single point for all $j = 0, \dots, m-1$: $M_j = \{p(j)\}$, $M_0 = \{p(0)\} = \{0\}$. The order of lowest terms at any point $p(j)$ does not exceed 2: more precisely, in the case 1 of Lemma 4

$$\nu_{p(j)} = 2, \quad j = 1, \dots, m-3, m-1; \qquad \nu_{p(0)} = \nu_{p(m-2)} = 1,$$

while in the cases 2 and 3 $\nu_{p(0)} = 0$ and the rest of ν_p's are equal to 2. Moreover, since all the blow-ups are nondiacritical, we have $\varkappa_{p(0)} = 0$ and $\varkappa_{p(j)} \leqslant 1$ for $j = 1, \dots, m-1$. Then

$$\sum_{j=0}^{m-1} \sum_{p\in M_j} \varkappa_p = \sum_{j=0}^{m-1} \varkappa_{p(j)} \leqslant m-1. \tag{28}$$

By virtue of the same Lemma 4 the number m of stages leading to the nice desingularization, does not exceed $\mu_0/2 + 2$, hence

$$m \leqslant \mu_0 + 1. \tag{29}$$

Inequalities (28) and (29) imply (25).

The proof of (25) in the case $\nu_0 \geqslant 2$ (when the linear part is identically zero) is carried over by induction in the number of stages.

The base of induction $m = 1$.

A. If the blow-up of the origin is not diacritical, then from (4′) it follows that

$$\sum_{p\in\mathbb{R}P^1} \mu_p + (\nu_0 - 1) \leqslant \mu_0 ,$$

since $\nu_0^2 \geqslant 2\nu_0$. Thus $\varkappa_0 \leqslant \mu_0$, since $\nu_0 - 1 = \varkappa_0$ and $\sum \mu_p \geqslant 0$.

B. If the blow-up is diacritical, then from (5′) it follows that

$$\sum_{p\in\mathbb{R}P^1} \mu_p + \sum_{p\in\mathbb{R}P^1} k_q + \nu_0 < \mu_0 ,$$

since $\nu_0^2 > \nu_0$ for $\nu_0 \geqslant 2$. Note that $\sum k_q = 0$ because the resolution goes in one step and hence there are no points of contact. Thus

$$\varkappa_0 < \mu_0 ,$$

since $\sum \mu_p \geqslant 0$ and $\nu_0 = \varkappa_0$ in the diacritical case. The base of induction is established.

Induction step. Assume that (25) is established for all nice desingularizations consisting of $m \leqslant l - 1$ stages, and prove it for $m = l$. As before, one has $\nu_0 \geqslant 2$. Consider the following cases.

A. The first blow-up is not diacritical. Then (4′) implies

$$\sum_{p\in M_1} \mu_p + \varkappa_0 \leqslant \mu_0. \tag{30}$$

For all points from M_1 by the induction assumption

$$\sum \varkappa_q^{(p)} \leqslant \mu_p , \qquad p \in M_1. \tag{31}$$

From (30) and (31) the required formula (25) follows.

B. If the first blow-up is diacritical, then from (5′) it follows that

$$\sum_{p\in M_1} \mu_p + \sum_{q\in M_1} k_q + \nu_0 < \mu_0. \tag{32}$$

Therefore

$$\sum_{p\in M_1} \mu_p + \varkappa_0 < \mu_0 , \tag{33}$$

since $\sum k_q \geqslant 0$, and blow-up of points of contact does not contribute to the left-hand side part of (25). Then repeating the above arguments, we conclude with (25). Lemma 8 is proved. □

LEMMA 9. *As before, let* μ_0 *stand for the multiplicity of a singular point at the origin for a real analytic vector field* v, m *is the number of stages of a nice desingularization and* μ *the maximal multiplicity of singularities of the field* $\tilde{v}$ *on the pasted in curve. Then the estimate*

$$m + \mu \leqslant \mu_0 + 2 \tag{34}$$

holds.

PROOF. First we establish (34) for the case when the order of lowest terms of the field v at the origin is 1 ($\nu_0 = 1$), and the linear part is a nilpotent Jordan block. Applying Lemma 4, one can see that in the case 1 the nice desingularization is achieved in $m = \mu_0/2 + 2$ steps and there appear m nondegenerate saddles on the pasted in curve, that is, $\mu = 1$.

One has $m + \mu = \mu_0/2 + 3 \leqslant \mu_0 + 2$, since $\mu_0 \geqslant 2$. Note that for $\mu_0 = 2$ $m + \mu = \mu_0 + 2$ and this is the only case when the equality in (34) is achieved.

In the case 2 of Lemma 4, $\mu \leqslant 2$ and $m = (\mu_0 + 1)/2$, therefore $m + \mu \leqslant (\mu_0 + 1)/2 + 2 \leqslant \mu_0 + 1 < \mu_0 + 2$, since $\mu_0 \geqslant 3$.

In the case 3, $m = n + 1$ and $\mu = \mu_0 - 2n$, therefore $m + \mu = \mu_0 - n + 1 \leqslant \mu_0 < \mu_0 + 2$, since $n \geqslant 1$.

Thus we proved the assertion of the lemma when the linear part is a nilpotent Jordan block. Now we assume that the linear part is identically zero, that is, $\nu_0 \geqslant 2$. The proof is carried on by induction in the number of stages of a nice desingularization.

Induction base. If a nice desingularization is achieved in one stage, $m = 1$, then in the nondiacritical case

$$\sum_{p \in \mathbb{R}P^1} \mu_p + 1 \leqslant \mu_0, \tag{35}$$

as this is implied by Lemma 3, hence

$$\begin{gathered} m + \mu = 1 + \mu \leqslant \mu_0 < \mu_0 + 2, \\ \mu = \max_{p \in \mathbb{R}P^1} \mu_p. \end{gathered} \tag{36}$$

In the diacritical case for $\nu_0 \geqslant 2$ Lemma 3 implies

$$\sum_{p \in \mathbb{R}P^1} \mu_p + \sum_{q \in \mathbb{R}P^1} k_q + 4 \leqslant \mu_0. \tag{37}$$

Since after a nice desingularization there are no points of contact on the pasted in curve, the second term in (37) vanishes. Together, (36) and (37) imply (34). The induction base is established.

Inductive step. Let (34) be established for all nice desingularizations with the number of stages less than l, and prove it for $m = l$. The following cases are possible.

1. The first stage of the desingularization is nondiacritical. Then (35) holds since $\nu_0 \geqslant 2$. Rewrite (35) in the form

$$\sum_{i=1}^{k} \mu_{p_i} + \sum_{i=1}^{r} \mu_{q_i} + 1 \leqslant \mu_0 , \tag{38}$$

where the first term corresponds to nonelementary points, while the second one to all elementary points. By the induction assumption, for each p_i the inequality

$$m(p_i) + \mu(p_i) \leqslant \mu_{p_i} + 2 \tag{39}$$

holds, where $m(p_i)$ is the number of stages for a nice desingularization of the point p_i, and $\mu(p_i)$ is the maximal multiplicity of singular points occurring on the pasted in curve after this decomposition. Then (38) and (39) imply that

$$\sum_{i=1}^{k} m(p_i) - 2k + \sum_{i=1}^{k} \mu(p_i) + \sum_{i=1}^{r} \mu_{q_i} + 1 \leqslant \mu_0. \tag{40}$$

Note that $\max m(p_i) = l - 1$, so that

$$\sum_{i=1}^{k} m(p_i) \geqslant (l-1) + (k-1). \tag{41}$$

Besides, one has

$$\mu = \max \{ \mu(p_i), \mu_{q_j}, \ i = 1, \dots, k, \ j = 1, \dots, r \} ,$$

hence

$$\sum_{i=1}^{k} \mu(p_i) + \sum_{j=1}^{r} \mu_{q_j} \geqslant \mu + (k + r - 1). \tag{42}$$

From (40), (41), and (42) it follows that

$$(l-1) + (k-1) - 2k + \mu + (k + r - 1) + 1 \leqslant \mu_0 ,$$

which is equivalent to

$$l + \mu + r - 2 \leqslant \mu_0. \tag{43}$$

Since $r \geqslant 0$ and $l = m$, we conclude from (43) that (34) holds.

2. If the first stage is diacritical, than (37) holds because $\nu_0 \geqslant 2$. Represent (37) as

$$\sum_{i=1}^{k} \mu_{p_i} + \sum_{i=1}^{n} \mu_{q_i} + \sum_{i=1}^{s} (\mu_{c_i} + k_{c_i}) + \sum_{i=1}^{t} k_{r_i} + 4 \leqslant \mu_0. \tag{44}$$

Here the first sum is over all nonelementary singularities, the second one over elementary points, the third (the fourth) sum is over singular (resp., nonsingular) points of contact.

By the induction assumption, for each nonelementary singular point p_i the estimate (39) holds. Next, a nonsingular point of contact r_i is completely eliminated after $m(r_i) = k_{r_i} + 1$ stages (Lemmas 5 and 6), and only nondegenerate singular points appear, hence $\mu(r_i) = 1$. Thus

$$m(r_i) + \mu(r_i) = k_{r_i} + 2. \tag{45}$$

At a singular point of contact c_i (recall that c_i is an elementary singularity) the contact is destroyed after $m(c_i) \leqslant k_{c_i}$ stages. The maximal multiplicity occurring in this process is $\mu(c_i) \leqslant \mu_{c_i} + 1$ (by virtue of Lemma 6). Thus

$$m(c_i) + \mu(c_i) \leqslant \mu_{c_i} + k_{c_i} + 1. \tag{46}$$

Combining the estimates (39), (45), (46), and (44), we conclude that

$$\begin{aligned} &\sum_{i=1}^{k}(m(p_i) + \mu(p_i)) - 2k + \sum_{i=1}^{n} \mu_{q_i} \\ &\quad + \sum_{i=1}^{s}(m(c_i) + \mu(c_i)) - s + \sum_{i=1}^{t}(m(r_i) + \mu(r_i)) - 2t + 4 \leqslant \mu_0. \end{aligned} \tag{47}$$

Rewrite (47) in the form

$$\sum m(\cdot) + \sum \mu(\cdot) - 2k - s - 2t + 4 \leqslant \mu_0. \tag{48}$$

The first sum contains $k + s + t$ terms, out of which at least one is equal to $l - 1$ while the remaining ones are $\geqslant 1$. The second sum of $k + n + s + t$ terms (each greater or equal to 1) contains at least one term equal to μ. Taking all of this into account, the inequality (48) takes the form

$$(l - 1) + (k + s + t - 1) + \mu + (k + n + s + t - 1) - 2k - s - 2t + 4 \leqslant \mu_0,$$

which is equivalent to $l + \mu + s + n + 1 \leqslant \mu_0$. Since s, n are nonnegative and $l = m$, one has

$$m + \mu + 1 \leqslant \mu_0,$$

whence the estimate (34) follows. Lemma 9 is proved. □

Completion of the proof of the principal theorem. Let σ be a nice desingularization of a singular point O for the polynomial vector field $P_N(x)$, see (20). Apply this desingularization to the smooth field

$$v(x) = P_N(x) + w(x), \qquad w(x) = o(|x|^N).$$

We obtain a field

$$\tilde{v}(y) = \tilde{P}_N(y) + \tilde{w}(y), \qquad y = \sigma(x),$$

defined in some neighborhood of the pasted in curve γ, where

$$\tilde{w}(y) = (F\sigma_* w) \circ \sigma^{-1}(y).$$

By Lemmas 7, 8, and 9, in a sufficiently small neighborhood U of the point O on $\mathbb{R}^2$ the following estimate holds,

$$|\tilde{w}(\sigma(x))| \leqslant c \cdot r(x)^{-\mu_0} \cdot r(x)^{-m} \cdot o(r(x)^N) = o(r(x)^{r+\mu_0-m}) = o(r(x)^{\mu}),$$

where μ as usual stands for the maximal multiplicity of singularities of $\tilde{P}_N(y)$ occurring on γ in the process of desingularization.

The presence of a characteristic orbit for P_N at the origin implies that $\tilde{P}_N(y)$ has at least one singular point on γ, therefore $\mu \geqslant 1$. Thus the assertion of Lemma 7 implies that in a sufficiently small neighborhood of γ

$$\tilde{w}(y) = o(d(y)^{\mu}). \tag{49}$$

It follows from (49) that all the singularities of $\tilde{P}_N$ on γ do not change their position after adding $\tilde{w}(y)$. Moreover, in sufficiently small neighborhoods of each of these points, the fields $\tilde{P}_N$ and $\tilde{P}_N + \tilde{w}$ are topologically equivalent. Indeed, for a nondegenerate point from the classification theorem list (§4), the corresponding field is topologically equivalent to its linearization [2], while the linear parts of $\tilde{P}_N$ and of $\tilde{P}_N + \tilde{w}$ are the same. For a degenerate elementary singularity from this list, the fields $\tilde{v}$ and $\tilde{P}_N$ are equivalent by the Shoshitaishvili reduction theorem [2, p. 63] and one-dimensionality of the center manifold.

Thus the desingularization schemes for P_N and v coincide, therefore the fields are orbitally topologically equivalent (see §5). This proves the principal theorem. □

REMARK. The estimate (34) in Lemma 9 for the case $\nu_0 \geqslant 2$ can be improved. Namely,

$$m + \mu \leqslant \mu_0.$$

The principal theorem can be thus formulated as follows: *the order of a topologically sufficient jet for a vector field possessing a characteristic orbit in a singular point of finite multiplicity does not exceed twice the multiplicity of the point, provided that the linearization is identically zero.*

The author wishes to express his deep gratitude to his advisor Professor Yu. Ilyashenko for continuous and attentive guidance and helpful advice.

References

1. F. Dumortier, *Singularities of vector fields on the plane,* J. Differential Equations **23** (1977), no. 1, 53–104.
2. V. I. Arnold and Yu. S. Ilyashenko, *Ordinary differential equations* Dynamical systems, I, Itogi Nauki. Sovremennye Problemy Matematiki: Fundamental′ nye Napravleniya., vol. 1, VINITI, Moscow, 1985; English transl., Encyclopaedia of Math. Sci., vol. 1, Springer-Verlag, Heidelberg, 1988.
3. Yu. I. Ilyashenko, *Algebraically and analytically solvable local problems in theory of ordinary differential equations,* Trudy Sem. Petrovsk., vol. 12, 1987, pp. 118–136; English transl. in J. Soviet Math. **47** (1989).

4. ______, *Dulac memoir "Sur les cycles limites" and related topics of local theory of ordinary differential equations*, Uspekhi Mat. Nauk **40** (1985), no. 6, 41–78; English transl. in Russian Math. Surveys **40** (1985).
5. J.-F. Mattei and R. Moussu, *Holonomie et integrales premiéres*, Ann. Sci. École Norm. Sup. (4) **13** (1980), 469–523.
6. A. Andronov, E. Leontovich, I. Gordon, and A. Mayer, *Qualitative theory of second order dynamical systems*, "Nauka", Moscow, 1967; English transl., Halsted Press, New York and Toronto, 1973.
7. A. Van den Essen, *Reduction of singularities of the equation* $A\,dy = B\,dx$, Equations differentielles et systemes de Pfaff dans le champ complexe (R. Gerard, J.-P. Ramis, eds.), Lecture Notes in Math., vol. 712, Springer-Verlag, Heidelberg, 1979, pp. 44-60.

Translated by S. YAKOVENKO

MOSCOW STATE UNIVERSITY, MOSCOW, RUSSIA

Amer. Math. Soc. Transl.
(2) Vol. **165**, 1995

On Few-Parameter Generic Families of Vector Fields on the Two-Dimensional Sphere

A. KOTOVA AND V. STANZO

Introduction

In this paper we investigate generic 2- and 3-parameter smooth families of vector fields on the two-dimensional sphere. The investigation is focused on determining the number of limit cycles that may appear in such families after small variations of the parameters. In other words, we study bifurcations of limit cycles in generic few-parameter families of vector fields on the sphere.

It is known (see, for example, the introductory paper [8] in this volume) that limit cycles are born from limit periodic sets that are closed invariant subsets of the sphere, eventually containing arcs of nonisolated singularities of vector fields. But since we study only *generic* families, with all singularities isolated, such limit periodic sets can be only *polycycles* (cyclically ordered collections of singular points together with trajectories connecting them in the specified order; the precise definition is given below). The question on the number of limit cycles which can be born from a polycycle occurring in a generic finite-parameter family, is closely related to the *Hilbert–Arnold problem* [8]. In order to solve this problem, at least for 2- and 3-parameter families, one needs to list all polycycles occurring in such families. This was our primary goal when we were writing this paper.

But in an attempt to compose such a list, we realized that it would be more natural to consider not only polycycles, but also their *ensembles*, unions of polycycles with some elements (arcs or vertices) shared by more than one polycycle; as soon as such ensembles occur in generic 2- and 3-parameter families, their bifurcations must be studied in the above context. It turns out that in 2-parameter families all ensembles are finite: they can be represented as finite unions of polycycles. But in the case of three parameters a completely new phenomenon was observed: there are 3-parameter families which for certain isolated values of parameters exhibit continuum of coexisting polycycles, and this effect cannot be destroyed by small perturbations of the families. Moreover, for any specified number $N \in \mathbb{N}$ one can construct a 3-parameter family of vector fields such that in this family and in all families sufficiently close to it, at least N limit cycles are born *simultaneously*.

1991 *Mathematics Subject Classification*. Primary 34C05, 34C20.

*The introductory §1 and the Appendix to §2 were written by S. Yakovenko.

The second author was supported in part by the Grant M 98000 of the ISF.

Thus there cannot be given any universal bound for the cyclicity of that ensemble in terms of the degree of its degeneracy (see below).

The paper consists of three relatively independent sections, with independent numeration of formulas, propositions and theorems. In §1 we establish the principles of classification of polycycles generically occurring in 1-, 2- and 3-parameter families and their ensembles.

The choice of the classification principles was motivated by the main goal of the first part of the paper: to compose a list of all polycycles which could generate limit cycles after perturbation. The first step in investigating bifurcations of a given polycycle is to represent the associated Poincaré map as a composition of *correspondence maps* near singular points and near regular arcs of that polycycle. The behavior of the correspondence maps near singular points is usually known (at least in small codimensions). So the main difficulty is to describe effects of taking compositions of such maps. The result depends on the combinatorial structure of the composition (the types of the correspondence maps and the order in which they occur in the composition). This way of reasoning leads naturally to the notion of the *combinatorial equivalence of rigged polycycles*, abbreviated to $\mathfrak{c}$-*equivalence*. In fact, it seems that all specialists working in this area, already think in terms of such equivalence when specifying "types" of polycycles. So the reader who understands the pictures shown in Figures 6A and 6B, may skip §1 and start reading from §2.

However, the formal definition tends to be rather cumbersome, mostly because with codimension increasing the singular points it may become rather complicated from the point of view of their local description. This formal definition of combinatorial equivalence does not necessarily make sense for generic k-parameter families for $k \geqslant 4$.

After introducing the notion of the combinatorial equivalence of polycycles, all equivalence classes are listed in §2. At the same time we give an alternative proof of the classical result for 1-parameter families from the book [7]. In the Appendix to this section, a list of cases of known cyclicity is given with a separate reference list.

The principal subject of §3 is an infinite ensemble of polycycles which occurs in generic 3-parameter families of vector fields. This ensemble is called *the lips*. We prove that an arbitrary number of limit cycles may be born from the lips after bifurcation. The bifurcation diagrams depend essentially on the transition map along flow curves between two smooth transversals. We investigate these diagrams using the notions of the standard and generalized Legendre transform of the graph of the transition function $y = f(x)$.

The introductory §1 and the Appendix to §2 were written by S. Yakovenko. All results from §2 and Theorem 2 from §3 (on the unboundedness of the number of limit cycles that may born from the lips) belong to A. Kotova. The bifurcation diagrams for the lips were constructed by V. Stanzo. He wrote §2 and §3; the proof given there was obtained using some ideas of S. Trifonov.

Acknowledgements. The authors are grateful to Yu. Ilyashenko and S. Trifonov for numerous and very helpful discussions. R. Roussarie and M. El Morsalani provided us with references to a large part of sources quoted in the Appendix to §2. D. Novikov made several remarks resulting in amelioration of the original text.

§1. Principles of classification of polycycles and their ensembles in small codimensions

The goal of the first two sections of the paper is to compose a list of polycycles occurring in generic few-parameter families (here and below "few" means "three or less"). This list, however, is not the ultimate goal, but rather an organizer for further studies of bifurcations occurring in generic few-parameter families: in the introductory paper to this volume [8] the notion of a *bifurcation number* $B(n)$ is introduced within the framework of the *Hilbert–Arnold problem*: speaking somewhat loosely, $B(n)$ is the maximal number of limit cycles that may be born from a polycycle in a generic n-parameter family. Complete investigation of bifurcations of polycycles from the lists obtained in the current paper would explicitly give the numbers $B(2)$ and $B(3)$.

In particular, various theorems on bifurcations of polycycles occurring in generic 2-parameter families yield the following result (see also [8]).

THEOREM. *We have*

$$B(2) = 2.$$

In other words, in a generic 2-parameter family of smooth vector fields on the sphere, any polycycle cannot generate more than 2 limit cycles.

To compose the list mentioned above, we need to introduce a certain equivalence relation between the polycycles. An ideal choice of that equivalence would mean that equivalent polycycles have homeomorphic bifurcation diagrams, but apparently the latter property can be established only after complete investigation of all polycycles. Thus we need some intermediate equivalence relation, which is introduced in §1 under the name of *𝔠-equivalence of rigged polycycles.* All definitions in §1 were chosen in order to incorporate as much of the information accumulated about the existing tools of investigation, as possible. But even this attempt was not completely successful: we still had to introduce a class of singularities which is not a single 𝔠-equivalence class. This class is labelled as *degenerate cusp* in Table 3 below. Its investigation is apparently much more difficult, compared to the other genuine 𝔠-equivalence classes.

1.1. Definition of 𝔠-equivalence of singular points. Unlike bifurcation diagrams for polycycles, bifurcation diagrams for singular points of vector fields are known, see [3], at least for degeneracies of small codimensions.

Recall that a ν-parameter unfolding of a vector field $v = v(x)$ with $x \in (\mathbb{R}^2, 0)$ is a local smooth family of vector fields $v(x, \varepsilon)$, $\varepsilon \in (\mathbb{R}^\nu, 0)$, such that $v(x, 0) \equiv v(x)$. A *bifurcation diagram* for an unfolding is a partition of the neighborhood $(\mathbb{R}^\nu, 0)$ of the parameter space into disjoint subsets such that on each subset the representatives $v(\cdot, \varepsilon)$ are topologically orbitally equivalent to each other.

DEFINITION 1A [3]. Two unfoldings $v_1(x, \varepsilon)$ and $v_2(y, \mu)$ of germs of vector fields v_1, v_2 are (strongly) *topologically orbitally equivalent*, if there exist germs of homeomorphisms

$$h\colon (\mathbb{R}^\nu, 0) \ni \varepsilon \mapsto \mu \in (\mathbb{R}^\nu, 0),$$
$$H\colon (\mathbb{R}^2, 0) \times (\mathbb{R}^\nu, 0) \ni (x, \varepsilon) \mapsto (y, h(\varepsilon)) \in (\mathbb{R}^2, 0) \times (\mathbb{R}^\nu, 0)$$

such that for any ε the restriction $H(\cdot, \varepsilon)$ conjugates the phase portraits of $v_1(\cdot, \varepsilon)$ and $v_2(\cdot, h(\varepsilon))$ in the domains of the unfoldings (we do not require preservation of orientation of phase curves).

An unfolding $v(x, \varepsilon)$ is a (topological orbital) *versal deformation* of the germ $v(\cdot, 0)$ if any other unfolding of that germ is topologically orbitally equivalent to an unfolding $v(\cdot, h(\varepsilon))$ *induced* from the unfolding $v(\cdot, \varepsilon)$ by an appropriate reparametrization.

Usually we omit the adjectives *strong orbital*, speaking about *topological equivalence* of unfoldings.

DEFINITION 1B [3]. Two unfoldings are *weakly topologically orbitally equivalent* if in the Definition 1A we drop the requirement that H is a homeomorphism, replacing it with the condition that for all $\varepsilon \in (\mathbb{R}^\nu, 0)$, the restrictions $H(\cdot, \varepsilon)$ are homeomorphisms that are defined in a certain *common* neighborhood of the origin $(\mathbb{R}^2, 0)$ and conjugate the phase portraits of $v_1(\cdot, \varepsilon)$ and $v_2(\cdot, h(\varepsilon))$.

DEFINITION 2A. Two germs v_1, v_2 of vector fields at singular points are called $\mathfrak{c}$-*equivalent* if their versal deformations are topologically orbitally equivalent, possibly after the time reversal. The number of the parameters of the versal deformation is called the *codimension* of the $\mathfrak{c}$-equivalence class of a germ.

REMARK. In many cases two germs of vector fields turn out to be $\mathfrak{c}$-equivalent if their jets of some low order belong to the same connected component of a certain semialgebraic subset of the corresponding jet space; examples are given below. One can see that in these cases the codimension of the (common) $\mathfrak{c}$-equivalence class as it is introduced in Definition 2A coincides with the usual codimension of this semialgebraic subset. This observation allows us to prove that a given class of germs is actually a single $\mathfrak{c}$-equivalence class in our sense (Theorem 1). The main part of the proof is simply a reference to the principal achievements of the local bifurcation theory of last decades. The remaining part of the proof is a standard application of Thom's transversality theorem.

DEFINITION 2B. Two germs are *weakly* $\mathfrak{c}$-*equivalent* if in Definition 2 the topological orbital equivalence is replaced by the weak topological orbital equivalence.

In other words, two germs are weakly $\mathfrak{c}$-equivalent, if their generic unfoldings with a sufficient number of parameters have homeomorphic bifurcation diagrams.

In this definition "generic unfoldings" means "unfoldings belonging to a certain residual subset from the (Banach) space of all C^∞-smooth unfoldings defined on a certain domain $(\mathbb{R}^2_x, 0) \times (\mathbb{R}^\nu_\varepsilon)$."

Thus to fix a $\mathfrak{c}$-equivalence class of a singularity means to specify the bifurcation pattern (in other words, to describe the bifurcation diagram up to a homeomorphism). In particular, $\mathfrak{c}$-equivalent germs are orbitally topologically equivalent themselves.

DEFINITION 3. The germ of a homeomorphism $H_0: (\mathbb{R}^2, 0) \to (\mathbb{R}^2, 0)$ is called a $\mathfrak{c}$-*equivalence* between two germs of vector fields, if it can be extended to a (strong) topological equivalence $H(\cdot, \varepsilon)$ between the versal unfoldings of those germs in the sense of Definition 1A, so that $H_0 = H|_{\varepsilon=0}$.

REMARK. There is no nontrivial weak counterpart for this definition.

EXAMPLE 1. Two germs of vector fields,

$$\begin{cases} \dot{x}_1 = x_1 , \\ \dot{x}_2 = -x_2^3 , \end{cases} \quad \text{and} \quad \begin{cases} \dot{y}_1 = y_1 , \\ \dot{y}_2 = -y_2^3 , \end{cases}$$

are c-equivalent, but any c-equivalence between them should map the x_i-axis into the y_i-axis for $i = 1, 2$, while the homeomorphism sending Ox_1 to Oy_2 and Ox_2 to Oy_1 and conjugating the phase portraits *is not a* c-*equivalence.*

1.2. Topological classification of singular points and their characteristic trajectories. The topological equivalence of singularities is a more rough, although a more explicit relation. We need it in order to describe types of orbits that tend to/from singular points.

The topological structure of a singularity satisfying the Łojasiewicz condition and having a characteristic orbit (a trajectory which tends to/from the origin with a certain limit slope), can be easily described in terms of *sectors*, see [7] for precise definitions. One of the basic results in [7] claims that an arbitrarily small neighborhood of a singularity can be partitioned into a finite union of *parabolic*, *elliptic*, and *hyperbolic* sectors.

Recall that *parabolic sectors* are filled with trajectories that tend to the singular point, possibly after the time reversal; the *hyperbolic* sectors are filled with trajectories that enter and then leave the neighborhood, and *elliptic sectors* are those filled by trajectories biasymptotic to the singularity. The boundaries between different sectors are phase curves of the vector field.

The partition of a neighborhood of the origin into sectors essentially depends on the choice of the neighborhood. In particular, if $U = \bigcup S_j$ is such a partition, and $\tilde{U} \subset U$ is a smaller neighborhood partitioned as $\tilde{U} = \bigcup \tilde{S}_j$, then in general it is not true that $\tilde{S}_j = S_j \cap \tilde{U}$. However, the latter assertion holds if S_j is a hyperbolic sector. On the contrary, if S_j is an elliptic sector and $\tilde{U}$, together with its closure, belongs to U, then $\tilde{S}_j \subsetneq S_j \cap \tilde{U}$: the difference $S_j \cap \tilde{U} \setminus \tilde{S}_j$ consists of two parabolic sectors that can be eventually joined to other parabolic sectors.

Thus the boundary between a hyperbolic and any other (in particular, another hyperbolic) sector is defined independently of the choice of the small neighborhood of the singularity. This is not the case for boundaries between elliptic and parabolic sectors: after shrinking the neighborhood, a boundary may stop to be one. Clearly, two parabolic sectors cannot have a common boundary unless their union is itself a parabolic sector.

DEFINITION 4. We say that a trajectory is a *p-curve* (with respect to a certain singular point) if after a suitable choice of the neighborhood of this singular point the trajectory belongs to the interior of a parabolic sector of this singular point. The trajectory is called an *h-curve* if it is a common boundary for two hyperbolic sectors. Finally, we say that a trajectory is a *b-curve* if it is a boundary between a hyperbolic and a parabolic sector.

In what follows we will mostly deal with the following classes of topological equivalence of vector fields on the (x, y)-plane near the origin (the classes are specified by their simplest representatives):

(1) Saddle: $\dot{x} = x$, $\dot{y} = -y$;

(2) Sink: $\dot{x} = -x$, $\dot{y} = -y$;
(3) Source: $\dot{x} = x$, $\dot{y} = y$;
(4) Saddle-node: $\dot{x} = x^2$, $\dot{y} = \pm y$; we call a saddle-node *contractive* if the sign is $-$ and *dispersive*, if the sign is $+$;
(5) Cusp: $\dot{x} = y$, $\dot{y} = x^2$.

The phase portraits of the corresponding vector fields are shown on Figure 0.

Note that p-curves exist for the sink, source and saddle-node; b-curves exist only for the saddle-node, while h-curves exist for the saddle, saddle-node and the cusp: the cusp has two hyperbolic sectors separated by two h-curves, while the saddle has four hyperbolic sectors separated by two pairs of h-curves.

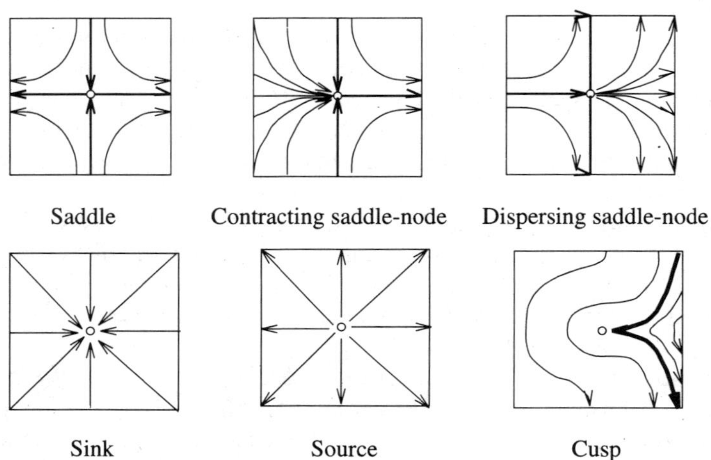

FIGURE 0. Topology of phase portraits and distinguished trajectories for simple singularities.

1.3. Description of $\mathfrak{c}$-equicalence classes in codimensions $\leqslant 3$. The four tables below give an almost complete $\mathfrak{c}$-classification of singular points of small codimension. Let $v(x)$, $x \in (\mathbb{R}^2, 0)$, be the germ of a smooth planar vector field, $v(0) = 0$, and let its linear part be Ax, where A is a 2×2-matrix with real coefficients. Denote by λ_j, $j = 1, 2$, the eigenvalues of this matrix.

Recall that codimension 0 means that the corresponding singularities are generic. It is known that they are structurally stable [7]. In codimension 0 there are two $\mathfrak{c}$-equivalence classes, listed in Table 0. Recall that in the definition of $\mathfrak{c}$-equivalence we do allow for the orientation reversal, which means in particular that we disregard the difference between stable and unstable singular points.

TABLE 0. Classes of $\mathfrak{c}$-equivalence in codimension 0

Class	Description	Topology
Hyperbolic saddle	$\lambda_i \in \mathbb{R}$, $\lambda_1\lambda_2 < 0$	Saddle
Foci and nodes	$\operatorname{Re}\lambda_1 \cdot \operatorname{Re}\lambda_2 > 0$	Sink or source

All other types of singularities have codimension $\geqslant 1$. They can be divided into four groups:

- *elliptic points* with a pair of purely imaginary eigenvalues $\pm i\omega$ with $\omega \neq 0$ (we will always assume that $\omega > 0$),
- *semihyperbolic points* with one zero and one nonzero eigenvalue,
- *cuspidal singularities*, whose linearization is a nonzero nilpotent matrix J, and
- *singularities with zero linearization matrix.*

REMARK. The last case means degeneracy of codimension 4 (all elements of the 2×2-matrix A must vanish), hence this case will never occur in our study.

LEMMA E [12], [13]. *An elliptic singularity without nontrivial formal first integral is C^∞-smoothly orbitally equivalent to the vector field on the plane $z = x+iy \in \mathbb{C} \simeq \mathbb{R}^2$ of the form*

$$\dot{z} = z(i\omega + a_\nu |z|^{2\nu} + \alpha |z|^{4\nu}), \qquad \nu \geqslant 1, \quad a_\nu, \alpha \in \mathbb{C}, \quad \operatorname{Re} a_\nu \neq 0. \tag{1}$$

LEMMA S [12], [13]. *A semihyperbolic singularity satisfying the Łojasiewicz condition is C^∞-smoothly orbitally equivalent to*

$$\begin{cases} \dot{x} = \pm x^{\mu+1} + \beta x^{2\mu+1}, \\ \dot{y} = \pm y, \end{cases} \qquad \mu \geqslant 1, \ \beta \in \mathbb{R}. \tag{2}$$

REMARKS. A singularity of the type (1) topologically is a source if $\operatorname{Re} a_\nu > 0$ and a sink if $\operatorname{Re} a_\nu < 0$. We call such a singularity a *slow focus* if $\nu = 1$, an *ultra-slow focus* if $\nu = 2$, and an *ultra-ultra-slow focus* if $\nu = 3$. The codimension of the ultra $^{\nu-1}$-slow focus is ν.

Semihyperbolic singularities with μ odd are topological saddle-nodes (independently of the choice of the combination of signs), and for μ even they are either saddles or nodes, depending on the signs. The simplest case $\mu = 1$ is called the *double saddle-node.*

LEMMA C [1]. *The 2-jet of a cuspidal point can be put into the form*

$$\begin{cases} \dot{x} = y, \\ \dot{y} = a x^2 + b xy, \end{cases} \tag{3}$$

by a polynomial transformation. □

DEFINITION 5. A cuspidal point is called a *Bogdanov–Takens cusp* if the 2-jet of the vector field has the form (3) with $ab \neq 0$. Otherwise a cuspidal point is called a *degenerate cusp.*

REMARK. The Bogdanov–Takens cusp is topologically equivalent to the cuspidal point from the above list, $\dot{x} = y$, $\dot{y} = x^2$, possibly after symmetry $x \mapsto -x$ and/or time reversal.

Conversely, the topological type of a degenerate cusp depends on whether $a = 0$ or $b = 0$ in the form (3), and also on the sign of the nonzero coefficient. Thus degenerate cusps *do not constitute a single* $\mathfrak{c}$*-equivalence class.* Still, such points do not occur in generic ν-parameter families of vector fields for $\nu < 3$, and in 3-parameter values they may occur only for isolated values of the parameters.

DEFINITION 5A. A *nonstandard class* is the class of singularities with 2-jet polynomially equivalent to the form (3) with $ab = 0$. We define the *codimension* of this class to be 3.

Introducing this artificially created class will simplify the formulation of the main result below, Theorem 1.

Description of $\mathfrak{c}$-equivalence classes in codimension $\leqslant 3$, is given below in the form of a series of tables. In these tables we refer to the parameters occurring in the normal forms (1) and (2). The nonstandard class appears in Table 3; both Table 1 and Table 2 contain only true $\mathfrak{c}$-equivalence classes.

TABLE 1. Classes of $\mathfrak{c}$-equivalence in codimension 1

Class	Description	Topology
Slow foci	Elliptic points with $\nu = 1$	Sink or source depending on the sign of $\operatorname{Re} a_1$
Double saddle-node	Semihyperbolic points with $\mu = 1$	Saddle-node, contractive or dispersive depending on the sign of $\dot{y}/y$

TABLE 2. Classes of $\mathfrak{c}$-equivalence in codimension 2

Class	Description	Topology
Ultra-slow foci	Elliptic points with $\nu = 2$	Sink or source, depending on the sign of $\operatorname{Re} a_2$
Degenerate saddle	Semihyperbolic points with $\mu = 2$, different signs in (2)	Saddle
Degenerate node	$\mu = 2$, coinciding signs in (2)	Sink or source, depending on the (common) sign
Bogdanov–Takens cusp	Cuspidal singularity with generic nonlinear terms: 2-jet equivalent to the normal form $\dot{x} = y$, $\dot{y} = x^2 \pm b\, xy$ with $b \neq 0$	Cusp

THEOREM 1. 0. *Each standard class from the Tables* 0–3 *is a single* $\mathfrak{c}$*-equivalence class in the sense of Definition* 2.

1. *Singular points which may occur in generic* k*-parameter families of smooth vector fields on the sphere for* $k \leqslant 3$, *belong to one of the eleven* $\mathfrak{c}$*-equivalence classes listed in Tables* 0–3 (*including the nonstandard class which consists of all degenerate cusps*).

2. *Singularities from a* $\mathfrak{c}$*-equivalence class of codimension* $\nu \leqslant k$ *generically occur on submanifolds of codimension* ν *in the parameter space.*

3. *Partition of the parameter space by submanifolds corresponding to different* $\mathfrak{c}$*-equivalence classes, is a regular stratification of the parameter space.* For a given $\mathfrak{c}$-equivalence type $\mathfrak{c}$ we call the corresponding submanifold the stratum $\mathfrak{c} = \text{const}$.

TABLE 3. Classes of $\mathfrak{c}$-equivalence in codimension 3

Class	Description	Topology
Ultra-ultra-slow foci	Elliptic points with $\nu = 3$	Sink or source, depending on the sign of $\operatorname{Re} a_3$
Degenerate saddle-nodes	Semihyperbolic points with $\mu = 3$	Saddle-node, contractive or dispersive depending on the sign of $\dot{y}/y$
Nonstandard $\mathfrak{c}$-equivalence class:		
Degenerate cusp	A cuspidal point with the 2-jet polynomially equivalent to $\dot{x} = y$, $\dot{y} = a\,x^2 + b\,xy$ with $ab = 0$	Many different cases, some of them not from the list of five simplest types mentioned in §1.2. In particular, a pair of parabolic sectors may occur for the so called elliptic type, see [11, §3]

4. *For any standard $\mathfrak{c}$-equivalence class $\mathfrak{c}$ with h- or b-curves, those curves depend smoothly on the parameters along the stratum* $\mathfrak{c} = \text{const}$.

REMARKS. 1. The smoothness assertion means that if we choose a transversal to an h- or b-curve of a singularity occurring for $\varepsilon = 0$, then for all small values of $\varepsilon \in (\mathbb{R}^k, 0)$ belonging to the corresponding stratum $\mathfrak{c} = \text{const}$, there will be again an h- (resp., a b)-curve intersecting this transversal, and the intersection point will smoothly depend on ε.

2. The regular stratification is a partition of a manifold into a locally finite number of submanifolds (strata) of different dimensions such that closures of each stratum consist of the stratum itself and some other strata of strictly smaller dimensions, and there must be a certain regularity of behavior of tangent subspaces on the boundaries. In fact, without loss of generality one may describe regular stratifications as preimages of real algebraic subsets by generic mappings.

SKETCH OF THE PROOF. One can easily verify that the conditions defining each $\mathfrak{c}$-equivalence class from Tables 0–3 (those conditions are given in the second column) determine semialgebraic subsets C_j, $j = 1, \ldots, 11$, in, say, 3-jet space of vector fields. The sets C_j together constitute a semialgebraic partition of the 3-jet space modulo an algebraic subset of codimension at least 4: there exists an algebraic subset C^* of codimension 4 such that

$$\left(\bigcup_{j=1}^{11} C_j\right) \cup C^* = (\text{the whole } 3\text{-jet space}).$$

The most difficult (though a well-known one) part of Theorem 1 is to show that all germs with jets belonging to any C_j really constitute a single $\mathfrak{c}$-equivalence class. For all standard classes this assertion is known, although sometimes rather difficult to prove, e.g., for nondegenerate cusps. We do not want to enter into details here: for references see [1], [3]. In fact, the formal proof for the case of ultra $^\nu$-slow foci for $\nu = 1, 2$ is not published yet. Moreover, for $\nu = 3$ the corresponding result is not true: as was discovered by R. Roussarie, unfoldings of ultra ^{3}-slow focus have functional moduli described in [3].

As soon as the first assertion is established, the remaining three are relatively easy. Since

$$\operatorname{codim} C_j = (\text{codimension of the corresponding } \mathfrak{c}\text{-equivalence class } \mathfrak{c}_j),$$

and all C_j together with C^* constitute a semialgebraic stratification of the entire jet space, standard arguments involving Thom's transversality theorem prove assertions 2 and 3 of the theorem (see also §4 from [10]).

We give only a brief explanation on how to prove the smoothness assertion for distinguished curves. For hyperbolic saddles this fact is well known. For elliptic points there is nothing to prove. For semihyperbolic points the result follows from theorems establishing a smooth normal form for unfoldings of elementary singularities [5], see also §1 from [10].

For the (nondegenerate) cusp we make a desingularization: the scheme of successive blowing-ups remains the same as soon as we are within the same $\mathfrak{c}$-equivalence class. After desingularization h-curves become separtrices of hyperbolic saddles occurring on the pasted in configuration of projective lines, and the assertion on smooth dependence on parameters follows from the standard results on saddle separatrices. □

REMARK. Partitioning the nonstandard class into true $\mathfrak{c}$-equivalence classes constitutes a very difficult problem; some partial results related to unfoldings of nilpotent singularities in generic 3-parameter families, were obtained on the heuristic level by A. Bazykin, F. Berezovskaya, and A. Khibnik. An intensive study was carried out by F. Dumortier, R. Roussarie, and J. Sotomayor, see [11, §§2,3] and references therein. Still, some problems remain open [11].

1.4. Oriented polycycles. Limit cycles can be born after perturbations of *limit periodic sets* of vector fields; we do not need the formal definition of the latter object (see the introductory paper [8] to this volume), since they can occur in generic finite-dimensional families only in the form of *(oriented) polycycles*.

DEFINITION 6. An oriented polycycle of a vector field on a sphere $\mathbb{S}^2$ is a cyclically ordered union of points (vertices) $A_0, A_1, \ldots, A_n = A_0$ (some of them may coincide) and different phase curves $\gamma_0, \ldots, \gamma_{n-1}$ endowed with the natural orientation, such that for any $i = 0, \ldots, n-1$ the curve $\gamma_i = \gamma_i(t)$ tends to the point A_i as $t \to -\infty$ and tends to A_{i+1} as $t \to +\infty$ (such curves are called *connections*, or *arcs of the polycycle*). In order to avoid trivial exceptions, we require that there is at least one singular point (vertex) on the polycycle.

In what follows we consider only polycycles that can generate limit cycles after bifurcations. We distinguish two subcases:

a) polycycles that consist of a single vertex, and
b) polycycles with at least one connection.

DEFINITION 7. An *ensemble* of oriented polycycles is a union of different oriented polycycles such that any two of them have at least one common point. The ensemble is *maximal* if it is not a proper subset of another (larger) ensemble.

REMARK. The notion of an ensemble (especially the maximal one) assumes that the domain of the vector field is explicitly or implicitly specified. For example, one may talk about maximal ensembles either on the sphere or in a certain (closed) annulus.

DEFINITION 8. We say that an arc of a polycycle is a $\sigma_1\sigma_2$-curve, where σ_i are symbols from the alphabet $\{p, b, h\}$ if the germs of that arc at the two endpoints are σ_i-curves in the sense of Definition 4.

There are nine possible types of oriented connections and six nonoriented types: pp, pb, ph, bb, bh, hh.

1.5. Definition of c-equivalence of polycycles and their ensembles. Now we give a definition of c-equivalence of polycycles: for polycyles which consist of just one singularity, it coincides with the above definition of c-equivalence of singular points.

DEFINITION 9. Two oriented polycycles for two vector fields v and v',

$$A_0 \xrightarrow{\gamma_0} A_1 \xrightarrow{\gamma_1} \cdots \xrightarrow{\gamma_{n-1}} A_{n-1} \xrightarrow{\gamma_n} A_0$$

and

$$A'_0 \xrightarrow{\gamma'_0} A'_1 \xrightarrow{\gamma'_1} \cdots \xrightarrow{\gamma'_{n-1}} A'_{n-1} \xrightarrow{\gamma'_n} A'_0$$

are *combinatorially equivalent*, or c-equivalent, if after a proper cyclical re-enumeration and/or possible direction reversal for one of the fields, say, $v \mapsto -v$,

(1) their vertices (singular points) and arcs (connections) are in one-to-one correspondence (repetitions of vertices taken into account); and
(2) there exist germs of homeomorphisms $H_i: (\mathbb{S}^2, A_i) \to (\mathbb{S}^2, A'_i)$, $i = 0, \dots, n-1$, that are c-equivalences between the germs of v at A_i and v' at A'_i (see Definition 3) and take the germs at the points A_i of all connections γ_j containing A_i in their closure to the germs of γ'_j at A'_i.

This definition can be immediately generalized for the case of ensembles of polycycles. Recall that even an infinite ensemble may contain only a finite number of singular points if it occurs in a generic finite-parameter family.

DEFINITION 10. Two ensembles of polycycles are c-equivalent if after an appropriate cyclical re-enumeration and/or direction reversal

(1) their vertices and arcs are in one-to-one correspondence, the correspondent vertices are c-equivalent, and the correspondence between connections is continuous in the sense of the Hausdorff metric;
(2) the germs of c-equivalencies H_i may be chosen in such a way that the germ of the union of all arcs of the first ensemble at the point A_i is taken by H_i into the analogous germ for the second ensemble, preserving the above correspondence.

EXAMPLE 2. The polycycles shown on Figure 1 are c-equivalent, although the flows in the small neighborhoods of the polycycles are not topologically equivalent.

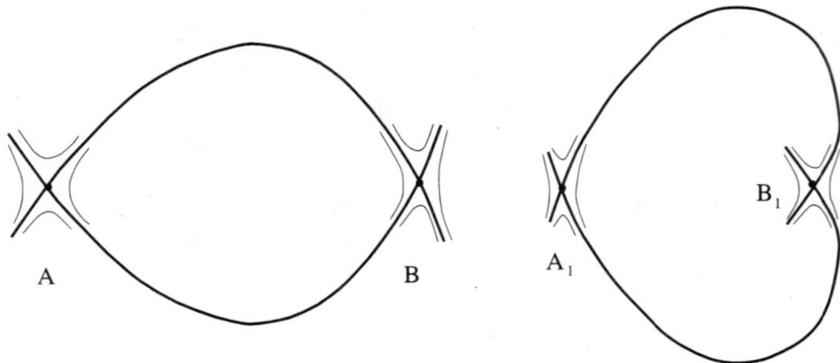

FIGURE 1. Equivalence of the *lentil* and *heart*: singular points are nondegenerate saddles.

EXAMPLE 3. Polycycles with a homoclinic saddle-node connection, shown in Figure 2A, are not equivalent, since in one case the loop is an hp-curve, while the other loop is of hb-type.

In Figure 2B there are three nonequivalent maximal ensembles, *the malignant frown*, the *plump lips* and the *paresis* (the *spadesuit*). However, all the three contain as a proper subensemble the configuration $\mathfrak{c}$-equivalent to the *lips*; the latter configuration is not maximal. The detailed investigation of the lips will be given in §3 below.

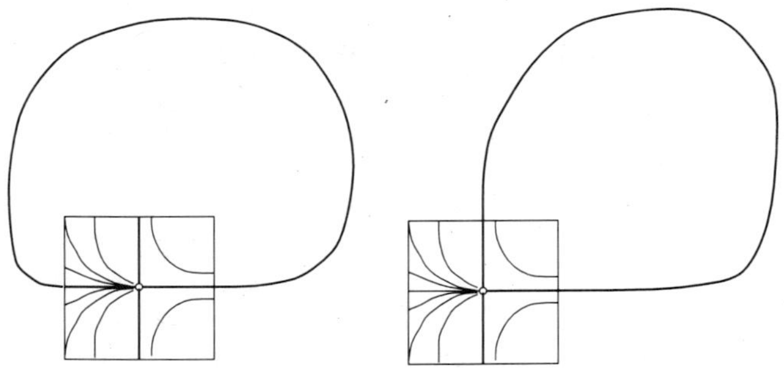

FIGURE 2A. $\mathfrak{c}$-equivalence of singularities is insufficient for $\mathfrak{c}$-equivalence of polycycles.

1.6. Codimension of polycycles and ensembles. In this section we formalize an evident assertion that occurrence of certain types of connections increases the degeneracy degree of a polycycle/ensemble as a whole.

DEFINITION 11. Let $\operatorname{cod}(\cdot)$ be an integer-valued function defined on singular points, connections between singular points, and their unions (finite or infinite) as follows:

(1) For a singular point A the value $\operatorname{cod}(A)$ is the codimension of the corresponding $\mathfrak{c}$-equivalence class;
(2) For a connection of $\sigma_1\sigma_2$-type, where $\sigma_i \in \{p, h, b\}$, the value $\operatorname{cod}(\gamma)$ is given the following table:

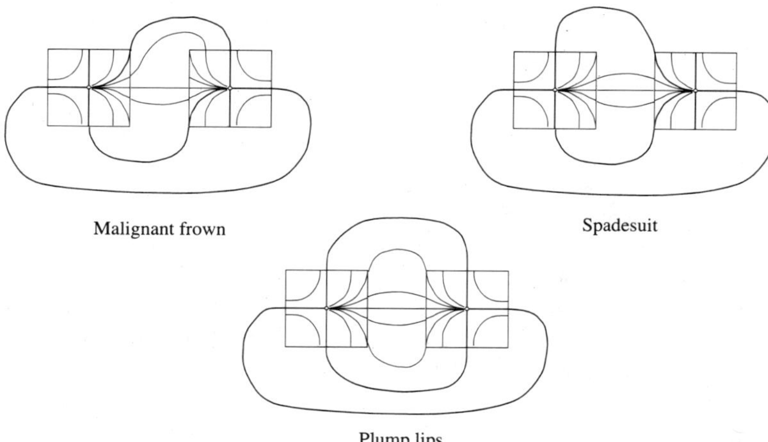

FIGURE 2B. Pairwise not $\mathfrak{c}$-equivalent ensembles from the *lips* family: malignant frown, plump lips, paresis (spadesuit).

Type of γ	pp	pb	ph	bb	bh	hh
$\operatorname{cod}(\gamma)$	0	0	0	1	1	1

(3) For a union (finite or infinite) of singular points and their connections the function cod extends in an additive way: for example, if P is an ensemble of polycycles, then

$$\operatorname{cod}(P) = \sum_{\text{vertices}} \operatorname{cod}(A_i) + \sum_{\text{connections}} \operatorname{cod}(\gamma_\alpha).$$

The function cod is called the *degeneracy codimension* of a polycycle or an ensemble.

PROPOSITION. *The function* cod *takes equal values on* $\mathfrak{c}$*-equivalent ensembles.* □

The characteristic property of this function is the following one.

THEOREM 2. *In generic k-parameter families of vector fields on the sphere with $k \leqslant 3$, ensembles belonging to a $\mathfrak{c}$-equivalence class with the codimension of degeneracy equal to ν, occur for values of the parameters belonging to an immersed submanifold of codimension ν.*

SKETCH OF THE PROOF. We already know that partition of the parameter space by submanifolds corresponding to different $\mathfrak{c}$-types of singularities is a regular stratification (see Theorem 1). Next, the simultaneous occurrence of several singular points corresponds to independent events: for generic families the stratifications are mutually transversal, hence codimensions of all strata corresponding to such occurrences can be obtained by adding the values of the function cod. It remains only to analyze how additional types of connections affect the codimension. By Theorem

1, we may restrict all considerations to submanifolds (strata) corresponding to fixed $\mathfrak{c}$-types of all singularities. Denote such a stratum by S.

Take an arc γ from an ensemble and choose another arc Σ transversal to it at a point Q. If we move along the stratum S, then by the same Theorem 1, an h- and/or b-curve would move differentiably with respect to the parameters, while p-curves intersect Σ at a set that contains an open neighborhood of Q on Σ. Thus we see that the occurrence of pp-, ph- and pb-connections is an open property on the stratum S, while for bb-, hb- and hh-connections one may define a *splitting function* $\varphi_\gamma: S \to \mathbb{R}$, the distance between two points of intersection of h- or b-curves with Σ, measured along Σ. The connection occurs if and only if $\varphi_\gamma = 0$. By changing slightly the family of vector fields in a small neighborhood of Q, we may assume that the function φ_γ is generic on the stratum S, hence the same singular points plus an additional hh-, hb- or bb-connection is an event occurring on a subset of codimension 1 *on the stratum* S. Evidently, multiple connections of that type may be perturbed independently, thus any number of splitting functions can be put in a generic position on the stratum. These arguments give the proof of the theorem. □

Remark. There are simple examples of generic 1-parameter families of vector fields on the sphere such that saddle connections in those families occur for nondiscrete subsets of the parameter space.

1.7. The classification problem: preliminary version. Now we can formulate the goal of the first section of the paper as follows.

Classification problem for ensembles. *Give a list of $\mathfrak{c}$-equivalence classes of polycycles and their ensembles in codimensions $\leqslant 3$.*

Note that this problem is now of almost purely combinatorial nature: we have to show how singularities from different $\mathfrak{c}$-equivalence classes can be connected together to form a polycycle or ensemble; we need to take into consideration also the types of endpoints of the connections.

1.8. Additional units of codimension. Consider a $\mathfrak{c}$-equivalence class of polycycles/ensembles of codimension ν occurring in a generic k-parameter family of vector fields with $k > \nu$. By Theorem 2, such polycycles would occur on a $(k-\nu)$-dimensional immersed submanifold of the parameter space. This in turn would mean that for certain isolated values of the parameters there may occur additional degeneracies, either for the vertices of the polycycle, or for its connections. Since we fixed the $\mathfrak{c}$-equivalence class of the polycycle, those degeneracies cannot change the types of vertices and/or connections. However, they may change the bifurcation pattern of a polycycle.

Example 4. Consider a separatrix loop of a nondegenerate saddle, that is, the $\mathfrak{c}$-equivalence class of a polycycle (hyperbolic saddle, hh-connection). Denote by λ_j, $j = 1, 2$, the eigenvalues of the saddle. Then it is known, see for example [5], that:

(1) if $\lambda_1 + \lambda_2 \neq 0$, then the cyclicity of such a loop is 1;
(2) if $\lambda_1 + \lambda_2 = 0$, but the improper integral taken along the loop $\gamma = \gamma(t)$

parameterized by the natural time,

$$\beta = \int_{-\infty}^{+\infty} \operatorname{div} v(\gamma(t))\, dt$$

is nonzero, then the cyclicity is 2;

(3) if $\lambda_1 + \lambda_2 = \beta = 0$, but the nonlinear terms of the Taylor expansion of the vector field at the saddle point are generic, then the cyclicity is equal to 3, *etc.*

Note that all conditions influencing cyclicity, do not change the $\mathfrak{c}$-equivalence class of the loop.

In order to take into consideration the possibility of such phenomena, we introduce the notion of a *rigged polycycle/ensemble*.

DEFINITIONS. A *rigging* of a polycycle/ensemble is a set of additional equality type constraints imposed on jets of the vector field at singular points and/or integrals of some functions over connections of the polycycle. The *codimension of a rigging* is the number of independent equality-type constraints imposed on both vertices and the arcs.

A *rigged polycycle/ensemble* is an equivalence class of polycycles/ensembles for vector fields satisfying the constraints imposed by the rigging: two polycycles/ensembles are equivalent if they are $\mathfrak{c}$-equivalent and the same rigging is imposed on the corresponding vertices and/or arcs.

The *degeneracy codimension* of a rigged polycycle/ensemble is the degeneracy codimension of the corresponding $\mathfrak{c}$-equivalence class plus the codimension of the rigging.

We call a rigging *essential* if the bifurcation diagram for a rigged polycycle/ensemble differs from that for nonrigged polycycles/ensembles from the same $\mathfrak{c}$-equivalence class. Otherwise a rigging is called *apparent*.

REMARK. Equivalence classes of rigged polycycles form a refinement of $\mathfrak{c}$-equivalence classes of (nonrigged) polycycles.

Since we do not know the bifurcation diagrams for all equivalence classes occurring in codimension 3, it may happen that in the classification table from §2 some $\mathfrak{c}$-equivalence classes should be further refined by imposing additional riggings. However, in the cases when we already know which rigging is essential, we include corresponding equivalence classes in the classification.

§2. Classification of polycycles and their ensembles in small codimensions

In this section we compose lists of $\mathfrak{c}$-equivalence classes of rigged polycycles occurring in generic few-parameter families. For apparent reasons we exclude from consideration polycycles consisting of just one single point that cannot generate limit cycles after bifurcation (such as a node or a saddle). From Table 1 it follows that there are no 0-degenerate polycycles.

2.1. 1-degenerate polycycles. Throughout this section and below we use the numeration for different classes of $\mathfrak{c}$-equivalent rigged polycycles in the form (c, m), where $1 \leqslant c \leqslant 3$ is the codimension and m is the number within the corresponding list.

THEOREM 1. *There exist only three 1-degenerate classes of rigged oriented polycycles, see Figure 4:*

(1.1) *an elliptic single point (slow focus);*
(1.2) *a double saddle-node with an hp-loop;*
(1.3) *a separatrix hh-loop of a hyperbolic saddle.*

REMARK. This result is a reformulation of a classical result from bifurcation theory [7]. We give its proof for completeness of the exposition. The proofs of Theorems 2 and 3 below will be modelled after this pattern.

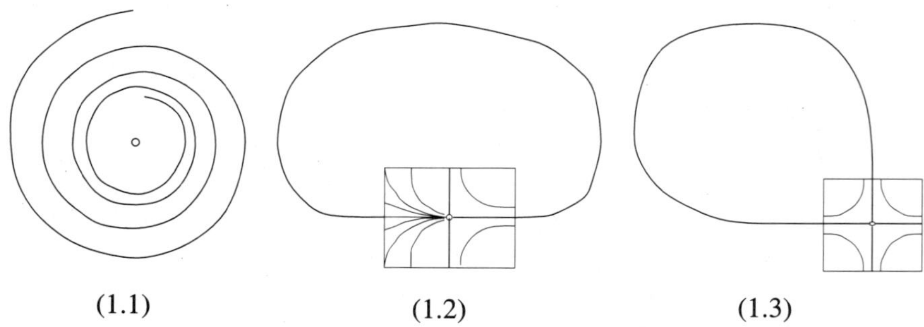

FIGURE 4. Gallery of 1-degenerate polycycles.

PROOF. In generic 1-parameter families only nondegenerate elementary singularities and double saddle-nodes may occur. So we can restrict ourselves to consideration of polycycles carrying those types of singular points only. Denote such a polycycle by P.

1. If the polycycle P has no connections, then the degeneracy condition must be imposed on the unique singular point A_0 itself, and the latter cannot be a saddle-node or saddle (this would mean zero cyclicity). Thus it should be either an elliptic singular point or a focus. The condition $\operatorname{cod}(A_0) = 1$ leaves only one possibility: A_0 is the slow focus (1.1).

2. If the polycycle carries a saddle-node A, then

$$1 = \operatorname{cod}(P) = \operatorname{cod}(A) \implies \sum_{\text{arcs}} \operatorname{cod}(\gamma_i) = 0.$$

In particular, if there are no other singularities on the polycycle, then the connection should be an hp-connection, and we have the case (1.2).

In fact, there cannot be other singular points on a polycycle carrying a saddle-node. Indeed, those points should be nondegenerate saddles with $\operatorname{cod} = 0$, hence there should be at least one separatrix connection between the saddle-node and one of the saddles, which is forbidden by the codimension arguments: such connection would have positive codimension $\operatorname{cod}(\gamma) > 0$.

3. If there are only hyperbolic saddles on the polycycle, then all connections should be of hh-type, and since

$$\sum_{\text{arcs}} \operatorname{cod}(\gamma_i) = 1,$$

there can be only one such connection, the separatrix hh-loop (1.3). □

COROLLARY. *Ensembles do not occur in generic* 1-*parameter families.* □

2.2. 2-degenerate polycycles.

THEOREM 2. *There are only nine equivalence classes of* 2-*degenerate rigged oriented polycycles and their ensembles, shown in Figure* 5:

(2.1) *an ultra-slow focus*;
(2.2) *a Bogdanov–Takens cusp*;
(2.3) *two double saddle-nodes connected by two* hp-*curves*;
(2.4) *a double saddle-node and a hyperbolic saddle, connected by one* hh-*curve and one* hp-*curve* (*the half-apple*);
(2.5) *a double saddle-node and a hyperbolic saddle, connected by an* hh-*curve and two* hp-*curves* (*the apple*);
(2.6) *an* hb-*loop of a saddle-node*;
(2.7) *a hyperbolic saddle with two separatrix* hh-*loops* (*the eight-figure*);
(2.8) *two hyperbolic saddles connected by two* hh-*curves* (*the heart and the lentil*);
(2.9) *a separatrix loop of a hyperbolic saddle with a nontrivial local rigging* (*the zero divergence at the singular point*).

PROOF OF THEOREM 2. By definition of the degeneracy codimension, only singularities from Tables 0–2 may occur on ensembles of codimension 2. To list all ensembles, we consider several cases.

2.2.1. *Polycycles consisting of a single point.* If the point is elliptic, then the polycycle may only be of the type (2.1).

Next, there is only one nonelementary singularity which has codimension 2, the cuspidal point. There cannot be any other connection, since the cusp has two hyperbolic sectors separated by h-curves. Hence a polycycle that carries the cusp and is 2-degenerate must consist of only one point, as in (2.2).

2.2.2. *Polycycles carrying a saddle-node.* A degenerate elementary singularity of codimension $\leqslant 2$ is either a double saddle-node, or a degenerate saddle, or node. One can easily see that the latter two cases are forbidden. Indeed, a node cannot occur on an oriented polycycle for topological reasons. On the other hand, a polycycle carrying a degenerate saddle must have at least one hh-connection which contributes another unit of codimension, so in 2-degenerate families degenerate saddles cannot occur as parts of polycycles. Thus the only remaining possibility for a polycycle is to carry double saddle-nodes and nondegenerate saddles.

Coexistence of two saddle-nodes is an event of the codimension 2, hence all connections can be of zero codimension, that is, of hp-type. The only possibility for that is (2.3).

If a polycycle carries one saddle-node A_0 with $\operatorname{cod}(A_0) = 1$ and several saddles A_i, $i \geqslant 1$, then there must be at least one hh-connection γ, a common separatrix for a saddle and the saddle-node, with $\operatorname{cod}(\gamma) = 1$. This implies that all other connections are of zero codimension, hence they must be of hp-type. Depending on whether there is one or two such connections, we have either (2.4) or (2.5). In both cases the codimension is 2, with no additional units left for further degeneracies.

If the saddle-node is the only singularity on the polycycle, then for the unique (by topological reasons) connection γ we must have

$$\operatorname{cod}(\gamma) + \operatorname{cod}(\text{rigging}) = 1.$$

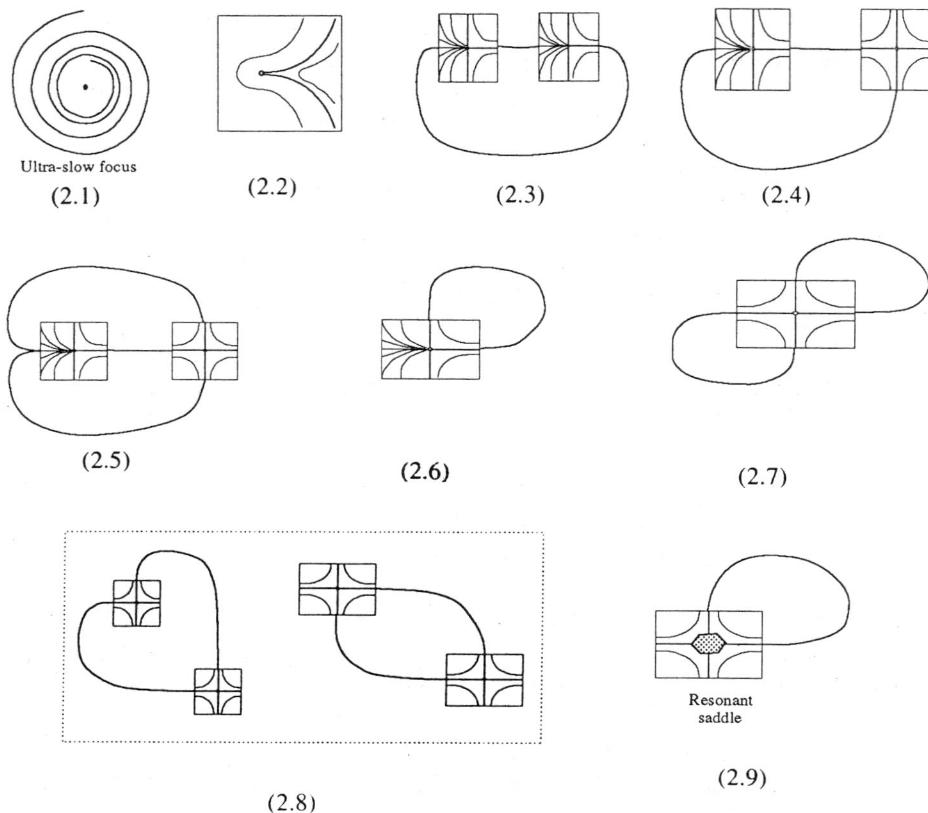

FIGURE 5. List of 2-degenerate polycycles and ensembles.

If $\operatorname{cod}(\gamma) = 1$, then the rigging is trivial, and γ must be an hb-curve, as in (2.6). If $\operatorname{cod}(\gamma) = 0$, then γ is an hp-curve, and the subcase is of an effective codimension 1, see Theorem 1.

2.2.3. *Polycycles carrying only hyperbolic saddles.* In this case there can be only hh-connections, all of them of positive codimension, hence their number may be 2 or 1, and in the latter case a nontrivial rigging is possible.

The number of saddle points participating in the polycycle can be also 1 or 2, since at most 4 endpoints of connections should land on the saddles, at least 2 endpoints at each saddle. This argument shows also that for 1 connection at most 1 saddle is admissible.

Thus we have the following possibilities:

(1) 2 saddles and 2 connections: the lentil or the heart (2.8);
(2) 1 saddle and 2 connections: the eight-figure (2.7);
(3) 1 saddle and 1 connection with a nontrivial rigging.

In the latter case the cyclicity and the bifurcation diagram are known, see the Appendix to §2. The additional rigging may affect cyclicity of the separatrix loop only if the corresponding unit of codimension is used to impose the equality $\lambda_1 + \lambda_2 = 0$ on the eigenvalues of the saddle point, see [5]. □

REMARK 1. In the case (2.8) (the *lentil* subcase) the bifurcation diagram depends on whether the divergence of the vector field has the same sign at the saddle points (and then cyclicity of the lentil is 1), or those signs are opposite (in which case cyclicity can be > 1). This remark is in fact of a universal character: within many $\mathfrak{c}$-equivalence classes the bifurcation diagram may vary from one connected component of the stratum $\mathfrak{c} = \text{const}$ to the other.

REMARK 2. In the case of two saddles and two hh-connections, there is one more admissible combination: two disjoint separatrix loops. We exclude this case from consideration, since being disjoint, these loops do not constitute an ensemble.

2.3. 3-degenerate polycycles and their ensembles. The description of polycycles occurring in generic 3-parameter families is roughly the same in spirit as the description of 2-degenerate polycycles, but much lengthier, see Figure 6. It makes sense to break all possible types of polycycles and their ensembles into three major groups as follows.

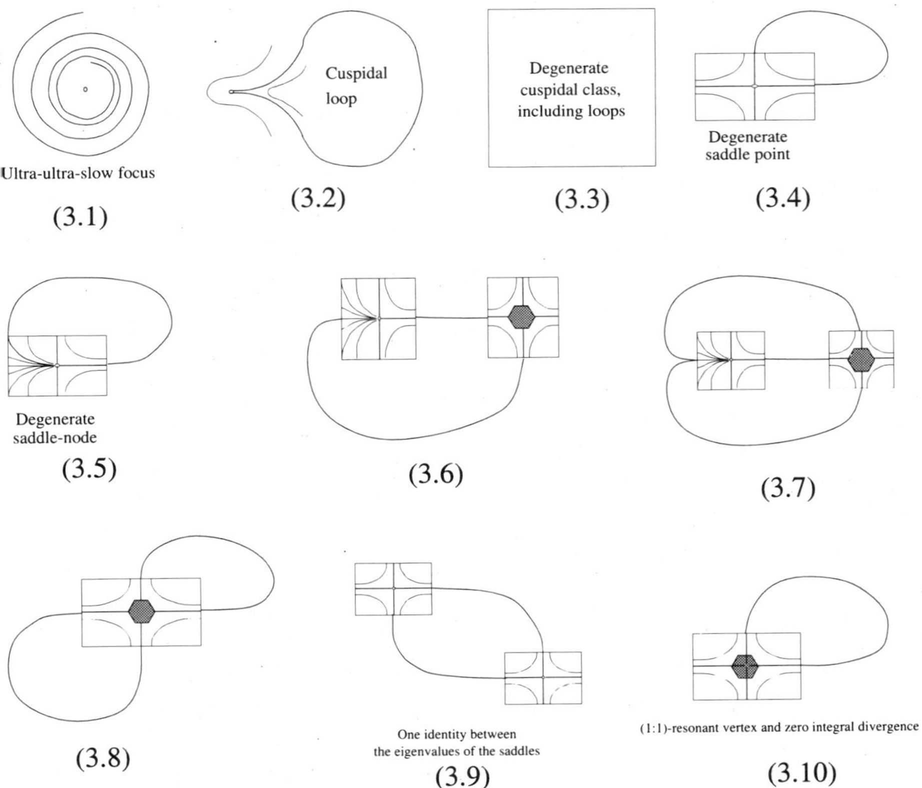

FIGURE 6A. Classification of 3-degenerate polycycles and their ensembles (beginning).

2.3.1. *Initial preclassification.*

(I) Elliptic points.

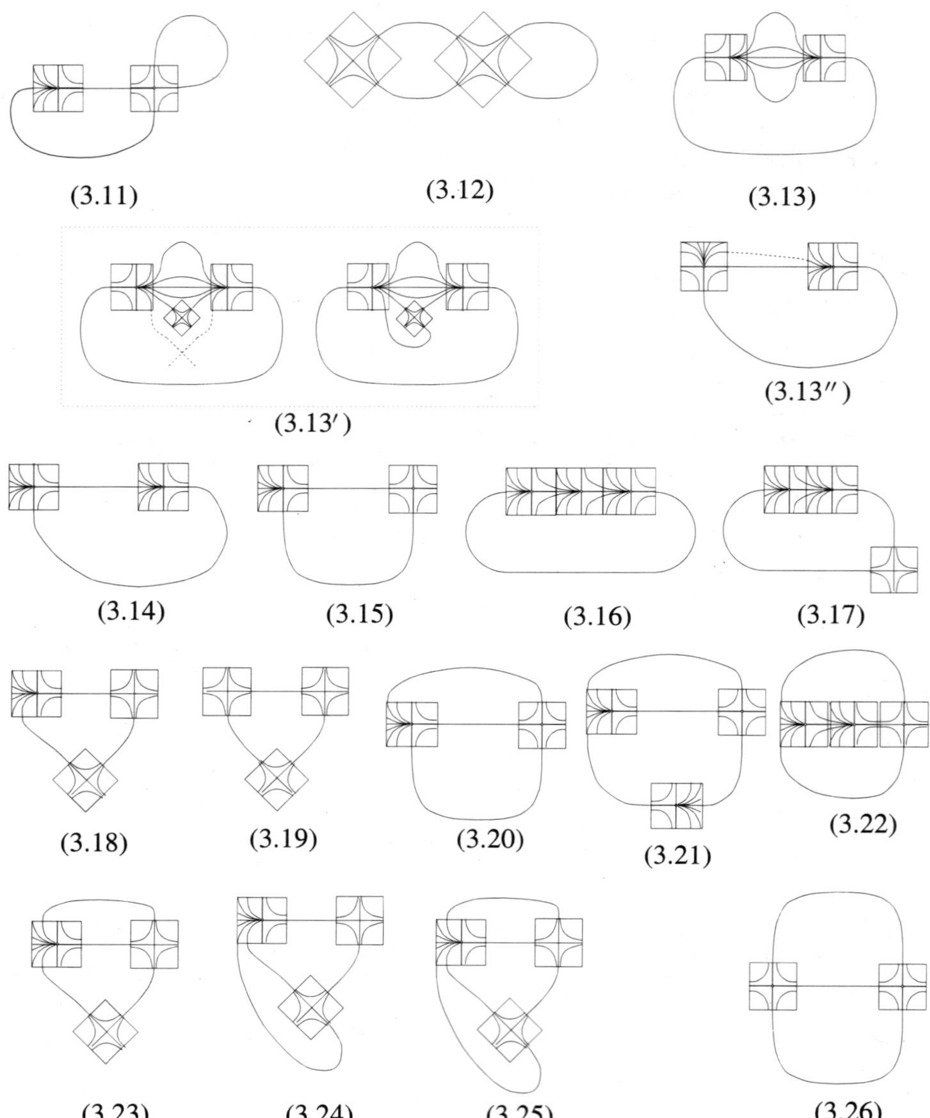

FIGURE 6B. Classification of 3-degenerate polycycles and their ensembles (continued).

(II) Polycycles and ensembles carrying cusps or singularities of codimension $\geqslant 2$ and polycycles/ensembles with a nontrivial rigging.

(III) Standard polycycles, that is polycycles carrying only double saddle-nodes and hyperbolic saddles without rigging, and ensembles of such polycycles.

2.3.2. *Formulation of the theorem.* In codimension 3 ensembles containing continuum of coexistent polycycles appears for the first time. All such ensembles can be derived from the one which we call *lips*.

DEFINITION 1. *Thin lips* is the ensemble of polycycles which consists of two saddle-nodes (one contractive and one dispersive) connected by one hh-curve and a continuum of pp-curves which form a 1-parameter connected family of smooth arcs.

THEOREM 3. *There are* 26 *principal classes of* 3*-degenerate maximal rigged polycycles and their ensembles with positive cyclicity, and an infinite series of pairwise nonequivalent classes, each of them containing the lips as a proper subensemble.*

Exact descriptions of the classes and the series is given in the proof. The schematic pictures of classes are shown on Figures 6A and 6B.

PROOF OF THEOREM 3. The proof is given in 2.3.3–2.3.9.

2.3.3. *Polycycles of the types* I *and* II *and their ensembles.* The first group consists of only one polycycle, ultra-ultra-slow focus (we exclude from consideration the cases of slow and ultra-slow foci with additional degeneracies since they have the effective codimension 1 and 2 and are already covered by Theorems 1 and 2). The corresponding case is numbered as (3.1).

Consider the group II. If a polycycle carries a cusp A_0, then there can be three possibilities:

(1) The cusp is nondegenerate, i.e. $\operatorname{cod}(A_0) = 2$ and there is only one hh-curve with $\operatorname{cod}(\gamma) = 1$: the codimension of this polycycle is exactly 3 and we have the case (3.2);

(2) $\operatorname{cod}(A_0) = 2 +$ (nontrivial rigging) without connections: as this is known from the Bogdanov–Takens analysis, within the specified $\mathfrak{c}$-equivalence class the bifurcation diagram is completely determined by the principal terms of the Taylor expansion of the vector field at the singular point. Hence any additional degeneration in higher order Taylor terms does not change the $\mathfrak{c}$-equivalence class, which has an effective codimension 2. The corresponding $\mathfrak{c}$-equivalence class was already described in Theorem 2.

(3) A_0 is the degenerate cusp, $\operatorname{cod}(A_0) = 3$: this case is the most difficult one. Two possible subcases can be distinguished:

(3a) The polycycle consists of a single point.

(3b) If the singularity is of an *elliptic type* in the terminology of [11], then the singular point has two parabolic sectors of opposite attractivity. Hence a pp-loop can occur without increasing the codimension of the polycycle.

Even the first subcase is not yet completely investigated, to say nothing about the much more difficult loop subcase. We simply label both subcases as (3.3), without going deeply into the subject.

If the polycycle carries a semihyperbolic singularity which is not a double saddle-node, then the possibilities are a topological saddle (of codimension 2) with the separatrix loop (3.4) and a degenerate saddle-node of codimension 3 with a homoclinic loop of zero codimension hence of hp-type (3.5). Any additional saddle carried by the polycycle would contribute at least one more hh-connection, hence the codimension limit would be exceeded.

The cases (3.6), (3.7), (3.8), and (3.9) are obtained by imposing additional degeneracy conditions (nontrivial rigging) on the $\mathfrak{c}$-equivalence classes of polycycles of original codimension 2 in the cases (2.4), (2.5), (2.7) and (2.8) respectively.

In fact, in all these cases the cyclicity (and sometimes the complete bifurcation diagram) is known. In particular, in the case (3.9) there can be two ways of imposing one degeneracy condition that affect the cyclicity and the bifurcation diagram in different ways. In the first subcase, labelled as (3.9a) in the Appendix, one of the saddles is resonant with the hyperbolicity ratio 1. The other subcase (3.9b) corresponds to two nondegenerate mutually inverse irrational hyperbolicity ratios. In the Appendix to §2 we list the degeneracy conditions and give the references to relevant publications.

The case (3.10) is obtained by imposing *two independent* degeneracy conditions on the *hh*-loop of a nondegenerate saddle, or (which is the same) by adding one more condition to the case described before as (2.4). Again the complete investigation of this case is available, see the Appendix.

2.3.4. *Standard polycycles and their ensembles: the program of investigation.* We arrange investigation of polycycles of the type III in the following order. In 2.3.5 we study polycycles with a separatrix loop. In 2.3.6 the connections admissible in standard polycycles are analyzed by means of *balance equations*. In 2.3.7 we describe all integer nonnegative solutions of those equations and construct the corresponding polycycles. The classification ends with a description of ensembles of standard polycycles.

2.3.5. *Polycycles with a homoclinic loop.* A homoclinic trajectory of a double saddle-node cannot be included into a bigger polycycle. Next, the occurrence of a separatrix loop is an independent event of codimension 1. Hence 3-degenerate polycycles with such a loop can be produced from 2-degenerate standard polycycles carrying saddles, by closing loops. More precisely, if a 2-degenerate polycycle has a saddle with only two separatrix semitrajectories belonging to the polycycle, then by closing another pair of such semitrajectories, one may produce an ensemble which is 3-degenerate. In such a way one obtains (3.11) from (2.4) and (3.12) from (2.8). Clearly, there are no more cases of 2-degenerate polycycles which would allow for such a construction.

2.3.6. *The balance equation.* In order to investigate polycycles/ensembles without homoclinic trajectories, consider all possible types of connections between singular points.

LEMMA 1. *Consider a* 3*-degenerate polycycle without homoclinic trajectories, carrying* N *saddle-nodes. Denote by* HP, PP, BP, HH, *and* HB *the number of connections of the types* $hp, \dots, hb$ *respectively. Then the following balance equations hold*:

$$HH + HB + N = 3, \tag{1}$$

$$HP + 2\,PP + 2\,BP + HB = N. \tag{2}$$

PROOF. Nondegenerate saddles and *pp*-connections do not contribute to the codimension of a polycycle. Double saddle-nodes have codimension 1, as well as *hh*- and *hb*-connections. Connections of *bb*-type are forbidden, since they require at least 2 saddle-nodes, plus there must be at least one more *hh*-arc, and the overall codimension would be at least 4. This observation proves (1).

The number of saddle-nodes on a polycycle is equal to the number of endpoints of connections of the types p and b. This implies the equality (2). □

Thus the problem of classification of standard polycycles is reduced to solving (1), (2) in nonnegative integers and a subsequent analysis of topological admissibility of solutions.

2.3.7. *Solution of the balance equations and construction of polycycles.* The following lemma gives a description of all standard polycycles (but not their ensembles).

LEMMA 2. *There exist exactly* 10 *integer nonnegative solutions of the system* (1), (2). *The cases when a solution corresponds to a polycycle, are given in Table* 4.

PROOF OF LEMMA 2. Choose the variables PP, BP, HB, and N as the free variables. They must satisfy the inequalities implied by (1), (2) and the nonnegativity conditions:

$$2\,PP + 2\,BP + HB \leqslant N \leqslant 3 - HB\,,$$
$$PP + BP + HB \leqslant 1.$$

As a result, we have 10 admissible choices yielding 10 solutions of the balance equations. Consider now the possibility of constructing the corresponding polycycles. The results are given in Table 4.

The number of nondegenerate saddles (which does not affect the degeneracy codimension) is in all cases determined by the number of h-type endpoints, after taking into account the number of saddle-nodes.

Two combinations are impossible for topological reasons. □

What remains to do in order to achieve the complete proof of Theorem 3 is to analyze 3-degenerate ensembles of polycycles. This is done in the next two sections.

2.3.8. *Finite ensembles of standard polycycles.* Here we describe finite ensembles consisting of 1-, 2- or 3-degenerate standard oriented polycycles.

A 1-degenerate polycycle which can occur as an element from an ensemble is a separatrix loop of a hyperbolic saddle. Such ensembles were already described in §2.3.5.

If an ensemble contains a 3-degenerate polycycle, then additional connections should not increase the codimension, thus only hp-, pb- and pp-connections are admissible. As for pb- and pp-connections, they can be added only if there are two saddle-nodes with different orientations, one contractive and one dispersive. There are no such patterns which would allow the addition of extra connections, besides those already considered.

As for hp-connections, they can be added only in the cases (3.15), (3.17), and (3.18). The case (3.15) yields the ensemble (3.20), out of the case (3.17) one may produce (3.21) and (3.22), and finally the addition of hp-connections to the case (3.18) gives rise to the ensembles (3.23), (3.24), and (3.25).

Now consider the case when a 3-degenerate ensemble consists of $\leqslant$ 2-degenerate polycycles. Such an ensemble can be produced by adding connections of the types hh or bh to a 2-degenerate polycycle with two singular points with nonclosed separatrices. The latter property holds only for polycycles (2.4) and (2.8).

TABLE 4. Integer nonnegative solutions of the balance equations and the corresponding c-equivalence classes of polycycles

No.	PP	BP	HB	N	HP	HH	Cases	Remarks
1	1	0	0	3	1	0	None	Two connections and three singular points: impossible to create a polycycle
2	1	0	0	2	0	1	(3.13)	Two saddle-nodes connected by hh- and pp-arcs: the lips arise
3	0	1	0	3	1	0	None	The same as in No. 1
4	0	1	0	2	0	1	(3.13″)	Occurrence of the polycycle (3.13″) implies occurrence of a polycycle (3.13) in an arbitrary small neighborhood of the former; the maximal ensemble containing (3.13″) must contain the lips
5	0	0	1	2	1	0	(3.14)	The limit case of (2.3)
6	0	0	1	1	0	1	(3.15)	There must be exactly one more saddle, since both hb- and hh-connections are present
7	0	0	0	3	3	0	(3.16)	3 saddle-nodes connected by hp-arcs
8	0	0	0	2	2	1	(3.17), (3.13′)	There must be a saddle. If the hh-arc connects a saddle-node and the saddle, then (3.17) occurs. If the hh-arc connects the two saddle-nodes, then (3.13′) is realized, and together with it the lips
9	0	0	0	1	1	2	(3.18)	Two hyperbolic saddles; one hh-arc connects the saddle-node with one of the saddles, the other is between the two saddles
10	0	0	0	0	0	3	(3.19)	Three saddles connected by hh-arcs

Addition of an extra hb-connection to (2.4) results in the 3-degenerate polycycle in ensemble (3.20). An extra hh-connection in (2.8) is possible only in the case of the lentil; as a result, the new ensemble (3.26) appears.

2.3.9. *Infinite ensembles of standard polycycles.* Occurrence of infinite ensembles of standard polycycles is due to p- and b-trajectories of saddle-nodes, which may be closed up directly or through hyperbolic points. If such a closing is direct (which means the presence of pp- or bp-connections), then one obtains the polycycle (3.13) "surrounded" by a continuum of polycycles of the same type.

Consider the case when parabolic sectors of two saddle-nodes are connected by a chain of heteroclinic trajectories of one or more singular points. As it was shown in Lemma 2, the only possibility for such a chain to be a part of a 3-degenerate polycycle, is the case

$$(ph\text{-connection}) \to (\text{hyperbolic saddle}) \to (hp\text{-connection}).$$

Near such a chain there exists an infinite number of pp-connections, hence polycycles of the type (3.13) appear in the form of the lips. All ensembles containing chains of such form, are included into the infinite series (3.13), (3.13′), (3.13″).

Thus we finished the description of all 3-degenerate polycycles and their ensembles and achieved the complete proof of Theorem 3. □

Appendix to §2: List of cases of known cyclicity

In this appendix we list $\mathfrak{c}$-equivalence classes of polycycles in codimension 1, 2, and, partly, in codimension 3 (when the cyclicity and/or bifurcation diagrams are known).

1-degenerate polycycles. The absolute cyclicity of polycycles from each of the three $\mathfrak{c}$-equvalence classes is known since the thirties; the bifurcation of the slow focus (1.1) was investigated by A. Andronov and H. Hopf, the other two cases are described in [7]. In each case the cyclicity is equal to 1.

2-degenerate polycycles. For the nine cases (in fact, when speaking about bifurcation diagrams, one should distinguish two subcases in (2.8), the heart and the lentil), the cyclicity of all polycycles is known and does not exceed 2. More detailed information is given in Table 5. We mention only briefly the classical results, focusing more on the recent achievements.

TABLE 5. Cyclicity of 2-degenerate polycycles: synopsis of results

Type	Cycl.	Remarks
(2.1)	2	F. Takens
(2.2)	1	R. Bogdanov, F. Takens
(2.3)	1	Simple
(2.4)	2	T. Grozovskiĭ [G], including the bifurcation diagram; the cyclicity estimate follows also from results of F. Dumortier, R. Roussarie, C. Rousseau [DRR]
(2.5)	3	T. Grozovskiĭ [G]
(2.6)	1	Cyclicity estimate is simple, the bifurcation diagram was constructed by V. Lukyanov [L] and S. Schechter [S]
(2.7)	1 ; 2	Cyclicity of the eight-figure as a polycycle is 1 (simple), but when considered as an ensemble, it has cyclicity 2 (D. Seregin, E. Malgina, in preparation, 1993)
(2.8)	2	L. Cherkas [C], the cyclicity estimate; V. Roitenberg (1985), V. Nozdracheva (1981), bifurcation diagrams for the lentil (preprints), the formulation is given in [3, chap. 3, §2.8]; A. Khibnik, M. Druzhkova (unpublished) for the heart
(2.9)	2	E. Leontovich (1951), cyclicity estimate; V. Nozdracheva [N], the bifurcation diagram

3-degenerate polycycles. Bifurcations of polycycles in generic 3-parameter families are not yet completely studied, and the main difficulties here are expected to occur when investigating the two wild cases (3.2) and (3.3). Yet the information on the cyclicity of the remaining cases is also incomplete, especially in the part

TABLE 6. Cyclicity of 3-degenerate polycycles and ensembles: synopsis of results

Type	Cycl.	Remarks
(3.1)	3	F. Takens
(3.2)	4?	Conjectured by R. Roussarie
(3.3)	?	Cyclicity is unknown even for the polycycle consisting of one degenerate cuspidal points, though some partial results are known [DRS]; the case of the loop is completely unknown
(3.4)	1	Cyclicity is simple; Li Weigu e.a. [LLZ], bifurcation diagram
(3.5)	1	Simple
(3.6)	2	F. Dumortier, R. Roussarie, C. Rousseau [DRR]
(3.7)	$\leqslant 4$	Exact cyclicity of the ensemble is unknown
(3.8)	5	Li Weigu (1993, unpublished), together with bifurcation diagram; the cyclicity estimate $\leqslant 7$ is due to M. Jebrane and A. Mourtada [JM]
(3.9a)	2	The case when one of the saddles is a resonant saddle with the hyperbolicity ratio equal to 1, A. Mourtada [M1]
(3.9b)	3	Two saddles with irrational hyperbolicity ratios $r_i \notin \mathbb{Q}$, $r_1 r_2 = 1$, A. Mourtada [M2]
(3.10)	3	E. Leontovich (1951), R. Roussarie (1986)
(3.11)		Ensemble of 2-degenerate polycycles
(3.12)		Ensemble of 2-degenerate polycycles
(3.13)	3	Stanzo (see §3 below); cyclicity of the ensemble is unbounded
(3.13′)	?	Cyclicity of the ensemble is unbounded
(3.13″)	?	Cyclicity of the ensemble is unbounded
(3.14)	1	Simple
(3.15)	2	F. Dumortier, R. Roussarie, C. Rousseau [DRR]
(3.16)	1	Simple
(3.17)	2	Formally does not follow from Theorem 2 in [DRR], but the proof remains the same
(3.18)	?	Unknown
(3.19)	3	A. Mourtada (thesis)
(3.20), ..., (3.26)		Ensembles of polycycles occurring before; exact cyclicity unknown

concerned with sharp estimates for cyclicity of ensembles. The known results are given in Table 6, where the main emphasis is on cyclicity of polycycles.

Simple cases. In Tables 5 and 6 several cases are listed as simple; the cyclicity of those polycycles does not exceed 1. This follows from one of the results obtained in [DRR]. Below we give an independent proof.

DEFINITION. A polycycle is called *contractive* if the divergence of the (unperturbed) vector field is negative at all vertices of this polycycle. The polycycle is called *dispersive* if the divergence is positive at all vertices.

THEOREM. *Cyclicity of any elementary contractive polycycle (independently of the number of vertices) does not exceed* 1.

COROLLARY. *Cyclicity of an elementary dispersive polycycle is* $\leqslant 1$.

PROOF OF THE THEOREM. The proof follows immediately from the two assertions:

(1) any limit cycle born from a contractive polycycle is stable, and
(2) two or more stable limit cycles cannot be simultaneously born from an elementary contractive polycycle.

To prove the first claim, note that the logarithm of the multiplicator of a periodic trajectory is equal to the integral of the divergence of the vector field over the trajectory parametrized by the natural time [7]. Any periodic trajectory born from a polycycle has a very large period, and a large part of this period the trajectory spends close to the places were the vertices of the polycycle originally were. Since the divergence of the vector field depends continuously on the parameters, we conclude that the integral in question takes very large negative values. Hence the multiplicator is very small, and the trajectory itself is stable by linear approximation.

The second assertion follows from the theorem on invariant manifolds. From the classical theorems on existence of invariant hyperbolic or center manifolds it follows that any singular point born from an elementary singular point after perturbation, has an invariant cross, the union of two segments parallel to the coordinate axes in the normalizing coordinates. The size of this cross in the original coordinates (the diameter of the union of local invariant manifolds) is not small, that is, the diameter of the union of invariant manifolds containing the singular point does not tend to zero as the parameters of perturbation tend to zero. Hence two periodic orbits sufficiently close in the sense of Hausdorff metric to the unperturbed polycycle, cannot contain singular points of the perturbed field in the thin annulus bounded by those two trajectories. Now the second assertion is evident: if two closed orbits together bound an annulus without singular points or periodic orbits inside, then those trajectories should be of the opposite stability. This contradicts the first assertion. □

§3. Bifurcation diagram for the lips

3.1. Main concepts and results. In this section we study the bifurcation diagram for a generic deformation of the ensemble called the lips in §2.

3.1.1. *Statement of the problem. Reparameterization.* The polycycle (3.13) occurring in the classification of §1 consists of two double saddle-nodes O_1 and O_2 connected by a pp-arc and a saddle connection of the type hh. In what follows we assume that the orientation is chosen as in Figure 7.

As it was shown in §2, near a polycycle (3.13) there must be a continuous family of polycycles of the same type, their pp-arcs filling an open subset of the sphere. On the topological boundary of the set filled by pp-curves, other types of connections may occur. We do not investigate effects caused by that circumstance; instead we choose a segment transversally intersecting a certain pp-connection and consider its saturation by the flow curves, assuming that all of them are also pp-connections. As a result, we obtain an ensemble called the *thin lips*. When the parameters of the family change, we consider only the trajectories intersecting the same segment, which is assumed independent of the parameters. This restriction causes the bifurcation when a limit cycle leaves the domain of consideration.

In order to investigate bifurcations of the lips, we introduce the system of parameters $\overline{\varepsilon} = (\varepsilon, \delta, \lambda)$. If $\overline{\mu} = (\mu_1, \mu_2, \mu_3)$ are the original parameters and the lips

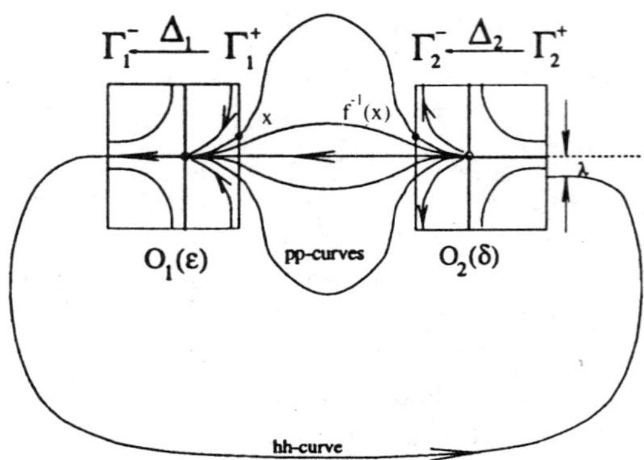

FIGURE 7. The lips.

occur for $\overline{\mu} = 0$, then, according to [3], [5], in a neighborhood of the saddle-node O_1 there exist local coordinates in which the family of vector fields has the form

$$\begin{cases} \dot{x} = -\dfrac{x^2+\varepsilon}{1+a(\overline{\mu})x}, \\ \dot{y} = -y, \end{cases} \qquad \varepsilon = \varepsilon(\overline{\mu}).$$

The smoothness of the normalizing chart in the phase variables and the parameters is arbitrarily high, but finite; the smoothness may be increased after shrinking the domain of the normalizing chart.

Let $\Gamma_1^{\pm}$ be two transversals to the flow that are given in the canonical chart by the equations $x = \pm 1$ respectively. Denote by x_1 the restriction of the coordinate y on Γ_1^+, and by y_1 the restriction $y|_{\Gamma_1^-}$. Then for $\varepsilon > 0$ we have the correspondence map

$$\Delta_1 : \Gamma_1^+ \to \Gamma_1^-$$

along the phase curves, which has the form

$$\Delta_1 : x_1 \mapsto y_1 = C_1(\varepsilon)\, x_1, \qquad C_1(\varepsilon) \to 0 \text{ as } \varepsilon \to 0.$$

More precisely,

$$C_1(\varepsilon) = \varphi(\varepsilon) \exp\left(-\frac{\pi}{\sqrt{\varepsilon}}\right),$$

$$\text{(1a)} \quad \varphi(\varepsilon) = \exp\left(\frac{2}{\sqrt{\varepsilon}}\left(-\frac{\pi}{2} + \arctan\frac{1}{\sqrt{\varepsilon}}\right)\right) = \exp\frac{2\arctan\sqrt{\varepsilon}}{\sqrt{\varepsilon}} = O(1).$$

Note that this map is defined only for $\varepsilon > 0$, although the transversals $\Gamma_1^{\pm}$ are well defined for all small ε. The representation (1a) is obtained by an explicit integration of the normal form, which has separated variables.

In the same manner, there exists a normalizing chart around O_2, in which the family has the form

$$\begin{cases} \dot{x} = \dfrac{x^2+\delta}{1+b(\overline{\mu})x}, \\ \dot{y} = y, \end{cases} \qquad \delta = \delta(\overline{\mu}).$$

We take the canonical transversals $\Gamma_2^{\pm}$ endowed with the local charts x_2, y_2 respectively so that the other correspondence map in these coordinates has the form

$$\Delta_2 \colon x_2 \mapsto y_2 = \Delta_2(x_2) = C_2(\delta)^{-1}x_2, \qquad C_2(\delta) \to 0 \text{ as } \delta \to 0,$$

or, more precisely,

$$C_2(\delta) = \varphi(\delta)\exp\left(-\frac{\pi}{\sqrt{\delta}}\right), \tag{1b}$$

where the function $\varphi(\cdot)$ is exactly the same as in (1a).

Besides the two maps (1a) and (1b), there are two regular correspondence maps along connections depending on parameters. In the normalizing charts they can be written as

$$F_{\overline{\mu}} \colon y_2 \mapsto x_1 = F_{\overline{\mu}}(y_2), \qquad G_{\overline{\mu}} \colon y_1 \mapsto x_2 = G_{\overline{\mu}}(y_1).$$

Since for $\overline{\mu} = 0$ the points O_i are connected by the saddle connection, $G_0(0) = 0$. Denote by λ the displacement (see Figure 7),

$$\lambda = G_{\overline{\mu}}(0).$$

In order to proceed further, we need additional genericity-type assumptions. We will require from now on the *Jacobian of the map* $(\mu_1, \mu_2, \mu_3) \mapsto (\varepsilon, \delta, \lambda)$ *to be nonvanishing*,

$$\det\left(\left.\frac{\partial\overline{\varepsilon}}{\partial\overline{\mu}}\right|_{\overline{\mu}=0}\right) \neq 0.$$

If this condition is satisfied, we will describe the bifurcation diagram for the lips in terms of the new parameters $\overline{\varepsilon}$ rather than $\overline{\mu}$: the above genericity assumption guarantees that the bifurcation diagram in the original parameter space will be *diffeomorphic* to the one we will construct below.

3.1.2. *Description of the bifurcation diagram outside the positive quadrant* $\varepsilon > 0, \delta > 0$. The complete description of the bifurcation diagram in the domain of parameters where there is at least one singular point of the vector field, is given by the following theorem.

THEOREM 1. *The bifurcation diagram in the intersection of a small neighborhood of the origin in the space* $\mathbb{R}^3_{\overline{\varepsilon}}$ *with the set* $M = \{\varepsilon \leqslant 0\} \cup \delta \leqslant 0$ *consists of twelve components corresponding to topologically nonequivalent phase portraits differing by the types of singular points and existence of connections between them:*

(1) $\varepsilon = \delta = \lambda = 0$—*two saddle-nodes connected by a separatrix,*

(2) $\varepsilon=\delta=0$, $\lambda\neq 0$—*two saddle-nodes without connection,*
(3) $\varepsilon<0$, $\delta<0$, $\lambda=0$—*two saddles connected by a separatrix, one stable and one unstable node,*
(4) $\varepsilon<0$, $\delta<0$, $\lambda\neq 0$—*two saddles without connection, one stable and one unstable node,*
(5) $\varepsilon<0$, $\delta=0$, $\lambda=0$—*saddle and saddle-node connected by a separatrix, one more stable node,*
(6) $\varepsilon<0$, $\delta=0$, $\lambda\neq 0$—*saddle, saddle-node and a stable node without connection,*
(7) $\varepsilon<0$, $\delta>0$—*saddle and stable node,*
(8) $\varepsilon=0$, $\delta<0$, $\lambda=0$—*saddle and saddle-node connected by a separatrix, and an unstable node,*
(9) $\varepsilon=0$, $\delta<0$, $\lambda\neq 0$—*saddle, saddle-node and an unstable node without connections,*
(10) $\varepsilon=0$, $\delta>0$—*one saddle-node,*
(11) $\varepsilon>0$, $\delta=0$—*saddle-node with a different orientation,*
(12) $\varepsilon>0$, $\delta<0$—*saddle, unstable node.*

The proof immediately follows from the local normal forms near the points $O_{1,2}$.

3.1.3. *The equation for determination of limit cycles. Preliminary remarks on the structure of the bifurcation diagram in the positive quadrant.* In the most interesting part of the bifurcation diagram, when there are no singular points, different strata of the diagram correspond to different numbers and position of limit cycles.

The Poincaré map $\Delta\colon \Gamma_1^+\to\Gamma_1^+$ for each $\overline{\varepsilon}$ is the composition,

$$\Delta=f_{\overline{\varepsilon}}\circ\Delta_2\circ g_{\overline{\varepsilon}}\circ\Delta_1\,,\qquad f_{\overline{\varepsilon}}=F_{\overline{\mu}(\overline{\varepsilon})}\,,\ g_{\overline{\varepsilon}}=G_{\overline{\mu}(\overline{\varepsilon})}\,,$$

considered as the composition of one-dimensional maps depending on the parameters. The equation determining limit cycles,

$$\Delta(x)=x\,,\quad \text{where } x=x_1 \text{ is the coordinate on } \Gamma_1^+\,,$$

can be rewritten in the form

$$\tilde{g}_{\overline{\varepsilon}}(x)=f_{\overline{\varepsilon}}^{-1}(x)\,,\qquad \text{where } \tilde{g}_{\overline{\varepsilon}}=\Delta_2\circ g_{\overline{\varepsilon}}\circ\Delta_1\,. \tag{2}$$

Without loss of generality one may assume that a part of the transversal Γ_1^+ chosen in 3.1.1 is simply the segment $[-1\,,\,1]$. Thus we need to investigate the equation (2) for $x\in[-1\,,\,1]$. When speaking about bifurcations of limit cycles, we in fact deal with bifurcations of real roots of (2) on $[-1\,,\,1]$ with the parameters changing. There are two possible types of bifurcation:

(1) splitting of multiple roots, and
(2) escaping of a root through the boundary points ± 1.

Thus the bifurcation surface of the equation (2), that is, the surface in the parameter space where the number of roots is changed, is the union of three surfaces, Σ_L, Σ_+, and Σ_-. On Σ_L we have multiple roots of the equation (2), on Σ_+ (resp.,

Σ_-) there is at least one root equal to $+1$ (resp., -1). In terms of bifurcations in the original system, the surface Σ_L corresponds to splitting of a multiple cycle, while the union $\Sigma_+ \cup \Sigma_-$ corresponds to cycles escaping from the domain where the system is considered. Bifurcations of such type were discovered by H. Zołądek in 1983, see [4].

The exact description of the surfaces Σ_L, Σ_+, Σ_- will be given in 3.1.7.

3.1.4. *Unboundedness of the number of limit cycles.*

LEMMA 1. *The equation for limit cycles* (2) *can be written in the form*

$$\begin{aligned} &\frac{\lambda}{C_2(\delta)} + \frac{aC_1(\varepsilon)}{C_2(\delta)}(x + r_1(x, \bar{\varepsilon})) = f(x) + r_2(x, \bar{\varepsilon}), \\ &f(x) = f_0^{-1}(x), \qquad a = \left.\frac{\partial g_{\bar{\varepsilon}}}{\partial y_1}\right|_{y_1=0}, \end{aligned} \tag{2'}$$

where the functions $r_i(\cdot, \bar{\varepsilon})$ *tend to* 0 *in the* C^k*-norm on* $[-1, 1]$ *as* $\bar{\varepsilon} \to 0$. *The smoothness* k *can be chosen arbitrarily high.*

PROOF. Since $g_{\bar{\varepsilon}}(y_1) = \lambda + ay_1 + O(y_1^2)$, we have

$$\Delta_2 \circ g_{\bar{\varepsilon}} \circ \Delta_1(x, \bar{\varepsilon}) = \frac{\lambda}{C_2(\delta)} + \frac{aC_1(\varepsilon)}{C_2(\delta)}(x + C_1(\varepsilon) \cdot O(x^2)).$$

Since the function $O(x^2)$ and its k derivatives are uniformly bounded on $[-1, 1]$, the assertion of the lemma holds for r_1. It holds for $r_2(x, \bar{\varepsilon}) = f_{\bar{\varepsilon}}^{-1}(x) - f_0^{-1}(x)$ since $f_{\bar{\varepsilon}}$ is C^k-smooth in $(x, \bar{\varepsilon})$. □

Introduce the new parameters

$$p = a\, C_1(\varepsilon)/C_2(\delta), \qquad q = -\lambda/C_2(\delta).$$

Then the equation (2′) can be written in the form

$$p(\bar{\varepsilon})x - q(\bar{\varepsilon}) + p(\bar{\varepsilon})r_1(x, \bar{\varepsilon}) - r_2(x, \bar{\varepsilon}) = f(x). \tag{3}$$

Isolated roots of the equation (3) on the segment $x \in [-1, 1]$ are in one-to-one correspondence with limit cycles intersecting the transversal Γ_1^+ that appear after perturbation of the lips. Moreover, simple roots correspond to hyperbolic cycles, double roots correspond to semistable cycles, etc. Thus from now on we will speak only about roots of the equation (3) and bifurcation diagrams for these roots; the reformulation of results for limit cycles is straightforward and will be omitted.

Fix $P > \max\limits_{x \in [-1, 1]} f'(x)$ and consider the subset in $\mathbb{R}^3_{\bar{\varepsilon}}$ for which $0 < p < P$. In this domain the equation (3) would be a small perturbation of the equation

$$px - q = f(x). \tag{4}$$

THEOREM 2. *The lips can generate an arbitrary number of limit cycles in generic families. More precisely, for any $N \in \mathbb{N}$ one can construct a 3-parameter family of vector fields on the sphere such that in this family and in any sufficiently close 3-parameter family there occurs an ensemble that is $\mathfrak{c}$-equivalent to the lips, and out of this ensemble at least N limit cycles are born when the parameters change.*

PROOF. Indeed, for any finite N one may choose the parameters p and q and a generic function $f(x)$ such that the unperturbed equation (4) would have N simple roots on $[-1, 1]$. Then the corresponding equation (3) would have at least N roots for arbitrarily small ε, δ, and λ.

The 3-parameter family of vector fields with the corresponding transition functions may be constructed using the standard glueing technique. □

COROLLARY. *The number of pairwise nonhomeomorphic bifurcation diagrams for the lips is infinite.*

3.1.5. *The second reparametrization. Shape of the surface* Σ_L. Consider a small neighborhood of the origin $V \subset \mathbb{R}^3_{\overline{\varepsilon}}$ and denote by V^+ the intersection $V \cap \{\delta > 0, \varepsilon > 0\}$. Without loss of generality we may put V^+ being a small cube with the edges parallel to the coordinate axes.

Consider the reparametrization map $\Phi: V^+ \to \mathbb{R}^{3+}_{\delta, p, q} = \mathbb{R}^3_{\delta, p, q} \cap \{p > 0\}$ defined by the formula

$$\Phi: (\delta, \varepsilon, \lambda) \mapsto (\delta, p, q) = \left(\delta, \frac{aC_1(\varepsilon)}{C_2(\delta)}, -\frac{\lambda}{C_2(\delta)}\right). \tag{5}$$

LEMMA 2 (on the geometry of the domain $\Phi(V^+)$ and the properties of Φ^{-1}).

1. *The domain $\Phi(V^+)$ is unbounded. The half plane $OPQ^+ = \mathbb{R}^{3+}_{\delta, p, q} \cap \{\delta = 0\}$ belongs to the boundary of $\Phi(V^+)$.*

2. *The inverse map Φ^{-1} is defined on $\Phi(V^+)$ and extends by continuity to OPQ^+. The extended map is continuously differentiable in δ at $\delta = 0$ and takes the half plane OPQ^+ into the point $\overline{\varepsilon} = 0$.*

3. *For any compact set $D \subset \mathbb{R}^{2+}_{p, q}$ there exists $\delta_D > 0$ such that the cylinder $Z_D = (0, \delta_D) \times D$ belongs to $\Phi(V^+)$.*

4. *For any point $A \in \mathbb{R}^{2+}_{p, q}$ the extended map Φ^{-1} takes the semi-interval $[0, \delta_A) \times A$ into a part of a curve tangent at $\overline{\varepsilon} = 0$ to the line $\delta = \varepsilon$, $\lambda = 0$.*

PROOF. Using the representations (1a), (1b), for the values C_i, we can obtain the explicit formula for Φ^{-1}:

$$\varepsilon = \frac{\delta}{\left(1 - \frac{\sqrt{\delta}}{\pi} \ln \frac{p\varphi(\varepsilon)}{a\varphi(\delta)}\right)}, \qquad \lambda = -q\varphi(\delta) \exp\left(-\frac{\pi}{\sqrt{\delta}}\right).$$

Here by (1a)

$$\varphi(\varepsilon) = \exp\left(-\frac{2 \arctan \sqrt{\varepsilon}}{\sqrt{\varepsilon}}\right)$$

is an analytic (after the natural extension to the origin) nonvanishing function, and $\varphi(\delta)$ is the same function considered as a function of δ.

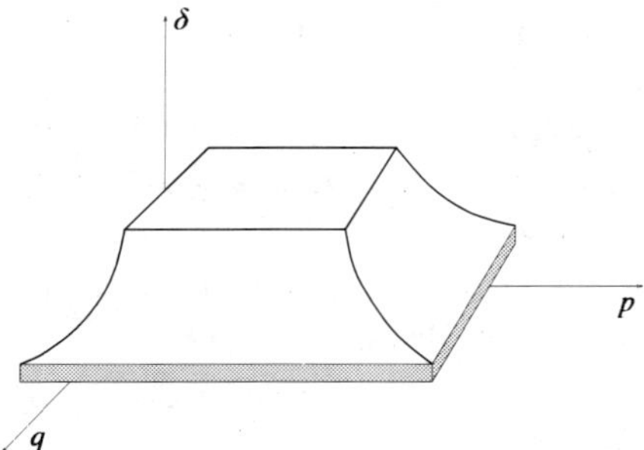

FIGURE 8. The domain $\Phi(V^+)$ for the case when V^+ is a cube.

These formulas imply all assertions of the Lemma. In the case when V^+ is a cube, the image $\Phi(V^+)$ has the shape shown in Figure 8.

COROLLARY 1. *For any compact $D \Subset \mathbb{R}^{2+}_{p,q}$, the map Φ^{-1} takes the cylinder Z_D into a narrow horn with the vertex at $\overline{\varepsilon} = 0$ tangent to the line $\delta = \varepsilon$, $\lambda = 0$; the projection of this horn to the plane $\delta O \varepsilon$ has the opening of the order $\sim \delta^{3/2}$, and the projection to the plane $\delta O \lambda$ has the exponentially small opening of the order $\sim \exp\left(-\frac{\pi}{\sqrt{\delta}}\right)$.*

REMARK. We say that the horn $\Phi^{-1}(Z_D)$ *corresponds* to the compact D.

COROLLARY 2. *There exists a compact $D \Subset \mathbb{R}^{2+}_{p,q}$ such that the surface Σ_L lies inside the corresponding horn.*

PROOF OF COROLLARY 2. Indeed, $\min_{x\in[-1,1]} f'(x) > 0$. Fix positive

$$p_0 < \min_{x\in[-1,1]} f'(x), \qquad p_1 > \max_{x\in[-1,1]} f'(x),$$

and

$$Q > \max_{x\in[-1,1]} |f(x)| + 2 \max_{x\in[-1,1]} f'(x).$$

Choose $D = [p_0, p_1] \times [-Q, Q]$. Then for $(p, q) \notin D$ the equation (3) has only simple roots, hence $\Sigma_L \subseteq \Phi^{-1}(Z_D)$. □

3.1.6. *Reparametrization of the equation for limit cycles.* After the second reparametrization $(\varepsilon, \delta, \lambda) \mapsto (\delta, p, q)$ the equation (3) will be transformed into the equation

$$(6) \quad px - q + r(x; \delta, p, q) = f(x), \qquad r = p\tilde{r}_1 - \tilde{r}_2, \ \tilde{r}_i = r_i \circ \Phi^{-1}, \ i = 1, 2.$$

From the assertions of the previous section and the properties of r_i the following properties of r easily follow:

(1) The function $r(x; \delta, p, q)$ can be extended continuously to the half plane OPQ^+; after such extension, $r(x; 0, p, q) \equiv 0$.
(2) For $\delta > 0$ the function r is smooth in the variables (x, p, q). Any finite number k of derivatives of r in these variables are uniformly bounded on

$[-1, 1]$ by a certain bound $M_k(\delta)$ as $p, q \in D$, and $M_k(\delta) = O(\delta)$ as $\delta \to 0$.

(3) For $\delta = 0$ the function r is C^1-differentiable in δ.

REMARK. Below we abbreviate the verbose expression "*any finite number of partial derivatives of a function* $\Psi(x; \delta, p, q)$ *in the variables* (x, p, q) *is uniformly bounded on* $x \in [-1, 1]$, *for* $(p, q) \in D$ *by a certain bound* $M(\delta)$ *which is of order of magnitude* $O(\delta)$ *as* $\delta \to 0$" to the condensed symbolic notation "$\Psi = O(\delta)$".

3.1.7. *Formulation of the main theorem and the beginning of the proof.* In this section we give the precise description of the bifurcation diagram for the equation (3) in the intersection V^+. As it was explained in 3.1.3, this diagram consists of three parts, Σ_L, Σ_+, and Σ_-. To describe the parts Σ_+ and Σ_- corresponding to escaping roots, it is sufficient to use the parameters $(\delta, \varepsilon, \lambda)$. The description of Σ_L is given in the (δ, p, q)-space.

THEOREM 3. 1. *The boundary surfaces* Σ_+ *and* Σ_- *are graphs of smooth functions* $\lambda = \lambda_\pm(\varepsilon, \delta)$ *continuously extendable to the boundary of the quadrant* $\varepsilon > 0$, $\delta > 0$. *The functions* $\lambda_\pm(\cdot)$ *are flat at the point* $\varepsilon = \delta = 0$.

2. *The surface* Σ_L *is a Legendre horn defined in the following way. Consider in the half plane* $\mathbb{R}^{2+}_{p,q} \cap \{p > 0\}$ *the Legendre transform* $L(\gamma)$ *of the graph of the function* $f(x)$. *Consider the embedding of* $\mathbb{R}^{2+}_{p,q}$ *into* $\mathbb{R}^3_{\delta,p,q}$ *as the part of the plane* $\delta = 0$. *Consider the closure* $\overline{\Phi(V^+)} \subset \mathbb{R}^3_{\delta,p,q}$ *and let* Z_L *be the cylinder in* $\Phi(V^+)$ *over* $L(\gamma)$ *with the axis parallel to* $O\delta$ *of the height* δ_0, *see Figure* 9.

Then for sufficiently small δ_0 *the "blown-up horn"*

$$Z = \Phi(\Sigma_L \cap V^+)$$

is diffeomorphic to Z_L. *The diffeomorphism* Δ *taking* Z *into* Z_L *preserves the foliation* $\delta = \text{const}$, *is* C^1-*smooth in* δ, *and its difference from the identity map on the fiber* $\delta = \text{const}$ *is* $O(\delta)$.

Regular points of Σ_L *correspond to double roots of the equation* (3). *Edges of return correspond to roots of multiplicity three of the equation* (3), *hence to nonhyperbolic limit cycles of the vector field* (*stable or unstable*).

3. *The intersection of the boundary of* Σ_L *with the layer* $0 < \delta < \delta_0$ *belongs to* Σ_+ *and* Σ_-. *At points of this intersection, the surface* Σ_L *is tangent to either* Σ_+ *or* Σ_-.

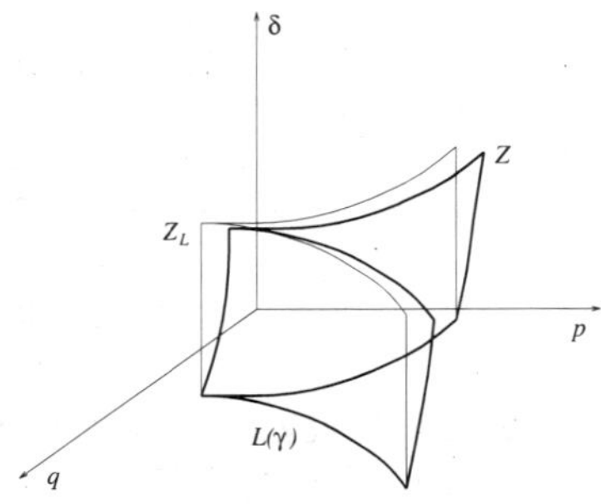

FIGURE 9. The Legendre horn.

The first assertion of the theorem can be easily deduced from Lemma 1. Indeed, these surfaces are defined by the equations

$$\begin{gathered}\lambda_+ + aC_1(\varepsilon)(1 + r_1(1, \overline{\varepsilon}_+)) = C_2(\delta)(f(1) + r_2(1, \overline{\varepsilon}_+)), \\ \lambda_- + aC_1(\varepsilon)(-1 + r_1(-1, \overline{\varepsilon}_-)) = C_2(\delta)(f(-1) + r_2(-1, \overline{\varepsilon}_-)), \\ \overline{\varepsilon}_+ = (\varepsilon, \delta, \lambda_+), \qquad \overline{\varepsilon}_- = (\varepsilon, \delta, \lambda_-),\end{gathered}$$

which, by the implicit function theorem, give the dependence of $\lambda_\pm$ on (δ, ε), since the terms r_i are small and smooth. Clearly those equations make sense also for the zero values of ε and δ, and their solutions $\lambda_\pm(\cdot)$ are flat at $\varepsilon = \delta = 0$.

The proof of the remaining assertions of the theorem occupies the rest of the paper, except for 3.3.4, where we prove the duality theorem for the generalized Legendre transform. The scheme of the proof is the following: the equation for limit cycles (6) is considered as a small perturbation of the equation (4) depending on two parameters p and q. The bifurcation diagram of roots of (4) on the plane $\mathbb{R}^2_{p,q}$ is the Legendre transform $L(\gamma)$ of the graph γ of the function $f(x)$. Bifurcations corresponding to different points of the curve $L(\gamma)$ are described in 3.2. A nonlinear perturbation occurring in (6) yields the *generalized Legendre transform* introduced in 3.3. Finally, in 3.4 we investigate properties of a family of generalized Legendre transforms depending on the parameter δ and close to the classical Legendre transform.

3.2. Investigation of the unperturbed equation. Consider in more detail the number n of different solutions of the unperturbed equation (4) on the segment $[-1, 1]$ for different p, q. Note that (4) is exactly the equation for limit cycles for the case when the transition map $g_{\overline{\varepsilon}}$ is linear and $f_{\overline{\varepsilon}}^{-1}$ does not depend on the parameters $\overline{\varepsilon}$. In this framework we can temporarily drop out the restriction $p > 0$.

3.2.1. *The classical Legendre transform.* The number of roots of equation (4) changes when the line $y = px - q$ is tangent to the graph of the function $f(x)$, or when this line passes through an endpoint of the graph. The set of points $(p, q) \in \mathbb{R}^2_{p,q}$ corresponding to tangent lines is the Legendre transform of the graph γ of the function f.

PROPOSITION 0 [2]. *Let γ be the graph of a generic smooth function $y = f(x)$ defined on a segment. Then the Legendre transform $L(\gamma)$ is a piecewise-smooth curve with a finite number of cuspidal points and without inflection points. Smooth components of the dual curve $L(\gamma)$ may have simple transversal self-intersections.* □

For brevity, below we call $L(\gamma)$ the *Legendre curve*.

LEMMA 3. *The points $(p, q) \in \mathbb{R}^2$ for which the equation (4) has a root ± 1, form the pair of straight lines $\ell_\pm$ tangent to the Legendre curve $L(\gamma)$ at the endpoints of the latter.*

PROOF. Substituting $x = \pm 1$ into (4), we obtain the equation of $\ell_\pm$:

$$\lambda_- = \{-p - q = f(-1)\}, \qquad \lambda_+ = \{p - q = f(1)\}.$$

Let $p = p(x)$, $q = q(x)$ be the canonical parameterization of the dual curve $L(\gamma)$. Then at the endpoints of $L(\gamma)$ we have the identities $dp/dq = x$, and at

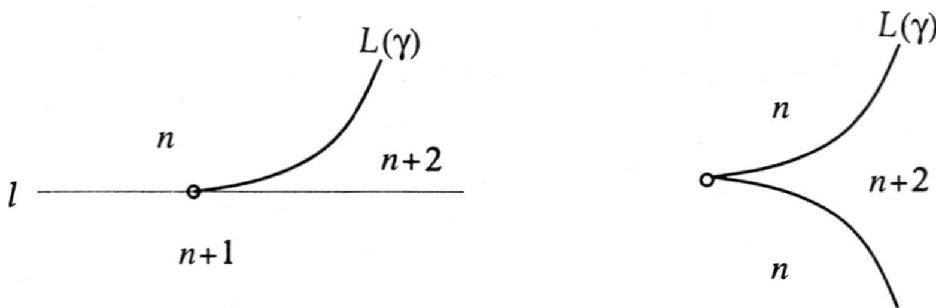

FIGURE 10. Bifurcation of zeros near cuspidal singularities of the Legendre curve and its endpoints.

the cuspidal points this identity holds for one-sided derivatives. The intersections $L(\gamma)\cap\ell_\pm$ correspond to $x=\pm1$. This implies the tangency conditions:

$$\left.\frac{dq}{dp}\right|_{x=\pm1}=\pm1. \qquad \square$$

3.2.2. *Bifurcation diagram for the nonperturbed equation.* A complete description of the bifurcation diagram for the equation (4) on the plane $\mathbb{R}^2_{p,q}$ is given by the following result, which can be established by simple arguments.

LEMMA 4. *The bifurcation diagram for the roots of the equation (4) is the union of the Legendre curve $L(\gamma)$ and two straight lines $\ell_\pm$ tangent to $L(\gamma)$ at the endpoints of the latter and parallel to the directions $p\pm q=0$:*

(1) *The number of roots is constant in each connected component of the complement to the union $L(\gamma)\cup\ell_-\cup\ell_+$.*
(2) *The number of roots in two components separated by the curve $L(\gamma)$ differs by 2; the domain which is locally convex has smaller number of roots.*
(3) *The number of roots in two components separated by the lines $\ell_\pm$, differs by 1.*
(4) *Transversal self-intersections of $L(\gamma)$ and intersections of $L(\gamma)$ with the lines $\ell_\pm$ describe simultaneous occurrence of two corresponding bifurcations.* $\square$

The local behavior of the number of roots near a cuspidal point and an endpoint of $L(\gamma)$ is uniquely determined by Lemma 4, see Figure 10: near a cuspidal point that corresponds to a root of multiplicity 3 of the equation (4), the number of roots decreases by 2 when passing from a "smaller" to a "larger" local connected component of $\mathbb{R}^2\setminus L(\gamma)$. Near the endpoint of the curve $L(\gamma)$ that corresponds to a double root at the point $x=\pm1$, there are three local connected components of $\mathbb{R}^2\setminus(L(\gamma)\cup\ell_\pm)$: one is a piece of half plane, the other is locally convex and the third one is a "thin" horn, see Figure 10. The number of roots is decreased by 1 each time when we move from the "thin" component to the half plane and then to the locally convex component.

3.3. Generalized Legendre transform. The classical Legendre transform described in the previous section, allows us to investigate the bifurcation diagram of limit cycles in the case when small nonlinear terms r_i are absent. In this case we study intersections of the graph of f with the 2-parameter family of lines $y = px - q$, $p, q \in \mathbb{R}$.

When the nonlinear terms in (3) are small but nonzero, the family of lines is replaced by a 2-parameter family of "almost straight" curves. The corresponding construction is called the *generalized Legendre transform.*

3.3.1. *The principal definition.*

DEFINITION 1. Consider a 2-parameter family of smooth curves on the plane (x, y) with the parameters (p, q), defined by the equation

$$\begin{gathered} F(x, y; p, q) = 0, \\ F'_y \neq 0, \quad F'_q \neq 0, \end{gathered} \tag{7}$$

and a curve γ on the plane $\mathbb{R}^2_{x,y}$. Choose the curves of the family (7) that are tangent to γ. The corresponding values of the parameters (p, q) constitute a subset $L_F(\gamma) \subset \mathbb{R}^2_{p,q}$. We call this subset the *generalized Legendre transform of the curve* γ.

REMARK. For the standard Legendre transform we have

$$F(x, y; p, q) = px - y - q.$$

Since the generalized Legendre transform depends on the function F, we sometimes refer to it as Legendre F-transform, when it would be necessary to distinguish between transforms generated by different functions.

3.3.2. *Genericity conditions.* Now we describe the genericity type conditions that must be be imposed on the curve γ, the graph of a function f, with respect to the family of curves F. Each point (x, y), $y = f(x)$, on this graph corresponds to a point $(p, q) = (p_x, q_x) \in L_F(\gamma)$ such that γ is tangent to the curve $F(\cdot, \cdot; p_x, q_x) = 0$ at the point $(x, f(x))$. By the implicit function theorem,

$$f'(x) = -\left.\frac{\frac{\partial F}{\partial x}(x, y; p_x, q_x)}{\frac{\partial F}{\partial y}(x, y; p_x, q_x)}\right|_{y=f(x)}.$$

Hence the point (p_x, q_x) is determined from the system of equations

$$\begin{gathered} F(x, y; p, q) = 0, \\ y = f(x), \\ \frac{\partial F}{\partial x}(x, y; p, q) + f'(x)\frac{\partial F}{\partial y}(x, y; p, q) = 0. \end{gathered} \tag{8}$$

Introducing the function

$$\varphi(x, p, q) = F(x, f(x); p, q), \tag{9}$$

we can rewrite (8) in the form

$$\varphi(x, p, q) = 0 \qquad (10)$$
$$\varphi'_x(x, p, q) = 0.$$

The solvability condition of the system (10) with respect to (p, q) is the inequality

$$\Delta = \det \begin{pmatrix} \varphi'_p & \varphi'_q \\ \varphi''_{xp} & \varphi''_{xq} \end{pmatrix} \neq 0 \qquad (11)$$

for all values of x, p, q. In that case the solutions

$$p_x = p(x), \qquad q_x = q(x) \qquad (12)$$

will be smooth functions.

Moreover, we require that there will be only a finite number of inflection points on φ, that is,

$$\#\left\{x : \frac{\partial^2 \varphi}{\partial x^2}(x, p_x, q_x) = 0\right\} < \infty. \qquad (13)$$

The exceptional values of x will be called F-critical, and we require in addition that for each F-critical value x

$$\frac{\partial^3 \varphi}{\partial x^3}(x, p_x, q_x) \neq 0. \qquad (14)$$

Definition. We say that a curve $\gamma = \{y = f(x)\}$ is in the general position with respect to the family F if the conditions (11), (13), (14) hold for the pair (F, f).

In particular, for $F = px - q - y$ we have $\Delta = 1$, and F-critical points are abscissas of inflection points of γ.

3.3.3. *The contact structure generated by the generalized Legendre transform.* Recall briefly some known constructions; for the detailed exposition see the book [2].

A *contact structure* on an odd-dimensional manifold is a field of hyperplanes in the tangent plane that satisfies certain nondegeneracy conditions (the so called maximal nonintegrability). Locally (and sometimes globally) such a field may be given as the distribution of zeros of a differential 1-form ω, the *contact form.* The nondegeneracy condition in terms of the form ω means that $d\omega|_{\omega=0}$ is everywhere nondegenerate antisymmetric bilinear form.

A *Legendre submanifold* is a submanifold of maximal dimension in a contact manifold such that $\omega|_{T\Lambda} = 0$. If the contact manifold has dimension $2n+1$, then the dimension of Λ is n.

A *Legendre foliation* (*bundle*) of a contact manifold M is a bundle with the total space M over some base B whose fibers are Legendre submanifolds. The projection $\pi: M \to B$ is the *Legendre projection.* The *Legendre singularities* are singularities of the Legendre projection restricted on a Legendre submanifold of M.

LEMMA 5. *Suppose that a function* $\varphi(x, p, q)$ *satisfies the inequality* (11). *Then the differential* 1*-form*

$$\omega = \varphi'_p\, dp + \varphi'_q\, dq \tag{15}$$

determines a contact structure in the space $\mathbb{R}^3_{x,p,q}$. *The curve* Λ *given by the equation* (10) *is a Legendre submanifold with respect to this structure.*

REMARK 1. The standard Legendre transform corresponds to the case $\omega = -dq + x\,dp$.

REMARK 2. A Legendre bundle in the space endowed with the contact structure ω of the form (15) may be defined by a family of lines parallel to the x-axis. The projection of the curve Λ along this direction coincides with the generalized Legendre transform of the curve $\gamma = \{y = f(x)\}$.

PROOF OF LEMMA 2. The form ω defines a contact structure on $\mathbb{R}^3_{x,p,q}$ if $\operatorname{rank}(d\omega|_{\omega=0}) = 2$. Since the rank of any antisymmetric form is always even, it is sufficient to verify that the restriction $d\omega|_{\omega=0}$ is nonzero. A simple computation shows that

$$d\omega = d\varphi'_p \wedge dp + d\varphi'_q \wedge dq = \varphi''_{px} dx \wedge dp + \varphi''_{qx} dx \wedge dq.$$

Consider the vector fields

$$v_1 = (1, 0, 0), \qquad v_2 = (0, -\varphi'_q, \varphi'_p). \tag{16}$$

One can easily see that

$$\omega(v_1) = \omega(v_2) = 0,$$

but

$$d\omega(v_1, v_2) = \Delta \neq 0$$

by the assumptions of the lemma. Thus ω is nondegenerate and defines a contact structure.

Now we show that the curve Λ is a Legendre submanifold. Indeed, by (10) on that curve we have

$$d\varphi|_\Lambda = (\varphi'_x\, dx + \varphi'_p\, dp + \varphi'_q\, dq)|_\Lambda = (\varphi'_p\, dp + \varphi'_q\, dq)|_\Lambda = 0. \quad \square \tag{17}$$

TERMINOLOGY. Here and below the direction on the (p, q)-plane parallel to the q-axis is called vertical.

COROLLARY 1. *The generalized Legendre transform of a smooth curve in the general position with respect to the family* F *is a piecewise smooth curve with cusps and without vertical tangent lines.*

PROOF. Indeed, for the function φ defined by (9), the assertion of the lemma holds true. The curve $L_F(\gamma)$ on the (p, q)-plane is obtained from Λ by projecting along the x-direction onto the base of the Legendre bundle. Let us verify that singularities of the projection $\Lambda \to L_F(\gamma)$ are generic Legendre singularities. The critical points of the projection are these points of Λ for which

$$\frac{\partial}{\partial x}\left(\frac{\varphi}{\varphi'_x}\right) = 0,$$

that is, $\varphi''_{xx} = 0$. By (14), this means a simple tangency of Λ with a fiber of the bundle, and this singularity cannot be destroyed by a small perturbation of Λ within the class of Legendre curves.

Next, from the classification of generic Legendre singularities [2] it follows that the parts of Λ between the critical points project as smooth curves on the (p, q)-plane, and the critical points themselves project as cuspidal points.

From (17) it follows that the tangent line at each point $(p, q) \in L_F(\gamma)$ is given by the equation

$$\varphi'_p(x, p, q)\, dp + \varphi'_q(x, p, q)\, dq = 0, \tag{18}$$

where (x, p, q) is the preimage of a point (p, q) on Λ. The nonverticality condition follows from the inequality (7), since

$$\varphi'_q(x, p, q) = F'_q(x, f(x), p, q) \neq 0. \quad \square$$

COROLLARY 2. *The parts of $L_F(\gamma)$ without cuspidal points are the graphs of smooth functions of the form $q = g(p)$, where*

$$g'(p) = -\frac{\varphi'_p(x, p, q)}{\varphi'_q(x, p, q)}. \tag{19}$$

This corollary is immediately implied by (18). $\square$

COROLLARY 3. *The germ of the projection of the curve Λ at a cuspidal point can be transformed into the standard semicubic parabola by a diffeomorphism of the (p, q)-plane. This diffeomorphism is called the normalization. If the contact structure, the Legendre bundle, and the curve depend smoothly on a parameter δ, then the normalization can be also chosen smoothly depending on δ.*

REMARK. This fact is known to specialists in the area. It is an elementary corollary to the following two facts: the normal form of a differential equation not resolved with respect to the derivative near a regular singular point [1, §4], and the Darboux theorem on the normal form of the contact structure. The authors are grateful to V. Arnold who pointed out this circumstance.

3.3.4. *Digression*: *Invertibility of the generalized Legendre transform. The dual transform.* Denote by Ω the set of all graphs of smooth functions $y = f(x)$ which are in a generic position with respect to the family F of curves and without F-critical points. As this was shown before, for any $\gamma \in \Omega$ its generalized Legendre transform $L_F(\gamma)$ is the graph of a smooth function $q = g(p)$. It turns out that the inverse transform which takes curves on the (p, q)-plane into their originals on the (x, y)-plane, is also the generalized Legendre transform generated by a family of smooth curves *dual* to the family $F = 0$.

DEFINITION. Consider the equation (7) as an equation defining a 2-parameter family of curves on the (p, q)-plane with the parameters x, y. We call this family the *dual family* to F and denote it by F^*.

REMARK. In [1], an invariant definition of the dual family is given, which does not depend on the choice of the function F.

The dual family F^* determines a generalized Legendre transform defined on curves in the (p, q)-plane. In particular, it is defined on graphs of smooth functions $q = g(p)$ that are in the general position with respect to the family F^* and have no F^*-critical points. The set of curves with such properties will be denoted by Ω^*.

THEOREM 4 (Duality theorem for generalized Legendre transform). *The generalized Legendre transform L_F defined on the set Ω and the generalized Legendre transform L_{F^*} defined on Ω^*, are mutually inverse.*

The proof is based on the following remarkably symmetric construction. The equation (7) defines a three-dimensional hypersurface E in the space $\mathbb{R}^4_{x,y,p,q}$. On the tangent bundle TE there we have an 1-form ω_E defined by one of the two equivalent formulas,

$$\omega_E = (F'_p\, dp + F'_q\, dq)|_{TE}, \tag{20a}$$

$$\omega_E = -(F'_x\, dx + F'_y\, dy)|_{TE}. \tag{20b}$$

The equivalence of (20a) and (20b) follows from the identity $dF|_{TE} = 0$. In the general case the nondegeneracy condition $\operatorname{rank} d\omega|_{\omega=0} = 2$ is satisfied, hence ω_E defines a contact structure on E.

Next, the projection operators $\Pi_{xy}: E \to \mathbb{R}^2_{x,y}$, $\Pi_{pq}: E \to \mathbb{R}^2_{p,q}$ define on E *two structures of Legendre bundle*. Indeed, by (20a), (20b) the fibers $(x, y) = \text{const}$ and $(p, q) = \text{const}$ are in both cases Legendre submanifolds. As before, we denote by γ the graph of the function $y = f(x)$.

LEMMA 6 (on one-to-one correspondence). 1. *Graphs of smooth functions on the plane $\mathbb{R}^2_{x,y}$ and Legendre submanifolds of the contact manifold E, projecting onto those graphs, are in one-to-one correspondence.*

2. *Points of the Legendre submanifold $\Lambda_\gamma \subset E \subset \mathbb{R}^4_{x,y,p,q}$ which projects onto the graph γ, satisfy the system of equations* (8).

3. *For any smooth function $f(x)$ the system of equations* (8) *determines a Legendre submanifold $\Lambda_\gamma \subset E$ which projects onto the graph $\gamma = \{y = f(x)\}$.*

PROOF. The first assertion of the Lemma follows from the second and the third ones. The last condition from (8) for points on the Legendre submanifold $\Lambda_\gamma \subset E \subset \mathbb{R}^4_{x,y,p,q}$ projecting onto γ, follows from the equality $\omega_E|_{T\Lambda_\gamma} = 0$ and (20b). All other conditions are trivially satisfied.

Now we proceed with the demonstration of assertion 3 of the Lemma. First we show that the equations (8) define a smooth submanifold for any choice of $f(x)$. To prove this, it is sufficient to establish the inequality (11) for the function $\varphi(x, p, q)$ defined by (9).

Consider the vector fields on TE,

$$v_1 = F'_y \frac{\partial}{\partial x} - F'_x \frac{\partial}{\partial y}, \qquad v_2 = -F'_q \frac{\partial}{\partial p} + F'_p \frac{\partial}{\partial q}.$$

From (20a), (20b) it follows that $\omega_E(v_1) = \omega_E(v_2) = 0$, hence $d\omega_E(v_1, v_2)$ is nonzero everywhere on E, since ω_E is nondegenerate. A straightforward computation with (9) taken into account, yields

$$d\omega_E(v_1, v_2) = F'_y \cdot \det \begin{bmatrix} \varphi'_p & \varphi'_q \\ \varphi''_{xp} & \varphi''_{xq} \end{bmatrix},$$

and we proved (11), thus showing that (8) determines a smooth curve in the space $\mathbb{R}^4_{x,y,p,q}$. One can easily see that this curve is a Legendre submanifold projecting onto γ. □

In a similar way the following result is proved.

LEMMA 7. 1. *Graphs of smooth functions on the plane $\mathbb{R}^2_{p,q}$ and Legendre submanifolds of the contact manifold E, projecting onto these graphs, are in one-to-one correspondence.*

2. *Points of the Legendre submanifold $\Lambda_{\tilde\gamma} \subset E \subset \mathbb{R}^4_{x,y,p,q}$ that projects onto the graph $\tilde\gamma = \{q = g(p)\}$ of a smooth function g, satisfy the system of equations*

$$\begin{gathered} F(x, y; p, q) = 0, \\ q = g(p), \\ \frac{\partial F}{\partial p}(x, y; p, q) + g'(p)\frac{\partial F}{\partial q}(x, y; p, q) = 0. \end{gathered} \tag{8'}$$

3. *For any smooth function $g(p)$ the system of equations (8′) determines a Legendre submanifold $\Lambda_{\tilde\gamma} \subset E$ that projects onto the graph $\tilde\gamma = \{q = g(p)\}$.* □

These two lemmas together imply the following result.

THEOREM 5. *Assume that the operator Π_{xy} projects a Legendre submanifold Λ onto the graph of a smooth function $y = f(x)$ and the operator Π_{pq} projects Λ onto the graph of $q = g(p)$. Then:*

(1) *the graph $\{q = g(p)\}$ is the generalized Legendre F-transform of the graph $\{y = f(x)\}$;*
(2) *the graph $\{y = f(x)\}$ is the generalized Legendre F^*-transform of the graph $\{q = g(p)\}$;*
(3) *there are no F-critical points on the x-axis;*
(4) *there are no F^*-critical points on the p-axis.*

PROOF. The first two assertions immediately follow from Lemmas 6 and 7. The third assertion is the inverse of Corollary 1 to Lemma 5. Namely, if there are F-critical points, then the generalized Legendre F-transform of the graph $y = f(x)$ has cusps. The fourth assertion is similar. □

PROOF OF THEOREM 4. Immediate corollary of Theorem 5. □

3.4. Construction of the surface Σ_L. In this section we complete the proof of assertions 2 and 3 of Theorem 3.

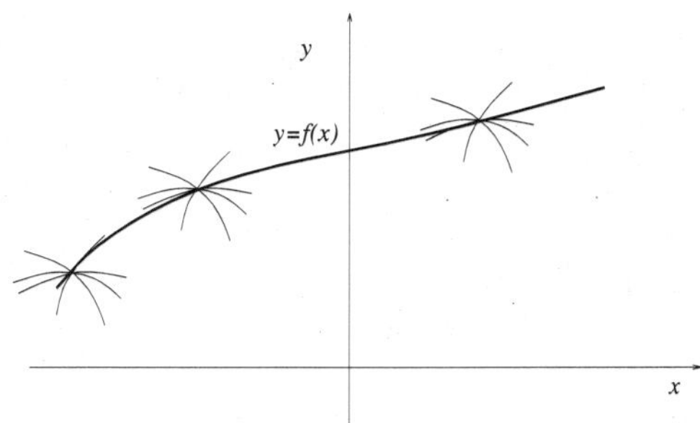

FIGURE 11. Generalized Legendre transform: the curve γ and the family of "almost straight lines" $F(x, y; p, q) = 0$.

3.4.1. *Preliminary considerations.* Recall that Σ_L is the bifurcation surface for limit cycles, that is, the surface in the parameter space on which the number of cycles is changed due to splitting and disappearance of multiple cycles or, what is the same, the surface where the number of roots of the equation (6) changes. Recall that the reparametrization $\Phi\colon (\delta, \varepsilon, \lambda) \mapsto (\delta, p, q)$ transforms the equation for limit cycles to the following equation equivalent to (6),

$$px - q + p\,r(x, \delta, p, q) = f(x), \qquad r = O(\delta).$$

For a fixed value of parameters the multiple roots of the equation (6) are the abscissas of the tangency points between the graph of the function

$$y = f(x) \tag{21}$$

and the "almost straight line"

$$y = px - q + r, \qquad r = r(x, \delta, p, q). \tag{22}$$

For a fixed δ, the graphs of the functions (22) constitute a 2-parameter family of curves, see Figure 11, thus in a natural way the generalized Legendre transform appears, which is close to the standard one. Below we prove that the generalized (for $\delta > 0$) and the standard (for $\delta = 0$) Legendre transforms of the graph of a generic function are diffeomorphic and the diffeomorphism is $O(\delta)$-close to the identity. These arguments allow us to construct the surface Σ_L and finish the proof of Theorem 3.

3.4.2. *Investigation of the bifurcation surface for $\delta \to 0$.* Denote the generalized Legendre transform associated with the family of curves (22) with a fixed value of δ by L_δ.

LEMMA 8. *The generalized Legendre transform $L_\delta(\gamma)$ of a generic graph γ is diffeomorphic to $L_0(\gamma)$ and the diffeomorphism smoothly depends on δ as $\delta \to 0$.*

PROOF. We split the proof into a number of steps.

PROPOSITION 1. *The set $L_\delta(\gamma)$ is a piecewise smooth curve with a finite number of cuspidal singularities and nowhere tangent to the vertical direction.*

Indeed, in this case the function $\varphi(x\,;\,p\,,\,q)$ defined in 3.3.2, becomes

$$\varphi_\delta(x\,;\,p\,,\,q) = px - q - f(x) + r(x\,,\,\delta\,,\,p\,,\,q). \tag{23}$$

We show that the curve γ is in the general position with respect to the family

$$F_\delta(x\,,\,y\,;\,p\,,\,q) = px - q - y + r(x\,,\,\delta\,,\,p\,,\,q)\,,$$

where r is the same as above. We need to check the conditions (11), (13), and (14).

The function r and its derivatives are uniformly bounded for $(p\,,\,q) \in D \Subset \mathbb{R}^2_{p\,,\,q}$ and tend to zero as $\delta \to 0$. In particular,

$$\varphi_0(x\,;\,p\,,\,q) = px - q - f(x).$$

Next, we have

$$\Delta_\delta = \det\begin{pmatrix} (\varphi_\delta)'_p & (\varphi_\delta)'_q \\ (\varphi_\delta)''_{x\,,\,p} & (\varphi_d)''_{xq} \end{pmatrix} = \det\begin{pmatrix} x + r'_p & -1 + r'_q \\ 1 + r''_{xp} & r''_{xq} \end{pmatrix} = 1 + O(\delta) \neq 0.$$

It follows that the curve $L_\delta(\gamma)$ is parametrized as

$$\begin{gathered} p = p_\delta(x)\,, \qquad q = q_\delta(x)\,, \\ p_\delta(x) = p_0(x) + O(\delta)\,, \quad q_\delta = q_0 + O(\delta)\,, \end{gathered} \tag{24}$$

where the functions $p_0(x)$, $q_0(x)$ parameterize the standard Legendre transform of γ.

Next,

$$\varphi''_{xx}(x\,;\,p\,,\,q) = -f''(x) + r''_{xx}.$$

By the general position assumption, the equation

$$f''(x) = 0 \tag{25}$$

has only simple roots. Hence the perturbed equation

$$-f''(x) + r''_{xx}(x\,;\,p_\delta(x)\,,\,q_\delta(x)) = 0. \tag{25$'$}$$

also has only simple roots, which are $O(\delta)$-close to the roots of (25).

Thus all conditions of the generic position of γ with respect to the family (22) are satisfied, and the first assertion follows from the Lemma 5.

PROPOSITION 2. *Parts of the curves $L_\delta(\gamma)$ and $L_0(\gamma)$ near the corresponding cuspidal points can be transformed into each other by a diffeomorphism $O(\delta)$-close to the identity.*

This immediately follows from Corollary 3 in 3.3.3.

PROPOSITION 3. *The self-intersection points of the curve* $L_\delta(\gamma)$ *are in one-to-one correspondence with the self-intersection points of the curve* $L_0(\gamma)$. *The parts of curves near the self-intersection points can be transformed into each other by a diffeomorphism that is* $O(\delta)$*-close to the identity.*

Since $f(x)$ is a generic function, the standard Legendre transform of its graph has only simple self-intersection points. From the formula (24) with the left-hand side smoothly depending on δ it follows that each self-intersection point $(\overline{p}_\delta, \overline{q}_\delta) \in L_\delta(\gamma)$ is a point of transversal simple self-intersection which is $O(\delta)$-close to the point $(\overline{p}_0, \overline{q}_0) \in L_0(\gamma)$ of the transversal self-intersection of $L_0(\gamma)$. The rest is evident. □

The fact that smooth pieces of $L_0(\gamma)$ and $L_\delta(\gamma)$ are diffeomorphic, is trivial. The global smooth equivalence between the entire curves is obtained using partitions of unity. Thus the proof of Lemma 8 is completed. □

COROLLARY. *There exists* $\delta_0 \in (0, \delta_D)$ *such that in the domain* $\Phi(V^+) \cap \{0 < \delta < \delta_0\}$ *the surface of multiple roots of the equation* (23) *is diffeomorphic to the cylinder* $Z_L = (0, \delta_0) \times L_0(\gamma)$. *The diffeomorphism extends onto the plane* $\delta = 0$ *and is continuously differentiable in* δ *at* $\delta = 0$. *The restriction of the extended diffeomorphism on the half plane* $\{\delta = 0, \varepsilon > 0\}$ *is the identity.* □

3.4.3. *Splitting and disappearance of limit cycles.* The bifurcation of limit cycles which occurs in the dynamical system when the parameters $(\varepsilon, \delta, \lambda)$ are on the surface Σ_L (that is equivalent to saying that the parameters (δ, p, q) are on the surface Z) is the same as the bifurcation of roots of the equation (3).

Smooth points of Z correspond to semistable cycles, which split into two hyperbolic cycles to one side of Z and disappear to the other. Edges of return correspond to limit cycles of multiplicity 3. Curves of self-intersection of Z are responsible for simultaneous occurrences of two semistable cycles.

3.4.4. *Tangency of the surface of multiple cycles with the boundary surfaces.* Recall that the boundary surfaces Σ_+, Σ_- correspond to the parameter values for which the equation of limit cycles has a root $x = \pm 1$. When a point moves across those surfaces, the limit cycles of the dynamical system leave the domain of consideration.

The values of the parameters for which a multiple root of the equation is equal to ± 1, belong both to the boundary of Σ_L and to one of the surfaces Σ_+, Σ_-. Let us show that at that point the surfaces are tangent to each other.

The latter claim is equivalent to saying that the surface Z is tangent to, say, Σ_+ (to simplify the exposition, we consider only the case of the boundary $x = +1$). We need to consider the values of δ different from 0 and δ_0. In turn, the assertion on tangency is equivalent to the fact that the generalized Legendre transform $L_\delta(\gamma)$ is tangent to the curve $\Phi(\Sigma_+) \cap \{\delta = \text{const}\}$ at the endpoint of $L_\delta(\gamma)$ corresponding to $x = 1$. The curve $\Phi(\Sigma_+) \cap \{\delta = \text{const}\}$ is described by the equation

$$\varphi_\delta(1; p, q) = 0,$$

where φ_δ is defined as in (23). Therefore everywhere on that curve we have

$$\frac{dq}{dp} = -\frac{(\varphi_\delta)'_p(1; p, q)}{(\varphi_\delta)'_q(1; p, q)}.$$

But, by (19), that same condition holds for the endpoint of the generalized Legendre transform $L_\delta(\gamma)$, corresponding to $x = 1$. This means that the two curves touch each other, hence the surfaces Σ_L and Σ_+ are tangent. The proof for Σ_- is the same. Thus the proof of Theorem 3 is complete. □

References

1. V. I. Arnold, *Geometric theory of ordinary differential equations*, Springer-Verlag, Berlin, Heidelberg, and New York, 1979.
2. ———, *Mathematical methods of classical mechanics* (2nd edition), Springer-Verlag, New York, 1989.
3. V. Arnold, V. Afraĭmovich, Yu. Ilyashenko, and L. Shilnikov, *Bifurcation theory*, Dynamical Systems–5, Itogi Nauki. Sovremennye Problemy Matematiki: Fundamental′ nye Napravleniya., vol. 5, VINITI, Moscow, 1985; English transl., Encyclopaedia of Math. Sci., vol. 5, Springer-Verlag, Heidelberg (to appear).
4. Kh. Zholondek [H. Zolądek], *On versality of a certain family of symmetric vector fields on the plane*, Mat. Sb. **120** (1983), no. 4, 473–499; English transl. in Math. USSR-Sb. **48** (1984).
5. Yu. Ilyashenko and S. Yakovenko, *Finitely-smooth normal forms for local families of diffeomorphisms and vector fields*, Russian Math. Surveys **46** (1991), no. 1, 1–43.
6. V. I. Arnold and Yu. S. Ilyashenko, *Ordinary differential equations* Dynamical systems, I, Itogi Nauki. Sovremennye Problemy Matematiki: Fundamental′ nye Napravleniya., vol. 1, VINITI, Moscow, 1985; English transl., Encyclopaedia of Math. Sci., vol. 1, Springer-Verlag, Heidelberg, 1988.
7. A. Andronov, E. Leontovich, I. Gordon, and A. Mayer, *Bifurcation theory of dynamical systems on the plane*, "Nauka", Moscow, 1967; English transl., Halsted Press, New York and Toronto, 1973.
8. Yu. Ilyashenko and S. Yakovenko, *Concerning the Hilbert sixteenth problem*, in this volume.
9. O. Kleban, *Order of the topologically sufficient jet of a smooth vector field on the real plane at a singular point of finite multiplicity*, in this volume.
10. Yu. Ilyashenko and S. Yakovenko, *Finite cyclicity of elementary polycycles in generic families*, in this volume.
11. F. Dumortier, *Techniques in the theory of local bifurcations: blow-up, normal forms, nilpotent bifurcations, singular perturbations*, Bifurcations and Periodic Orbits of Vector Fields (Dana Schlomiuk, ed.), NATO ASI Series C (Mathematical and Physical Sciences), vol. 408, Kluwer, Dordrecht, 1993, pp. 19–74.
12. F. Takens, *Normal forms for certain singularities of vector fields*, Ann. Inst. Fourier Grenoble **23** (1973), no. 2, 163–195.
13. G. Belitskiĭ, *Equivalence and normal forms of germs of smooth mappings*, Uspekhi Mat. Nauk **33** (1978), no. 1, 95–155; English transl. in Russian Math. Surveys **33** (1978).

Sources for Tables 5 and 6

[C] L. Cherkas, *Structure of a successor function in the neighborhood of a separatrix of a perturbed analytic autonomous system on the plane*, Differentsial′ nye Uravneniya **17** (1981), no. 3, 469–478; English transl. in Differential Equations **17** (1981).

[DRR] F. Dumortier, R. Roussarie, and C. Rousseau, *Elementary graphics of cyclicity.* 1, 2, Preprint, Université de Bourgogne (1993).

[DRS] F. Dumortier, R. Roussarie, and J. Sotomayor, *Generic 3-parameter families of vector fields on the plane, unfolding a singularity with nilpotent linear part. The cusp case*, Ergodic Theory and Dynamical Systems **7** (1987), 375–413; *Generic 3-parameter families of vector fields on the plane, unfoldings of saddle, focus and elliptic singularities with nilpotent linear parts*, Lecture Notes Math., vol. 1480, Springer-Verlag, Berlin, Heidelberg, and New York, 1991, pp. 1–164.

[G] T. Grozovskiĭ, *Bifurcation of polycycles of the types "apple" and "half-apple" in generic 2-parameter families*, Preprint (1993).

[JM] M. Jebrane and A. Mourtada, *Cyclicité finie des lacets doubles non triviaux*, Nonlinearity **7** (1994), 1349–1365.

[LLZ] Li Weigu, Li Chengzhi, and Zhang Zhifen, *Unfolding of the critical homoclinic orbit of a class of degenerate equilibrium points*, Preprint (1991).

[L] V. Lukyanov, *On bifurcations of dynamical systems with a saddle-node separatrix loop*, Differentsial′ nye Uravneniya **18** (1982), no. 9, 1493–1506; English transl. in Differential Equations **18** (1982).

[M1] A. Mourtada, *Degenerate and on-trivial polycycles with two vertices*, J. Differential Equations **113** (1994) (to appear).

[M2] ———, *Analytical unfolding of irrational and trivial* 2-*polycycle*, Preprint Université de Bourgogne **16** (1992), submitted to *Global analysis*.

[N] V. Nozdracheva, *Bifurcations of a structurally unstable separatrix loop*, Differentsial′ nye Uravneniya **18** (1982), no. 9, 1551–1558; English transl. in Differential Equations.

[S] S. Schechter, *The saddle-node separatrix loop bifurcation*, SIAM J. Math. Anal. (1987), 1142–1156.

Translated by S. YAKOVENKO

MOSCOW STATE UNIVERSITY, MOSCOW, RUSSIA

Amer. Math. Soc. Transl.
(2) Vol. **165**, 1995

A Geometric Proof of the Bautin Theorem

S. YAKOVENKO

ABSTRACT. We give a short geometric proof of the Bautin theorem on cyclicity of elliptic singular points of quadratic vector fields on the real plane. The key argument is $\mathbb{Z}_3$-symmetry of an auxiliary Hamiltonian system.

§1. Formulation of the result

A (planar) vector field is *quadratic* if its right-hand side part is given by two polynomials of degree 2, and a singular point is said to be *elliptic* if the corresponding eigenvalues are $\sigma \pm i\omega \notin \mathbb{R}$. In 1939, N. N. Bautin announced and, in 1952, published a proof of the following result.

BAUTIN THEOREM [1]. *If a quadratic vector field on the real plane has an elliptic singular point, then any small variation of coefficients of the field produces at most three limit cycles in a sufficiently small neighborhood of this point. There exist variations producing three cycles.*

In the terminology introduced in the introductory paper to this volume, this theorem can be reformulated as follows.

BAUTIN THEOREM (Reformulation). *Cyclicity, within the family of quadratic vector fields, of a polycycle consisting of just one elliptic point, equals* 3.

The original proof was achieved by lengthy computations with power series, integration of recurrent systems of equations, etc., with many omissions and several (unessential) errors. In this paper we give a complete proof which is based on a hidden symmetry of the problem, thus avoiding almost all computations. The most difficult part of the original proof is concerned with investigation of a certain auxiliary perturbed Hamiltonian system: unlike the standard situation, one needs to investigate the second variation of the Poincaré return map.

The proof consists of three parts. The first part, containing the arguments that are now completely standard, reduces the problem of determining the cyclicity to a certain assertion concerning Taylor coefficients of the Poincaré map. This assertion involves all coefficients, and the second part of the proof (the algebraic one) deals with few initial coefficients. This part involves heavy computations which

1991 *Mathematics Subject Classification.* Primary 34C05.

however can be relegated to a computer. The third part (the transcendental one) establishes some divisibility properties for *all Taylor coefficients*, hence it cannot be computerized in principle. This third part is the most dark and cumbersome fragment of the original proof, and it is this part which required simplification.

In its algebraic part the original proof was also partly erroneous. Bautin computed explicitly the first seven coefficients of the Poincaré map, but omitted the computations. Later it was found that he made an error in a numeric factor (which does not affect the validity of the proof), and a computer-aided computation of the coefficients was done, among other works, in [5]. But in fact we do not need the explicit values of the numeric factors: for the cyclicity estimate it is sufficient to establish only that those factors are nonzero. This demonstration (still computer-aided) can be organized in a much simpler way, and we supply a short code written for the `Mathematica` symbolic computations program, which does all necessary computations in seconds of the processor time.

Concluding this short introduction, we want to emphasize that the exposition below is self-contained except for the *center conditions for quadratic vector fields*: those were established by Dulac [2] in 1908 and (unlike some later versions) are widely recognized as being definitely correct. All the rest (including the computer-aided computations) is explicitly described or proved.

§2. Reductions of the problem

2.1. Dulac and Bautin ideals. Quadratic systems on the plane were investigated by H. Dulac in 1908 [2], who found necessary and sufficient conditions for such a field to possess a first integral in a neighborhood of an elliptic point.

Any quadratic vector field with an elliptic singular point on the real plane can be put (after an appropriate affine transformation) into the standard *Dulac–Kapteyn form*

$$\begin{cases} \dot{x} = \sigma x - y - \lambda_3 x^2 + (2\lambda_2 + \lambda_5)xy + \lambda_6 y^2 , \\ \dot{y} = x + \sigma y + \lambda_2 x^2 + (2\lambda_3 + \lambda_4)xy - \lambda_2 y^2 . \end{cases} \tag{1}$$

REMARK. The transformation taking a general quadratic field

$$\dot{x} = \sum a_{ij} x^i y^j , \quad \dot{y} = \sum b_{ij} x^i y^j , \ 0 < i + j \leqslant 2 , \qquad a_{ij} , \ b_{ij} \in \mathbb{R} ,$$

to the form (1), is *discontinuous* with respect to the coordinates a_{ij}, b_{ij}.

DULAC THEOREM [2]. *The system* (1) *has a singularity of the type center at the origin* (*is integrable*) *if and only if* $\sigma = 0$ *and one of the four sets of equalities holds:*

$$\{ \lambda_3 - \lambda_6 = 0 , \tag{D.1}$$

$$\begin{cases} \lambda_2 = 0 , \\ \lambda_5 = 0 , \end{cases} \tag{D.2}$$

$$\begin{cases} \lambda_4 = 0 , \\ \lambda_5 = 0 , \end{cases} \tag{D.3}$$

$$\begin{cases} \lambda_5 & = 0 , \\ \lambda_4 + 5\lambda_3 - 5\lambda_6 & = 0 , \\ \lambda_3\lambda_6 - 2\lambda_6^2 - \lambda_2^2 & = 0. \end{cases} \tag{D.4}$$

To an elliptic point of any vector field, we can associate the *Poincaré return map* Δ and the *displacement function* δ: if $(x, 0)$ is a point on the y-axis, then $\Delta(x)$ is defined as the x-coordinate of the next intersection of the phase trajectory emanating from the point $(x, 0)$, with the same ray of y-axis, and $\delta(x) = \Delta(x) - x$ is the displacement: $\delta(x) = 0$ if and only if the trajectory passing through $(x, 0)$ is closed. Both Δ and δ do depend on the parameters σ, λ_j. *Isolated roots of the displacement function correspond to limit cycles of the vector field.*

LEMMA 1 [1]. *The displacement map is analytic at $x = 0$, therefore expandable in a convergent Taylor series*

$$\delta(x, \sigma, \lambda) = \sum_{k=1}^{\infty} \mathbf{a}_k(\sigma, \lambda)\, x^k.$$

The coefficients $\mathbf{a}_k$ of the expansion are entire functions of the parameters σ, λ, which turn out to be homogeneous polynomials of degree $k-1$ in λ at $\sigma = 0$.

PROOF. The analyticity is implied by the fact that after blowing up the family (1) yields a line field which is nonsingular on the pasted in projective line for all values of the parameters. The assertion about the degrees is evident: the line field spanned by the vector field (1) with $\sigma = 0$ is invariant with respect to transformations of the form $(x, y, \lambda) \mapsto (\nu x, \nu y, \nu\lambda)$, therefore each $\mathbf{a}_k(0, \lambda)$ is a homogeneous entire function of λ of degree $k-1$, hence a homogeneous polynomial. □

DEFINITION. The *Dulac ideal* $\mathfrak{D}$ is the ideal of polynomials from $\mathbb{R}[\lambda]$, $\lambda = (\lambda_2, \dots, \lambda_6)$, that vanish on the union of the four loci (D.1)–(D.4) given by Dulac theorem.

One can easily see that the following three polynomials,

$$\begin{aligned} A_1 &= \lambda_5(\lambda_3 - \lambda_6), \\ A_2 &= \lambda_2\lambda_4(\lambda_3 - \lambda_6)(\lambda_4 + 5\lambda_3 - 5\lambda_6), \\ A_3 &= \lambda_2\lambda_4(\lambda_3 - \lambda_6)(\lambda_3\lambda_6 - 2\lambda_6^2 - \lambda_2^2), \end{aligned} \tag{2}$$

constitute a basis of the Dulac ideal $\mathfrak{D}$. This follows from the fact that different polynomials from the Dulac list (D) are mutually prime.

DEFINITION. The *Bautin ideal* is the ideal generated by the coefficients of the displacement function, evaluated at $\sigma = 0$:

$$\mathfrak{B} = \langle a_1, a_2, \dots, a_k, \dots \rangle \subseteq \mathbb{R}[\lambda],$$

where

$$a_k = a_k(\lambda) = \mathbf{a}_k(0, \lambda) \in \mathbb{R}[\lambda].$$

Apparently, $\mathfrak{B} \subseteq \mathfrak{D}$, since in the integrable case the displacement vanishes identically.

2.2. Two fundamental lemmas. The displacement function can be written in the form of a term divisible by σ and a term corresponding to $\sigma = 0$:

$$\delta(x, \sigma, \lambda) = \sigma h(x, \sigma, \lambda) + \delta(x, 0, \lambda) = \sigma x\, h_0(x, \sigma, \lambda) + \sum_{k\geqslant 2} a_k x^k,$$

with $a_k = a_k(\lambda)$, as before. Now we can formulate the basic algebraic fact concerning the first seven coefficients of the displacement map.

LEMMA 2. *The first seven Taylor coefficients of the displacement function $\delta(\cdot, \sigma, \lambda)|_{\sigma=0}$ as polynomials in λ satisfy the following identities, in which $A_j \in \mathbb{R}[\lambda]$, $j = 1, 2, 3$, are the generators of the Dulac ideal, and the asterisks stand for nonzero constant factors*:

$$\begin{aligned} a_2 &= 0, \\ a_3 &= *A_1, \\ a_4 &= 0 \mod A_1, \\ a_5 &= *A_2 \mod A_1, \\ a_6 &= 0 \mod \langle A_1, A_2\rangle, \\ a_7 &= *(\lambda_3 - \lambda_6)A_3 \mod \langle A_1, A_2\rangle. \end{aligned} \tag{3}$$

The proof of this result is given in §4 below. We establish the form (3) in a completely traditional way, and show how the computer-aided proof of the fact that the factors denoted by asterisks are indeed *nonzero*. The proof is independent of the rest of the paper.

REMARK. Note that if there were no factor $(\lambda_3 - \lambda_6)$ in the representation for a_7, then a_3, a_5, and a_7 would constitute a basis for $\mathfrak{D}$, hence there would be proved the coincidence of $\mathfrak{B}$ and $\mathfrak{D}$. In fact, the situation is more complicate.

LEMMA 3 [1].

$$\mathfrak{B} \subseteq \langle A_1, A_2, A_3^*\rangle \subsetneq \mathfrak{D},$$

where $A_3^* = (\lambda_3 - \lambda_6)A_3$.

This basic fact, which establishes certain properties of all coefficients of the displacement function, will be proved in §3. Lemma 3 is the core of the Bautin's proof: the rest in this section contains arguments that nowadays are rather standard.

Together, Lemmas 2 and 3 imply the following assertion.

COROLLARY.

$$\mathfrak{B} = \langle a_3, a_5, a_7\rangle = \langle A_1, A_2, A_3^*\rangle. \quad \square$$

2.3. Proof of the Bautin theorem. The above Corollary implies the required estimate on the number of isolated zeroes of the displacement function, as this was shown by R. Roussarie [3]. Indeed, the assertion of the corollary means that for any $j > 7$ we have

$$a_j = q_{j3}a_3 + q_{j5}a_5 + q_{j7}a_7,$$

where $q_{jk} \in \mathbb{R}[\lambda]$ are polynomials. On the formal level, this means that the Taylor expansion for $\delta(\cdot, \sigma, \lambda)|_{\sigma=0}$ can be re-expanded into the sum of three terms, and the displacement function δ itself can be expanded into the sum of four series

$$\delta(x, \sigma, \lambda) = \sigma x\, h_0 + a_3 x^3\, h_1 + a_5 x^5\, h_2 + a_7 x^7\, h_3,$$
$$h_j \in \mathbb{R}[\lambda]\,[[x]], \quad j = 1, 2, 3, \qquad h_j(0, 0) = 1.$$

In fact, this re-expansion can be achieved also within the class of convergent series, using the bounded division principle [3], so one can consider $h_0, \dots, h_3$ as being analytic functions of x, σ, λ which take nonzero constant values at $x = 0$.

As it is proved in [3], the number of isolated nonzero roots of such a combination does not exceed the number of terms minus 1, that is, 3 in our case. Roughly, the inductive proof goes as follows: the function h_0 is nonvanishing in a small neighborhood of the origin in the (x, σ, λ)-space, thus division by h_0 does not change the number of zeros, while the function δ becomes

$$\frac{\delta(x, \sigma, \lambda)}{h_0(x, \sigma, \lambda)} = \sigma + a_3(\lambda)\, x^3 \tilde{h}_1(x, \sigma, \lambda) + \cdots + a_7(\lambda)\, x^7 \tilde{h}_7(x, \sigma, \lambda).$$

Next, differentiating the last expression in x, we diminish the number of real roots on the small interval $(0, r)$ no more than by 1. On the other hand, after such differentiation we get rid of the constant term, obtaining the combination of the form

$$a_3\, x^2 h_1^* + \cdots + a_7\, x^6 h_3^*$$

with h_j^* taking nonzero constant values when $x = 0$. Division of this combination by x^2 *does not change the number of positive roots on the interval* $(0, r)$. But after such a division we obtain a combination of the same form as at the beginning of the inductive step, though with one term less.

Thus we proved that the displacement function, if not identically zero, can have at most 3 positive roots in a sufficiently small interval $(0, r)$. This proves the Bautin theorem on cyclicity. □

REMARK. Nontriviality of the Bautin result roots in the presence of the additional factor in A_3^* as compared to A_3: *the Bautin ideal is not a radical one*, rad $\mathfrak{B} = \mathfrak{D} \supsetneq \mathfrak{B}$. An explanation of such a phenomena was found recently: H. Żołądek [4] discovered that the coefficients of the displacement function are rotationally invariant with respect to a certain natural $\mathbb{S}^1$-action, eventually after an appropriate regrouping [4, p. 233]. Then the ideal spanned by the first several coefficients *in the ring of rotationally symmetric polynomials* turns out to be a radical one. However, this approach also involves lengthy computations in the spirit of those given below in §4. Another aspect of the rotational invariance of ideals of coefficients was studied by P. Joyal.

§3. Investigation of the Bautin ideal and proof of Lemma 3

3.1. Reduction to a perturbed Hamiltonian system. To prove Lemma 3, we need to show that all the terms a_k with $k > 7$ modulo the *radical* ideal $\langle A_1, A_2 \rangle$ are divisible by an *extra* factor $(\lambda_3 - \lambda_6)$. The locus $\{A_1 = A_2 = 0\}$ is the union of three linear subspaces. Two of them correspond to the zero displacement, and only the third one, $L = \{\lambda_5 = 0,\ \lambda_4 + 5(\lambda_3 - \lambda_6) = 0\}$, yields a nontrivial case:

$$\delta(x, \sigma, \lambda)\big|_{\sigma=0,\ \lambda\in L} \not\equiv 0.$$

Since $A_3|_L = -5(\lambda_3 - \lambda_6)^2\lambda_2(\lambda_3\lambda_6 - 2\lambda_6^2 - \lambda_2^2)$, and divisibility of $a_k|_L$ by $A_3|_L$ already follows from Dulac theorem, we have to *prove that all* $a_k|_L$ *are divisible by* $(\lambda_3 - \lambda_6)^3$, or, what is essentially the same, that the displacement δ restricted on L is divisible by $(\lambda_3 - \lambda_6)^3$.

PROPOSITION [1]. *The quadratic system* (1) *restricted on the subspace*

$$L = \{\sigma = 0,\ \lambda_5 = 0,\ \lambda_4 + 5(\lambda_3 - \lambda_6) = 0\}$$

in the parameter space, can be written as the perturbed Hamiltonian system

$$\begin{cases} \dot{x} = -H_y + \mu p, & p(x, y) = -x^2, \\ \dot{y} = \ \ H_x + \mu q, & q(x, y) = -3xy, \end{cases} \tag{4}$$
$$\mu = \lambda_3 - \lambda_6,$$
$$H(x, y) = \tfrac{1}{2}(x^2 + y^2) + \lambda_2(\tfrac{1}{3}x^3 - xy^2) - \lambda_6(\tfrac{1}{3}y^3 - x^2y).$$

From now on we consider μ as a new parameter replacing the combination $\lambda_3 - \lambda_6$. Thus the new parameters are $\mu, \lambda_2, \lambda_6$. The role played by μ is completely different from that played by λ_2 and λ_6, so we sometimes drop the latter parameters from the notation.

PRINCIPAL LEMMA (Hamiltonian version of Lemma 3). *The displacement function* $\delta = \delta(x, \mu, \lambda_2, \lambda_6)$ *for the perturbed Hamiltonian system* (4) *is divisible by* μ^3: *if we denote by* ∂_μ *the differential operator* $(\partial/\partial\mu)$ *and drop the parameters* λ_2 *and* λ_6 *from the notation, then*

$$\delta(\cdot, 0) = \partial_\mu\delta(\cdot, 0) = \partial_\mu^2\delta(\cdot, 0) \equiv 0.$$

3.2. Second variation of the displacement function. Let

$$\omega = -p\,dy + q\,dx$$

be the *perturbation form*: the family (4) with λ_2, λ_6 considered as fixed parameters, corresponds to the Pfaffian equation $dH + \mu\omega = 0$. We endow the transversal $y = 0,\ 0 < x < r$, with the new chart $h = H|_{y=0}$ and compute the displacement with respect to this new chart, denoting it by $\delta(h, \mu)$. Apparently, $\delta(h, 0) \equiv 0$ (the unperturbed system is integrable), and it is well known that the first variation of the displacement function is given by the Abelian integral:

$$\partial_\mu\delta(h, 0) = \oint_{H=h} \omega.$$

LEMMA 4. *If the first variation of the displacement function vanishes identically,* $\oint_{H=h} \omega \overset{h}{\equiv} 0$, *then*

$$\partial_\mu^2 \delta(h\,,\,0) = \oint_{H=h} \mathbf{K}[\omega\,,\,H] \cdot \omega\,, \tag{5}$$

where $G = \mathbf{K}[\omega\,,\,H]$ *is any function satisfying the equation*

$$d\omega = dG \wedge dH.$$

PROOF. Denote by $\Gamma(h\,,\,\mu)$ the segment of the integral trajectory of the (perturbed) Pfaffian system $dH + \mu\omega = 0$ between two subsequent intersections with the transversal at the points h and $\Delta(h\,,\,\mu)$ respectively, with the natural orientation: for $\mu = 0$ this segment becomes the *closed* oval $\{H = h\}$ of the Hamiltonian level curve. In what follows we use the symbol $\int$ for integration over $\Gamma(h\,,\,\mu)$, while $\oint$ means integration over $\Gamma(h\,,\,0)$. We write $\omega_1 \sim \omega_2$ if two forms are cohomologous, and $\omega_1 \overset{\circ}{=} \omega_2$, if their restrictions on any closed oval $\Gamma(h\,,\,0)$ coincide. The standard notations i_v and L_v are used for the inner antidifferentiation (substitution of a vector field as the first argument of a form) and the Lie derivative: e.g., $i_v dH = L_v H = dH(v)$.

The displacement map is given by the *exact* formula

$$\delta(h\,,\,\mu) = \int dH = -\mu \int \omega.$$

Consider a *vector field of normal variation* v on the level curve $\Gamma(h\,,\,0)$: by definition,

$$v(g_0^t(z_h)) = \left.\frac{\partial}{\partial\mu}\right|_{\mu=0} g_\mu^t(z_h)\,,$$

where $z_h = (x_h\,,\,0)$ is the intersection point of the transversal $\{y = 0\,,\ x > 0\}$ with the level curve $\Gamma(h\,,\,0)$, and g_μ^t is the flow of the (perturbed for $\mu \neq 0$) Hamiltonian vector field. In fact, the field v is defined up to the addition of a vector field tangent to the curve $\Gamma(h\,,\,0)$, which justifies the name of the normal variation.

Then $dH(v) = i_v\,dH$ is a function on $\Gamma(h\,,\,0)$ which is a primitive for the restriction of ω on this curve:

$$(i_v\,dH)(g_0^t z_h) = \left.\frac{\partial}{\partial\mu}\right|_{\mu=0} H(g_\mu^t(z_h)) = \left.\frac{\partial}{\partial\mu}\right|_{\mu=0} \left(\mu \int_{z_h}^{g_\mu^t z_h} \omega\right) = \int_{z_h}^{g_0^t z_h} \omega\,,$$

provided that the integration is carried along subarcs of $\Gamma(h\,,\,\mu)$ and $\Gamma(h\,,\,0)$ respectively. This function is well defined since $\oint \omega = 0$, and by construction,

$$d(i_v\,dH) \overset{\circ}{=} \omega.$$

Next,

$$\partial_\mu^2 \delta(h\,,\,0) = \partial_\mu|_{\mu=0} \int \omega = \oint L_v \omega.$$

Using the homotopy formula $L_v = i_v d + d i_v$, we obtain

$$L_v\omega = d(i_v\omega) + i_v\, d\omega \sim i_v(dG \wedge dH) = i_v\, dG \cdot dH - i_v\, dH \cdot dG.$$

The term $i_v\, dG \cdot dH$ vanishes identically on the curve $\Gamma(h, 0)$ since $dH \overset{\circ}{=} 0$. Therefore

$$L_v\omega \overset{\circ}{=} -(i_v\, dH)\, dG \sim G\, d(i_v\, dH) \overset{\circ}{=} G\omega.$$

Since any two forms with $\omega_1 \sim \omega_2$ or $\omega_1 \overset{\circ}{=} \omega_2$ yield the same integrals over the *closed loop* $\Gamma(h, 0)$, the lemma is proved. □

REMARK. Apparently formula (5) allows for an inductive application: if the second variation of the displacement map also vanishes identically, the same construction yields an explicit formula for the third variation *etc.*, until the first not identically zero variation occurs. Recently this formula was independently written by J.-P. Françoise [6].

3.3. Integration over time-parameterized curves. The assertion of Lemma 4 is fairly general, and we make some comments on the nature of the function $G = \mathbf{K}[\omega, H]$. This function depends linearly on the *cohomology class* of the form ω (rather than on the form itself). Moreover, the integration of a form $dG = d\mathbf{K}[\omega, H]$ over ovals of H can be reduced to the integration over time-parameterized solutions to the (unperturbed) Hamiltonian system: if

$$\Theta \in \Lambda^1(\Gamma(h, 0)), \Theta = \begin{cases} -\dfrac{dx}{H_y(x, y)}, & \text{for } H_y \neq 0, \\ \dfrac{dy}{H_x(x, y)}, & \text{for } H_x \neq 0, \end{cases}$$

is the time form, so that

$$\oint_{\Gamma(h, 0)} \Theta = \text{the period of the trajectory } \Gamma(h, 0),$$

and

$$*\colon \Lambda^2(\mathbb{R}^2) \to \Lambda^0(\mathbb{R}^2), \qquad *(f(x, y)\, dx \wedge dy) = f(x, y)$$

stands for the duality operator (division of a 2-form by the standard symplectic structure $\Omega = dx \wedge dy$), then $d\mathbf{K}[\omega, H] \overset{\circ}{=} (*d\omega) \cdot \Theta$.

3.4. The $\mathbb{Z}_3$-symmetry of the Hamiltonian. The key fact that makes possible computation of integrals is the following $\mathbb{Z}_3$-symmetry of the Hamiltonian: *if $T\colon \mathbb{R}^2 \to \mathbb{R}^2$ is rotation of the (x, y)-plane by $2\pi/3$, then $H \circ T = H$.* Indeed, H is the linear combination of three terms, each of them being $\mathbb{Z}_3$-invariant. This immediately implies that $T^*\Theta \overset{\circ}{=} \Theta$. As another corollary we obtain the following statement.

LEMMA 5. *Integral of any homogeneous 1-form of degree 2 over any oval* $\Gamma(h, 0)$ *is zero.*

PROOF. By the Stokes theorem,

$$\oint_{\Gamma(h,0)} \omega = \iint_{H\leqslant h} d\omega = \iint_{H\leqslant h} (\alpha x + \beta y)\, dx \wedge dy.$$

The latter integral is the momentum of the $\mathbb{Z}_3$-symmetric figure $\{H \leqslant h\}$ with respect to a certain axis passing through the origin, which is zero. □

3.5. Proof of the Principal Lemma. As a corollary to Lemma 5, we obtain that for the perturbation (4) written in the Pfaffian form $dH + \mu\omega = 0$, $\omega = x^2\, dy - 3xy\, dx$, the first variation $\partial_\mu \delta(h, 0)$ vanishes.

To compute the second variation $\oint \mathbf{K}[\omega, H] \cdot \omega$, we use the trick suggested by Bautin: find another homogeneous form $\tilde{\omega}$ in the cohomology class of ω such that the corresponding perturbation $dH + \mu\tilde{\omega}$ would be integrable for *any choice* of μ so that the corresponding displacement vanishes identically, $\tilde{\delta}(h, \mu) \equiv 0$. Then, according to Lemma 4,

$$0 = \partial_\mu^2 \tilde{\delta}(h, 0) = \oint \mathbf{K}[\tilde{\omega}, H] \cdot \tilde{\omega} = \oint \mathbf{K}[\omega, H](\omega + dS) =$$
$$\oint \mathbf{K}[\omega, H] \cdot \omega - \oint S\, d\mathbf{K}[\omega, H],$$

and we conclude that

$$(6) \qquad \begin{cases} \tilde{\omega} = \omega + dS, \\ dH + \mu\tilde{\omega} \text{ is integrable for all } \mu \end{cases} \implies \oint \mathbf{K}[\omega, H] \cdot \omega = \oint S\,(*d\omega)\,\Theta,$$

where the form Θ and the duality operator $*$ are the same as in §3.3.

The space of perturbations leaving the system $dH + \mu\tilde{\omega}$ integrable is rather ample: for example, the condition $\lambda_3 - \lambda_6 = 0$ written in the natural coordinates a_{ij}, b_{ij} for the general system

$$\dot{x} = -y + \sum a_{ij}\, x^i y^j, \qquad \dot{y} = x + \sum b_{ij}\, x^i y^j, \quad i + j = 2,$$

determines the codimension 2 linear subspace $a_{20} + a_{02} = 0$, $b_{20} + b_{02} = 0$. On the other hand, a form $\tilde{\omega} = \sum p_{ij} x^i y^j\, dx + q_{ij} x^i y^j\, dy$, $i + j = 2$, belongs to the cohomology class of ω if $2q_{20} - p_{11} = 5$, $q_{11} - 2p_{02} = 0$. In particular, for the form

$$\tilde{\omega} = -5xy\, dx$$

the perturbation $dH + \mu\tilde{\omega} = 0$ is integrable and

$$*d\tilde{\omega} = *d\omega = 5x, \quad \omega = \tilde{\omega} + d(x^2 y).$$

Applying (6), we conclude that

$$\oint \mathbf{K}[\omega, H] \cdot \omega = 5 \oint x^3 y\, \Theta.$$

But the function $f(x, y) = x^3 y$ has zero T-average, $f + f \circ T + f \circ T^2 \equiv 0$, while the form Θ is T-symmetric, $T^*\Theta \overset{\circ}{=} \Theta$, hence the last integral is zero. Thus we proved the Principal Lemma, and together with it Lemma 3. □

REMARK. In fact, the assertion proved in this section, is somewhat stronger then the claim of Lemma 3: we proved that *on the common locus* $A_1 = A_2 = 0$ *the displacement function is divisible by* $(\lambda_3 - \lambda_6)^3$. We will need this assertion later in §4 in the form of a divisibility assertion in $\mathbb{R}[\lambda]$.

LEMMA 6 (Corollary to the Principal Lemma). *For any* $j \in \mathbb{N}$ *we have*

$$a_j = 0 \mod \langle A_1, A_2, (\lambda_3 - \lambda_6)^3 \rangle. \quad \square$$

PROOF. As this was noted in 3.1, the locus $A_1 = A_2 = 0$ is the union of three linear subspaces. Two of them correspond to the identically zero displacement, and for the third one the displacement is divisible by $(\lambda_3 - \lambda_6)^3$. □

§4. Computation of the low order Taylor coefficients of the displacement map and proof of Lemma 2

The first seven coefficients of the displacement map must be computed in order to establish the formulas (3). The algorithm suggested in [1] is extremely inefficient: computation of explicit expressions for $a_1, \dots, a_7$ requires hours of computer time (though computation of a next coefficient on zeros of previous ones is less labor consuming). Actually the information about degrees of a_k in $\lambda_2, \dots, \lambda_6$ together with the knowledge of their zero locus is *sufficient to find* $a_1, \dots, a_7$ *up to constant factors,* denoted by asterisks in (3). The proof for a_6, a_7 uses the divisibility assertion of Lemma 6.

4.1. Computation up to constants. As this follows from Lemma 1, the coefficients a_j are homogeneous polynomials of degree j in the variables $\lambda = (\lambda_2, \dots, \lambda_6)$. Their zero locus is known from the Dulac theorem: it is the union of the hyperplane (D.1), two subspaces of codimension 2, (D.2) and (D.3), and a quadric (D.4) of codimension 3. Denote by $\ell(\lambda)$ the linear form and by $q(\lambda)$ the quadratic form occurring in the Dulac conditions,

$$\ell(\lambda) = \lambda_4 + 5(\lambda_3 - \lambda_6), \qquad q(\lambda) = \lambda_3\lambda_6 - 2\lambda_6^2 - \lambda_2^2,$$

so that the four loci (D. j), $j = 1, \dots, 4$ would have the form

$$\text{(D.1)} \qquad \lambda_3 - \lambda_6 = 0,$$

$$\text{(D.2)} \qquad \lambda_2 = \lambda_5 = 0,$$

$$\text{(D.3)} \qquad \lambda_4 = \lambda_5 = 0,$$

$$\text{(D.4)} \qquad \lambda_5 = \ell(\lambda) = q(\lambda) = 0.$$

We start with the following remark: all coefficients a_j should be divisible by $(\lambda_3 - \lambda_6)$, since they must vanish on the hyperplane (D.1). Hence we can represent them as

$$(7) \qquad a_j(\lambda) = (\lambda_3 - \lambda_6)\, b_j(\lambda), \qquad \deg b_j = j - 2.$$

Below we use asterisks for constants and in the next subsection will prove that these constants are in fact nonzero.

Case $j = 2$. We have $\deg b_2 = 0$, hence b_2 is a constant vanishing somewhere. Thus $b_2 = 0 \implies a_2 = 0$.

Case $j = 3$. For the same reasons b_3 is a linear form vanishing on (D.2) and (D.3). The only possibility is $b_3 = *\lambda_5 \implies a_3 = *A_1$.

Case $j = 4$. The quadratic polynomial b_4 restricted on $\lambda_5 = 0$ must vanish if $\lambda_2 = 0$ or $\lambda_4 = 0$, since a_4 vanishes on the union of (D.2) and (D.3). Thus we have

$$b_4 = *\lambda_2\lambda_4 \mod \lambda_5, \qquad a_4 = *(\lambda_3 - \lambda_6)\lambda_2\lambda_4 \mod A_1,$$

and since a_4 must vanish also on (D.4), where $\lambda_4 = 5(\lambda_6 - \lambda_3)$, we have

$$*(\lambda_3 - \lambda_6) \cdot 5\lambda_2(\lambda_3 - \lambda_6) = 0 \mod q(\lambda).$$

But since the quadratic polynomial $q(\lambda)$ is nondegenerate, this is possible only if the factor denoted by $*$ is zero. Hence $b_4 = 0 \mod \lambda_5 \implies a_4 = 0 \mod A_1$.

Case $j = 5$. For the same reasons as before,

$$a_5 = (\lambda_3 - \lambda_6)\lambda_2\lambda_4 r(\lambda) \mod A_1, \qquad \deg r = 1.$$

Note that the minimal linear subspace containing the locus (D.4) is the subspace $\Pi = \{\lambda_5 = \ell(\lambda) = 0\}$, since the quadratic form q is nondegenerate. The above linear form r must vanish on Π, hence it must be a linear combination of λ_5 and $\ell(\lambda)$. The first term of this combination yields an element divisible by A_1 after multiplication by $(\lambda_3 - \lambda_6)$, while the term proportional to ℓ gives an element $*A_2$, hence

$$a_5 = *A_2 \mod A_1.$$

Before analyzing the last two cases, we make the following trivial observation: if $p \in \mathbb{R}[\lambda_2, \lambda_3, \lambda_6]$ is a polynomial in three variables $\lambda_2, \lambda_3, \lambda_6$ that vanishes on the cone $q = 0$, then $p = rq$, where $r \in \mathbb{R}[\lambda_2, \lambda_3, \lambda_6]$ is a polynomial with $\deg r = \deg p - 2$. This follows from the irreducibility of the polynomial q.

Case $j = 6$. In the same manner as before, we write

$$a_6 = (\lambda_3 - \lambda_6)\, q_6, \qquad q_6 = u(\lambda)\, \lambda_5 + \lambda_2\lambda_4\, s_2(\lambda_2, \lambda_3, \lambda_4, \lambda_6),$$

where $\deg s_2 = 2$; this is a general form of a polynomial vanishing on (D.1), (D.2), and (D.3). Next, we can write

$$s_2 = \ell(\lambda)\, v(\lambda) + \tilde{s}_2(\lambda_2, \lambda_3, \lambda_6), \qquad \deg \tilde{s}_2 = 2, \ \ \tilde{s}_2 \in \mathbb{R}[\lambda_2, \lambda_3, \lambda_6].$$

Note that $\tilde{s}_2$ depends only on the variables $\lambda_2, \lambda_3, \lambda_6$. Since a_6 vanishes on (D.4), the quadratic polynomial $\tilde{s}_2$ must vanish on the cone $q = 0$, hence we have $\tilde{s}_2 = r\, q(\lambda)$, where $\deg r = 0$, which means that r is a constant. Putting everything together, we have

$$(8) \quad a_6(\lambda) = (\lambda_3 - \lambda_6)\lambda_5\, u(\lambda) + (\lambda_3 - \lambda_6)\lambda_2\lambda_4\ell(\lambda)\, v(\lambda) + \\ r\,(\lambda_3 - \lambda_6)\lambda_2\lambda_4\, q(\lambda_2, \lambda_3, \lambda_6), \qquad u, v \in \mathbb{R}[\lambda].$$

But by Lemma 6 the restriction of a_6 on the subspace $\lambda_5 = \ell(\lambda) = 0$ must be divisible by $(\lambda_3 - \lambda_6)^3$. Computing the restriction, we obtain

$$a_6(\lambda)\big|_{\lambda_5=\ell(\lambda)=0} = -5r\,(\lambda_3 - \lambda_6)^2 \lambda_2\, q(\lambda)\,, \tag{9}$$

and since q is irreducible, we conclude that the above divisibility is possible only if $r = 0$. Then (8) immediately implies that $a_6 = 0 \mod \langle A_1\,, A_2\rangle$.

Case $j = 7$. The same arguments as in the case $j = 6$ show that for a_7 one can write the decomposition analogous to (8) but with $r = r(\lambda)$ being now a linear homogeneous form rather than a constant. Then the divisibility assertion can hold only if $r = *(\lambda_3 - \lambda_6)$, where the asterisk stands for a constant. This means that

$$a_7 = *(\lambda_3 - \lambda_6)\,A_3 \mod \langle A_1\,, A_2\rangle\,,$$

and we proved the identities (3) modulo the assertion that the constants denoted by asterisks are *nonzero*.

Remark. Bautin in [1] justifies the form (3) for the first seven coefficients of the displacement function by the following argument: since all a_j are known to be in $\mathfrak{D}$, there exists a representation

$$a_j = h_{j1}A_1 + h_{j2}A_2 + h_{j3}A_3\,, \qquad h_{jk} \in \mathbb{R}[\lambda]\,,$$

and *for the degrees of the coefficients* h_{jk} *one has the equality*

$$\deg h_{jk} = \deg a_j - \deg A_k\,,$$

which implies that for $\deg a_j < \deg A_k$ *the corresponding coefficient vanishes.* However, the italicized assertion is in general wrong. For this assertion to be true for an arbitrary finite family A_k with $\deg A_k < \deg A_{k+1}$ and any polynomial a_j from the ideal $\langle A_1\,, A_2\,, \dots\rangle$, it is sufficient that A_{k+1} be not a zero divisor in the quotient ring $\mathbb{R}[\lambda]/\langle A_1\,, \dots\,, A_k\rangle$, which is *not the case* for the Dulac polynomials. Thus in the given specific case one needs specific considerations based on the known factor structure of the polynomials A_k. I am grateful to André Reznikov for explanations concerning the subject.

4.2. Lyapunov order of weak foci and its computation. In order to show that the asterisks in (3) stand for *nonzero* constants, it is sufficient to show that the displacement function does not have a zero of an unexpectedly large order at the origin.

Let $v = v(x\,, y)$ be a vector field with an elliptic singular point at the origin, and $\delta(x)$ be the displacement function for the Poincaré return map.

Definition. The Lyapunov order of an elliptic point is the order of zero of the displacement function at the origin:

$$\operatorname{ord}_0 v = \nu < \infty \iff \delta(x) = x^\nu \alpha(x)\,, \qquad \alpha(0) \neq 0.$$

By definition, the Lyapunov order for an integrable field is equal to $+\infty$.

From the general theory it follows that if the Lyapunov order is finite, it is an odd number. Evidently, the definition of this order is independent of the choice of the transversal and the coordinate system as well: the order $\operatorname{ord}_0 v$ is a geometric invariant of the germ of a vector field v.

To compute the Lyapunov order from the definition, one needs to find the lowest nonzero jet of the return map, in other words, to integrate the differential equation in the class of mixed Taylor–Fourier series of the form

$$R(\varphi) = \sum_{r=0}^{\infty} r^k \left(a_k \cos k\varphi + b_k \sin k\varphi\right),$$

where $r = R(\varphi)$ is a solution in the polar coordinates (r, φ). This is the procedure Bautin suggested to use for determining the first coefficients of the displacement map. However, this procedure from the computational point of view is very hard to implement.

There exists a simpler way to determine the Lyapunov order. Recall that for a k-jet $\hat{f} \in J^k(\mathbb{R}^2, 0)$ the Lie derivative along a vector field v is a well-defined k-jet denoted by $L_v \hat{f}$, provided that $v(0) = 0$. Indeed, we can define the Lie derivative of the k-jet $\hat{f}$ as the k-jet of the Lie derivative of an arbitrary function f with $j^k(f) = \hat{f}$; the result will not depend on the choice of f.

Definition. A *k-jet integral* of the vector field v at the elliptic point $(0, 0)$ is the k-jet $I^{(k)} \in J^k(\mathbb{R}^2, 0)$ whose Lie derivative $L_v I^{(k)}$ is the zero jet.

The k-jet integral is called *nondegenerate* if it is an extension of the 2-jet $\frac{1}{2}(x^2 + y^2)$.

Evidently, the property of having or not having a k-jet integral is independent of the choice of local coordinates.

Lemma 7. *If an elliptic point has a finite Lyapunov order $\nu < +\infty$, then it has a nondegenerate ν-jet integral and does not have nondegenerate $(\nu+1)$-jet integral.*

Proof. Since both the Lyapunov order and existence of jet integrals of a given order are independent of the choice of coordinates, one can choose the local coordinate system (x, y) in such a way that in the complex coordinate $z = x + iy$ the vector field would have the form

$$\dot{z} = z \sum_{j=0}^{N} c_j |z|^{2j} + O(|z|^{2j+3}), \qquad c_0 = i$$

(the Poincaré–Dulac normal form). Let k be the number of the first coefficient in the sequence $c_0, c_1, \ldots$ such that $\operatorname{Re} c_k \neq 0$.

Then, as one can easily see, for the polar radius $r = |z|$ one has the equation

$$\dot{r} = (\operatorname{Re} c_k)\, r^{2k+1} + O(r^{2k+3}),$$

which means that the displacement function has a root of multiplicity $\nu = 2k+1$ at the origin, hence $2k+1$ is the Lyapunov order.

On the other hand, the ν-jet $I^{(\nu)} = \frac{1}{2}|z|^2$ is a ν-jet integral. Moreover, there cannot be $(\nu+1)$-jet integrals. Indeed, if there would be such a jet $I^{(\nu+1)}$, then for any function $\tilde{I}(z)$ with such a $(\nu+1)$-jet, the variation of the function along a trajectory of v starting at $(x, y) = (r, 0)$ and ending at $\Delta(x, y) = (r+\delta(r), 0)$ would not exceed $O(r^{\nu+2})$, while the difference $|\tilde{I}(x, y) - \tilde{I}(\Delta(x, y))|$ is no less than

$$\inf_{x\in[r, r+\delta(r)]} |\tilde{I}'_x(x, 0)| \cdot |\delta(r)| \geqslant |\delta(r) \cdot \tfrac{1}{2} r| \geqslant \tfrac{1}{2} r^{2\nu+1}. \quad \square$$

The procedure of finding the maximal order of a jet integral is very simple for quadratic vector fields. Indeed, let

$$v(x, y) = \mathbf{R} + \mathbf{Q}, \qquad \mathbf{R}(x, y) = -y\tfrac{\partial}{\partial x} + x\tfrac{\partial}{\partial y},$$
$$\mathbf{Q}(x, y) = Q_1(x, y)\tfrac{\partial}{\partial x} + Q_2(x, y)\tfrac{\partial}{\partial y}, \qquad \deg Q_j = 2,$$

be the decomposition of v into the linear rotational and the quadratic homogeneous parts, and suppose that we look for a k-jet integral in the form of a polynomial represented as the sum of its homogeneous terms

$$I^{(k)} = I_2 + \cdots + I_k, \qquad \deg I_j = j.$$

Then the equation $L_v I^{(k)} = 0^{(k)}$ means that the chain of identities

$$L_{\mathbf{R}} I_{j+1} = L_{\mathbf{Q}} I_j, \qquad j = 2, \ldots, k-1, \tag{10}$$

is satisfied. Each identity is a system of linear equations for the coefficients of $(j+1)$ st homogeneous term I_{j+1}, and for solvability of this system it is necessary and sufficient that average value of the function

$$\psi_{j+1} = L_{\mathbf{Q}} I_j \tag{11}$$

over the unit circle vanishes (then its average value over any other concentric circle also vanishes and the equations can be solved). Thus the algorithm of computing the Lyapunov order can be organized as follows: start with $j = 2$ and $I_2 = \frac{1}{2}(x^2+y^2)$, then compute ψ_3 which, being a polynomial of the third order, necessarily has zero average, so the term I_3 can be found. Then compute ψ_4; if its average is nonzero, then the Lyapunov order is 3, otherwise proceed further by determining I_4, etc. The number of the step on which a nonzero average occurs for the first time, is the Lyapunov order of the singularity.

This algorithm applied to the system in the Dulac–Kapteyn form, is shown on Figure 1 (we give the script for the `Mathematica` system).

Comments to the program. After termination of the program, the variables `i[k], k=2,3,...,j-1`, contain the homogeneous terms (in the polar coordinates) of the expression

```
I[r,f]=i[2]+i[3]+...+i[j-1]
```

which is the jet of the highest possible order. The variable `j` after termination of the algorithm contains the number of the first term which cannot be obtained, hence the Lyapunov order of the vector field is equal to `j` -1.

```
(* Initial data, to be supplied by user or left unassigned *)
lambda[2]=1; lambda[3]=1; lambda[4]=-5; lambda[5]=0; lambda[6]=0;

(* Quadratic vector field *)
p[x_,y_]:= -lambda[3]x^2 +(2 lambda[2]+lambda[5])x y + lambda[6] y^2;
q[x_,y_]:= lambda[2]x^2 + (2 lambda[3]+lambda[4])x y - lambda[2] y^2;

(* Output formats *)
Format[lambda[i_]]:= Subscripted[l[i]];
Format[Iterm[d_]] :=  Subscripted[I[d]];

                        (* ROUTINES *)

(* Radial component of the vector field *)
RadialComponent[r_,f_]:=
   Simplify[
     Expand[( x p[x,y] + y q[x,y]
           //. {x -> r Cos[f], y -> r Sin[f]} ),
           Trig -> True]]

(* Angular component of the vector field *)
AngularComponent[r_,f_]:=
   Simplify[
     Expand[((x^2+y^2)^(-1) ( q[x,y] x - p[x,y] y)
           //. {x -> r Cos[f], y -> r Sin[f]}),
           Trig -> True]]

(* Right Hand Side for the j-th Equation *)
rhs[j_]:= D[i[j-1],r]r^(-1) * RadialComponent[r,f] +
          D[i[j-1],f]       * AngularComponent[r,f];

(* Averaging: Solvability condition for a LPDE *)
Average[fun_]:=Integrate[fun,{f,0,2Pi}];

(* Solution of LPDE *)
Primitive[fun_]:= - Integrate[fun,f];

(* Exit routine *)
ExitRoutine[w_,j_]:=
  Print[
    StringForm[
       "Nonzero average = `` is an obstacle on the step = `` ",w,j]];

                        (* MAIN BODY *)

w=0; j=3; i[2]=r^2;      (* head of the loop *)
While[w==0,              (* while the average is zero, do computations *)
     z=rhs[j];           (*Evaluating Right Hand Side*)
     w=Average[z];       (* Computing the average *)
     If[w==0,
        (* if the next term is available, then *)
        i[j]=Primitive[z];
        term=Collect[i[j],r];
        Print[StringForm["\n`` = ``", Iterm[j],Short[term]]];
        j=j+1,
        (* otherwise cyclicity determined! *)
         ExitRoutine[w,j],
         ExitRoutine[w,j]
     ]; (* End If *)
     ] (* End While *)
```

FIGURE 1. Script for Mathematica

4.3. End of the proof of Lemma 2. Denote by c_j the constant occurring in the formula for a_j in (3), $j = 3, 5, 7$. Our goal is to prove that $c_j \neq 0$.

Assume that $c_3 = 0$ and choose a value $\lambda^{(3)} \in \mathbb{R}^5$ such that $A_1\left(\lambda^{(3)}\right) = 0$.

Let ν_3 be the Lyapunov order of the corresponding quadratic vector field. Since $c_3 = 0$, then ν_3 must be at least 4 for any such $\lambda^{(3)}$. On the other hand, the algorithm for computing the Lyapunov order (see the previous section) allows to compute this order independently. If we choose $\lambda^{(3)}$ to be a rational point in $\mathbb{R}^5$, then all computations can be performed in the symbolic form over the rational multiples of π, and the answer is exact. In Table 1 we specify the choice of $\lambda^{(3)}$ and give the result of the computation. It turns out that no unexpected degeneracy occurs, and the Lyapunov order corresponding to the chosen value of $\lambda^{(3)}$ is 3. The contradiction proves that $c_3 \neq 0$.

TABLE 1. Computation of Lyapunov orders

j	$\lambda = (\lambda_2, \dots, \lambda_6)$	Locus	ν_{comp}	T
3	$(0, 1, 0, 1, 0)$	$A_1(\lambda) \neq 0$	3	$18''$
5	$(1, 1, 1, 0, 0)$	$A_1(\lambda) = 0,\ A_2(\lambda) \neq 0$	5	$27''$
7	$(1, 1, -5, 0, 0)$	$A_1(\lambda) = A_2(\lambda) = 0,\ A_3(\lambda) \neq 0$	7	$54''$

NOTES TO TABLE 1. In this table the fourth column ν_{comp} shows the results of the computation of the Lyapunov order for the quadratic system (1) with the corresponding values of λ. The last column indicates the computation time T in seconds of the processor time.

The computations were performed with `Mathematica 2.0` on the personal computer with Intel 80486 processor operating at 25 MHz. On DEC 5000/200 workstations these computations are approximately five times faster, so that even the symbolic computation of the zero locus of $A_1 = A_2 = A_3 = 0$ takes 113 seconds of the processor time.

In the same manner, choosing $\lambda^{(5)}$ such that $A_1\left(\lambda^{(5)}\right) = 0$ and $A_2\left(\lambda^{(5)}\right) \neq 0$, we use a computer to prove that $c_5 \neq 0$. The last computation for $\lambda^{(7)}$ satisfying $A_1\left(\lambda^{(7)}\right) = A_2\left(\lambda^{(7)}\right) = 0$, $A_3\left(\lambda^{(7)}\right) \neq 0$ shows that $c_7 \neq 0$. This completes the proof of Lemma 2. □

REMARKS. In fact, the same algorithm (see Figure 1) allows us to find explicitly the Dulac loci, but this involves computations in the ring of polynomials $\mathbb{R}[\lambda]$, which is much more time-consuming. The advantage of the approach suggested here can be explained as follows: *to evaluate a constant function, it is sufficient to compute it at just one point*. We stress once again the fact that for rational values of $\lambda^{(j)}$ all computations are exact hence demonstrative.

References

1. N. N. Bautin, *On the number of limit cycles appearing with variations of coefficients from an equilibrium state of the type of a focus or a center*, Mat. Sb. (N. S.) **30** (1952), 181–196; English Transl., Amer. Math. Soc. Transl., 1954; Reprinted in: Stability and Dynamical Systems, Amer. Math. Soc. Transl. Series 1, vol. 5, Providence, RI, 1962, pp. 396–413.
2. H. Dulac, *Détermination et intégration d'une certaine classe d'équations différentielles ayant pour point singulier un centre*, Bull. Soc. Math. France **32** (1908), no. 2, 230–252.
3. R. Roussarie, *Cyclicité finie des lacets and des points cuspidaux*, Nonlinearity **2** (1989), 73–117.
4. H. Żołądek [Kh. Zholondek], *Quadratic systems with center and their perturbations*, J. Differential Equations **109** (1994), no. 2, 223–273.
5. W. Farr, Li Chengzhi, I. Labouriau, and W. Langford, *Degenerate Hopf bifurcation formulas and Hilbert's 16th problem*, SIAM J. Math. Anal. **20** (1989), no. 1, 13–30.

6. J. P. Françoise, *Successive derivatives of a first return map, application to the study of quadratic vector fields*, Preprint, October 1993, Université de Paris VI.

Translated by THE AUTHOR

Department of Theoretical Mathematics, The Weizmann Institute of Science, Rehovot 76100, Israel

E-mail address: yakov@wisdom.weizmann.ac.il

Recent Titles in This Series

(Continued from the front of this publication)

126 **S. A. Akhmedov et al.,** Eleven Papers on Differential Equations
125 **D. V. Anosov et al.,** Seven Papers in Applied Mathematics
124 **B. P. Allakhverdiev et al.,** Fifteen Papers on Functional Analysis
123 **V. G. Maz′ya et al.,** Elliptic Boundary Value Problems
122 **N. U. Arakelyan et al.,** Ten Papers on Complex Analysis
121 **V. D. Mazurov, Yu. I. Merzlyakov, and V. A. Churkin, Editors,** The Kourovka Notebook: Unsolved Problems in Group Theory
120 **M. G. Kreĭn and V. A. Jakubovič,** Four Papers on Ordinary Differential Equations
119 **V. A. Dem′janenko et al.,** Twelve Papers in Algebra
118 **Ju. V. Egorov et al.,** Sixteen Papers on Differential Equations
117 **S. V. Bočkarev et al.,** Eight Lectures Delivered at the International Congress of Mathematicians in Helsinki, 1978
116 **A. G. Kušnirenko, A. B. Katok, and V. M. Alekseev,** Three Papers on Dynamical Systems
115 **I. S. Belov et al.,** Twelve Papers in Analysis
114 **M. Š. Birman and M. Z. Solomjak,** Quantitative Analysis in Sobolev Imbedding Theorems and Applications to Spectral Theory
113 **A. F. Lavrik et al.,** Twelve Papers in Logic and Algebra
112 **D. A. Gudkov and G. A. Utkin,** Nine Papers on Hilbert's 16th Problem
111 **V. M. Adamjan et al.,** Nine Papers on Analysis
110 **M. S. Budjanu et al.,** Nine Papers on Analysis
109 **D. V. Anosov et al.,** Twenty Lectures Delivered at the International Congress of Mathematicians in Vancouver, 1974
108 **Ja. L. Geronimus and Gábor Szegő,** Two Papers on Special Functions
107 **A. P. Mišina and L. A. Skornjakov,** Abelian Groups and Modules
106 **M. Ja. Antonovskiĭ, V. G. Boltjanskiĭ, and T. A. Sarymsakov,** Topological Semifields and Their Applications to General Topology
105 **R. A. Aleksandrjan et al.,** Partial Differential Equations, Proceedings of a Symposium Dedicated to Academician S. L. Sobolev
104 **L. V. Ahlfors et al.,** Some Problems on Mathematics and Mechanics, On the Occasion of the Seventieth Birthday of Academician M. A. Lavrent′ev
103 **M. S. Brodskiĭ et al.,** Nine Papers in Analysis
102 **M. S. Budjanu et al.,** Ten Papers in Analysis
101 **B. M. Levitan, V. A. Marčenko, and B. L. Roždestvenskiĭ,** Six Papers in Analysis
100 **G. S. Ceĭtin et al.,** Fourteen Papers on Logic, Geometry, Topology and Algebra
99 **G. S. Ceĭtin et al.,** Five Papers on Logic and Foundations
98 **G. S. Ceĭtin et al.,** Five Papers on Logic and Foundations
97 **B. M. Budak et al.,** Eleven Papers on Logic, Algebra, Analysis and Topology
96 **N. D. Filippov et al.,** Ten Papers on Algebra and Functional Analysis
95 **V. M. Adamjan et al.,** Eleven Papers in Analysis
94 **V. A. Baranskiĭ et al.,** Sixteen Papers on Logic and Algebra
93 **Ju. M. Berezanskiĭ et al.,** Nine Papers on Functional Analysis
92 **A. M. Ančikov et al.,** Seventeen Papers on Topology and Differential Geometry
91 **L. I. Barklon et al.,** Eighteen Papers on Analysis and Quantum Mechanics
90 **Z. S. Agranovič et al.,** Thirteen Papers on Functional Analysis
89 **V. M. Alekseev et al.,** Thirteen Papers on Differential Equations
88 **I. I. Eremin et al.,** Twelve Papers on Real and Complex Function Theory
87 **M. A. Aĭzerman et al.,** Sixteen Papers on Differential and Difference Equations, Functional Analysis, Games and Control

(See the AMS catalog for earlier titles)